European Union Foreign Policy and the Global Climate Regime

P.I.E. Peter Lang

Bruxelles · Bern · Berlin · Frankfurt am Main · New York · Oxford · Wien

Simon Schunz

European Union Foreign Policy and the Global Climate Regime

"College of Europe Studies"
No.18

Cover picture: "Enviest", © Hélène Le Borgne.

All manuscripts in this series are subject to a critical review by the book series editors as well as external peer review.

No part of this book may be reproduced in any form, by print, photocopy, microfilm or any other means, without prior written permission from the publisher. All rights reserved.

© P.I.E. PETER LANG S.A.
Éditions scientifiques internationales
Brussels, 2014
1 avenue Maurice, B-1050 Brussels, Belgium
info@peterlang.com; www.peterlang.com

ISSN 1780-9665
ISBN 978-2-87574-134-9
D/2014/5678/24

CIP available from the Library of Congress, USA, and the British Library.

Bibliographic information published by "Die Deutsche Nationalbibliothek".

"Die Deutsche Nationalbibliothek" lists this publication in the "Deutsche National-bibliografie"; detailed bibliographic data is available on the Internet at <http://dnb.d-nb.de>.

Contents

Acronyms and Abbreviations 11

Foreword and Acknowledgements 15

INTRODUCTION. Rationale of the Study 17
    The EU, Climate Change and Global Climate Politics 21
    The Structure of the Study 23

CHAPTER 1. Analytical Framework: Studying the European Union's Influence on the Global Climate Regime 27
    Building the Key Concepts: Influence Attempts and Influence 27
    Setting the Theoretical Scene: Insights from EU Foreign Policy Analysis and Regime Theory 33
    Methodological Bases: Analysing and Determining Influence 42

CHAPTER 2. Historical Foundations (1980s–1995): EU Influence on the Set-up of the Global Climate Regime 47
    The Pre-negotiation Phase: From Scientific Circles to First Political Negotiations 47
    The EU's Influence on the UN Framework Convention on Climate Change (1991–1992) 52
    The Road to COP 1 (1992–1995) 64

CHAPTER 3. From the Berlin Mandate to the Kyoto Protocol (1995–1997): EU Influence on the First Development of the Global Climate Regime 67
    The Context: Major Developments in Global Politics and Climate Science 67
    Key Actors in the Global Climate Regime and their Positions 68
    The Negotiation Process and the EU's Influence Attempts 76
    The Outcome: the Kyoto Protocol 95
    The EU's Influence Attempts: Extracting Patterns 98

The EU's Influence in the Kyoto Protocol Negotiations 101
Explaining the EU's Influence during the Period
1995 to 1997 109

**CHAPTER 4. From the Buenos Aires Action Plan
to the Year 2007 (1998–2007): EU Influence on
the Consolidation of the Global Climate Regime** 113
COP 4 to COP 7: From the Buenos Aires Action Plan
to the Marrakech Accords (1998–2001) 113
After the Marrakech Accords: Ensuring Ratification
of the Kyoto Protocol (2002–2004) 127
Towards a Post-2012 Regime: Loose Talks
on the Way to Bali (2005–2007) 134
Determining and Explaining the EU's Influence
during the Period 1998 to 2007 147

**CHAPTER 5. From the Bali Roadmap to the Copenhagen
Accord (2007–2009): EU Influence on the Post-2012
Global Climate Negotiations** 151
The Context: Major Developments in Global Politics
and Climate Science 151
Key Actors in the Global Climate Regime
and their Positions 154
The Negotiation Process and the EU's Influence Attempts 170
The Outcome: the Copenhagen Accord 222
The EU's Influence Attempts: Extracting Patterns 225
The EU's Influence in the Post-2012
Climate Negotiations until 2009 231
Explaining the EU's Influence during
the Period 2007 to 2009 244

**CHAPTER 6. Gradually "Back on Track" (2010–2012):
EU Influence on the Resumed Post-2012 Global
Climate Negotiations** 251
The Context: Major Developments in Global Politics
and Climate Science 251
Key Actors in the Global Climate Regime
and their Positions 252
The Negotiation Process and the EU's
Influence Attempts 257

The Outcomes: the Cancun Agreements,
the Durban Package and the Doha Gateway 265
Determining and Explaining the EU's Influence
during the Period 2010 to 2012 269

**CHAPTER 7. Explaining EU Influence
on the Global Climate Regime** 275
Patterns of EU Influence across Time 275
Comparing EU Influence Attempts to its
Actual Influence: the "Goodness of Fit" Puzzle 283
Determinants of EU Influence over Time: Propositions
on Causal Mechanisms and their Scope Conditions 287

Conclusion 301
Major Findings of the Study and their Significance 301
Research and Normative Implications of the Study 303

References 311

Annexes 361

Index 365

Tables
Table 1: Toward a Typology of Influence 33
Table 2: How the EU Can Exert Influence –
EU Foreign Policy Acts 36
Table 3: Establishing EU Influence – Constitutive
Dimensions 45
Table 4: The Negotiations of the Copenhagen Accord 217
Table 5: EU Influence on the Global Climate
Regime over Time 276
Table 6: EU Influence Attempts and their Success
in the Global Climate Regime (1991-2012) –
a Compilation of Instruments 285

Acronyms and Abbreviations

ADP	Ad Hoc Working Group on the Durban Platform for Enhanced Action
AGBM	Ad Hoc Group on the Berlin Mandate
AOSIS	Alliance of Small-Island States
APP	Asia-Pacific Partnership on Clean Development and Climate Change
APEC	Asia-Pacific Economic Cooperation
Art.	Article
AWG	Ad Hoc Working Group
AWG-KP	Ad Hoc Working Group on the Kyoto Protocol
AWG-LCA	Ad Hoc Working Group on Long-Term Cooperative Action
BAPA	Buenos Aires Plan of Action
BASIC	Brazil, South Africa, India, China
BRIC	Brazil, Russia, India, China
CA	Copenhagen Accord
CBDR	Common but differentiated responsibilities
CDM	Clean Development Mechanism
CFSP	Common Foreign and Security Policy
COP	Conference of the parties
COW	Committee of the Whole
DG	Directorate-General
EC	European Community
EGFA	Expert Group on Further Action
ENB	Earth Negotiations Bulletin
ENGOs	Environmental non-governmental organisations
EP	European Parliament
ETS	Emissions Trading System
EU	European Union
FAR	Fourth Assessment Report

G-77/China	Group of 77 and China
GDN	Green Diplomacy Network
GHG	Greenhouse gas
HFC	Hydrofluorocarbon
INC	Intergovernmental Negotiating Committee
IPCC	Intergovernmental Panel on Climate Change
IR	International Relations
JU(S)SCA(N)NZ	Japan, United States, Canada, Australia, New Zealand (Switzerland, Norway, Iceland)
KP	Kyoto Protocol
LDC	Least Developed Country
LULUCF	Land-use, land-use change and forestry
MEF	Major Economies Forum
MEM	Major Economies Meeting
MOP	Meeting of the parties
MRV	Measurement, reporting and verification
NAFTA	North American Free Trade Agreement
NAMAs	Nationally appropriate mitigation actions
NCCCC	National Coordination Committee for Climate Change
NDRC	National Development and Reform Commission
NGOs	Non-governmental organisations
OECD	Organisation for Economic Cooperation and Development
OPEC	Organisation of Petroleum Exporting Countries
PCA	Partnership and Cooperation Agreement
PFC	Perfluorocarbon
PMs	Policies and Measures
ppm	Parts per million
QELROs	Quantified Emission Limitation and Reduction Objectives
REIO	Regional Economic Integration Organisation
RGGI	Regional Greenhouse Gas Initiative
SAR	Second Assessment Report
SBI	Subsidiary Body for Implementation

Acronyms and Abbreviations

TAR	Third Assessment Report
TEC	Treaty establishing the European Community
TFEU	Treaty on the Functioning of the European Union
TEU	Treaty on European Union
UN	United Nations
UNCHE	United Nations Conference on the Human Environment
UNFCCC	United Nations Framework Convention on Climate Change
UNGA	United Nations General Assembly
UK	United Kingdom
US(A)	United States of America
WEIS	World Event Interaction Coding Scheme
WG	Working Group
WPIEI-CC	Working Party on International Environmental Issues (– Climate Change)

Foreword and Acknowledgements

This book is the fruit of a long-term research project on a fascinating topic: climate change and the role of the EU in the global attempts to combat it. It comprises the significantly revised version of my doctoral dissertation, defended in October 2010 at the University of Leuven (KULeuven). Neither this book nor the thesis would have seen the light of day without the intellectual input of my supervisors, Prof. Hans Bruyninckx (KULeuven, European Environment Agency) and Prof. Stephan Keukeleire (KULeuven, College of Europe), to whom I feel deeply indebted. I am equally thankful to Prof. Bart Kerremans, Prof. Jan Wouters (KULeuven, Leuven Centre for Global Governance Studies) and the external assessors, Prof. Philipp Pattberg (Free University of Amsterdam) and Prof. John Vogler (Keele University), for serving on the doctoral jury that assessed the thesis. The study would not have been written without the financial support provided through the interdisciplinary KULeuven Impuls project "The European Union and multilateral governance" (2006–2010). I would also like to express my gratitude to the editors of the College of Europe Studies series, particularly Prof. Sieglinde Gstöhl, as well as to the team at Peter Lang's Brussels office.

Many of my colleagues affiliated with the Institute for International and European Policy and the Leuven Centre for Global Governance Studies made valuable contributions to this research project over the years. I am particularly grateful to Dr. Rouba Al-Fattal, Sue Basu, David Belis, Dr. Sofie Bouteligier, Dr. Tim Corthaut, Prof. Tom Delreux, Dr. Sijbren de Jong, Dr. Ana Maria Dobre, Prof. Edith Drieskens, Dr. Sander Happaerts, Björn Koopmans, Dr. Montserrat Gonzalez-Garibay, Dr. Kolja Raube, Dr. Karoline Van den Brande and Dr. Louise van Schaik. Special mention goes to Jed Odermatt, who provided support to improve the linguistic quality of this study.

The research project also benefitted greatly from numerous exchanges with policy-makers and observers of the intra-EU and global climate negotiations. I am indebted to my interview partners and many other professionals involved in processes of global and European climate politics, who generously shared their thoughts with me.

Last but not least, I would like to warmly thank my family and friends for their support throughout the process of researching for and writing up this study. Above all, I feel enormously indebted to Hélène.

INTRODUCTION
Rationale of the Study

This book combines the analysis of two topis that have immensely gained in political importance over the past two decades: the foreign and external policies of the European Union (EU) and global climate change. The EU is still a relatively recent player on the global scene, even when it comes to the environment, arguably *the* domain – beyond trade – in which it has made the first and most visible steps to become acknowledged as a foreign policy actor in its own right (Bruyninckx 2005: 213–214). Yet, especially since the entry into force of the Maastricht Treaty in 1993, the EU's capacity and ambitions to shape global politics have grown considerably. This is especially true in an area that has equally obtained ever-increasing attention in the past twenty years: climate change – one, if not "the defining challenge of our generation" (United Nations Secretary General Ban Ki-moon, Reuters 2007a). Following the first compelling natural scientific insights into the risks associated with anthropogenic interference with the global climate, this collective action challenge was for the first time politically tackled at a global level in the early 1990s. Initial negotiations under United Nations (UN) auspices led to the adoption of the UN Framework Convention on Climate Change (UNFCCC, in force since 1994). Since then, attempts to complement the soft legal framework convention so as to foster durable global solutions to the climate problematique have been ongoing in the UN regime, with the intermediate results embodied in the Kyoto Protocol (in force since 2005) and the Cancun Agreements of 2010.

By bringing these two topics together, the study intends to provide an in-depth understanding and explanation of how the European Union behaves, and what effects its behaviour yields, in global climate politics. In so doing, it conceives of climate change as an ever more "important foreign policy issue" (Ott 2001a). It consequently treats the EU's activities targeted at the global climate regime not simply as the external dimension of intra-EU climate and energy policies – and thus as a part of its external relations – but as genuine foreign policy.[1] EU foreign policy is understood as "that area of [EU] politics which is directed at the external

---

[1] This approach is particularly justified by the fact that the EU was a *foreign* climate policy player even before it had a domestic climate policy acquis (Pallemaerts 2004).

environment with the objective of influencing that environment and the behaviour of other actors within it, in order to pursue interests, values and goals" (Keukeleire/MacNaughtan 2008: 19). Indeed, the primary objective of EU activities in the global climate regime has been, rather than maintaining *external relations* with third parties, that of *influencing* this regime and the behaviour of other actors within it for the purpose of protecting the climate in line with EU interests, values and goals. If one considers the EU thus as a foreign policy player in global climate politics and intends to scrutinize its activities, influence becomes a key measure for its effectiveness.

Closely accompanying the evolution of the EU's external and foreign environmental policies, the political and academic debates about the EU's role on the global scene have regularly observed that the Union was "recognized as a leader" (Sbragia 2000: 312; Zito 2005) with "extensive influence in the politics of the global environment" (Vogler 2005: 848). In the academic debate on the EU's role in global climate politics more specifically (see, above all, Bäckstrand/Elgström 2013; Wurzel/Connelly 2010; Oberthür *et al.* 2010; Parker/Karlsson 2010; Lindenthal 2009; Costa 2009; Schreurs/Tiberghien 2007; Harris 2007; Groenleer/van Schaik 2007; Pallemaerts/Williams 2006; Pallemaerts 2004), the scrutiny of its proactive approach in this domain has led to claims that "EU leadership in international climate policy over the past 15 years or so has remained largely unrivalled" (Oberthür 2007: 79). Such claims have regularly been based on studies that employ the analytical concept of "leadership" (Gupta/Grubb 2000; Gupta/Ringius 2001). Most importantly, the notion of *directional* leadership has repeatedly been used to describe how the EU attempts to show the way, employing "perceptions and solutions developed domestically as a 'model' to diffuse internationally" (Grubb/Gupta 2000: 21). The Union's "model" first took shape in the late 1990s and early 2000s, partially as a result of internal regime creation aimed at strategy-building in reaction to international developments (Pallemaerts 2004: 42–56). A flagship initiative in this regard was the establishment of an Emissions Trading System (ETS) that has been in operation since 2005 (Skjaerseth/Wettestad 2008). In 2008/2009, a major climate and energy package was then adopted, lifting the EU's acquis to a new level of harmonization in these domains (Morgera *et al.* 2011). In the face of these evolutions, claims about EU "leadership by example" have resonated well with popular intra-EU political discourse about the Union as a "green power" and global climate leader, notably prior to the 2009 Copenhagen climate summit (e.g. Barroso 2008).

In stark contrast to much of these debates and to the European Union's apparently persistently high and, at least until the 2009 Conference of the Parties in Copenhagen (COP 15), even steadily increased level of

Introduction

proactivity as a global climate player, the effects of its activities have, at first sight, been limited across time. For the early 1990s, it had been observed that "the EU had a comparatively limited impact on the UNFCCC and the Kyoto Protocol" (Oberthür/Roche-Kelly 2008: 36). After years of stalemate in the global regime, it then seemingly suffered a severe setback regarding both its reputation and objectives when it became partially sidelined in the final stages of the 2009 COP 15 where none of its major proposals made it into the "Copenhagen Accord". The most striking observation from these debates, and *prima facie* from the evolution of EU participation in the global climate regime more generally, is thus a strong discrepancy between an almost linear increase in EU activity as a global climate player and the apparently limited impact it has had over time. This observation forms the major puzzle that this book addresses. To do so in a systematic manner, the study responds to three closely intertwined research questions:

- **Question 1**: *How did/does the European Union attempt to exert influence on the multilateral negotiations pertaining to the development[2] of the global climate regime?*
- **Question 2**: *Did the European Union actually exert influence on the multilateral negotiations pertaining to the development of the global climate regime?*
- **Question 3**: *Why did/does the European Union exert influence on the multilateral negotiations pertaining to the development of the global climate regime?*

The responses to these questions contribute to the current political and academic debates in four main ways. *First*, the study adds to these debates by providing comprehensive *empirical* knowledge about what the Union has done and does in global climate negotiations and what effects this has (had), especially in answer to questions 1 and 2. To this end, a longer, discontinued time frame is considered necessary so as to overcome the "presentism bias" of many EU foreign policy analyses (Jørgensen 2007). Cross-time comparisons allow for a clearer understanding of the EU's influence on the global climate regime, which has gone through several phases. After negotiations on the Framework Convention itself (1991–1992), the regime awaited its formal confirmation (1992–1995). Ratification of the UNFCCC was followed by a novel negotiation phase resulting in the Kyoto Protocol (1995–1997). Subsequently, several conferences of the parties had to prepare for the ratification of the Protocol (1997–2005). Finally, post-2012 negotiations were started, first loosely (2005–2007) and then more intensely, with the intermediate outcome of the Copenhagen

---

[2] Regime development refers to the formation of the regime and its evolution.

Accord (2007–2009), later formally integrated into the UN regime by the Cancun Agreements (2010). The 2011 Durban Package and the 2012 Doha Gateway then re-started negotiations on a legally binding global agreement. These time periods are regrouped into five phases that are analysed in depth (1991–1995, 1995–1997, 1998–2007, 2007–2009, 2010–2012). *Second*, the study contributes to the debates in *conceptual-theoretical* terms by advancing the understanding of why the EU does or does not have an impact on global politics. Within the discipline of EU foreign policy analysis, the study thus inserts itself into emergent debates on the EU's performance/external effectiveness as a foreign policy actor by developing the concept of influence as a measure of effectiveness and by providing, in answer to questions 2 and 3, a better understanding and explanation of the Union's impact in the studied regime (Jørgensen *et al.* 2011; Dee 2013). To this end, explanatory factors from the EU and international levels of analysis are considered and combined. Moreover, a close scrutiny of the instruments and resources the Union utilizes as a foreign policy player advances the state of the art on the link between EU foreign policy tools, influence and, ultimately, foreign policy effectiveness. This closely ties in with the *third* contribution made by this work, which concerns *methodology*. By developing a method that integrates the mapping of EU activity and the assessment and explanation of its impact, the study refines the toolbox of the discipline of (EU) foreign policy analysis. A *fourth* contribution results from the *normative and political-practical relevance* of the research. The study produces insights into the EU's performance in global politics that allow for an appreciation of whether it actually lives up to the expectations it creates by evoking certain conceptions – such as "leadership" – of its own global role, notably in the significant policy domain of climate change.

By precisely tracing the EU's activities and their effects on the global climate regime across time, the study challenges and nuances some of the claims made in current debates. It demonstrates how the EU has gradually made itself the champion of the global fight against climate change, trying hard, and through various means, to get a grip on the regime. It also shows, however, how and why the Union has oftentimes failed to effectively do so. Although the study demonstrates that the EU has, at least since the mid-1990s, indeed been a very, if not the most proactive foreign climate policy player, its activities qualify best as *attempted* – but regularly unsuccessful – leadership. A number of reasons related to both the external context and internal prerequisites for EU activities, but especially the often underestimated interplay between these two, conceptually embodied in the very notion of foreign policy, can account for this lack of success.

The EU, Climate Change and Global Climate Politics

Analysing the contents and effects of the EU's foreign policy represents a general interest of many EU foreign policy specialists, as illustrated by Karen Smith (2007: 12–13, 2010):

> Much more research needs to be done on the EU's influence in the wider world, and particularly on the EU's impact on the international system (…), and its actual impact on outsiders (…) (does the EU influence them and how?). Too often, we lapse into assertions that the EU has either considerable or little influence, without the backing of clear, substantial evidence for such influence. 'Proving' the EU has influence (or not, and what sort and why) requires considerable empirical research (…) but unless we try to get to the bottom of this, we are left with unsubstantiated assertions about the EU's place/role/influence in the world. (…) Debates about whether the EU is or is not a civilian power, a normative power, a superpower and so on, are not really leading us anywhere right now. (…) We should instead engage in a debate about what the EU does and why it does it and with what effect, rather than what it is.

In striving to address this interest, the present study focuses on an emblematic concern of EU (foreign) policy, a key domain in which the EU has – following the principle of precaution – the long-standing intention to influence its environment and other actors: global climate politics (Van Schaik/Schunz 2012). Its foreign policy activities and the global politics in this area cannot be understood without a basic understanding of the issue of climate change itself.

When it comes to this issue, the Fourth Assessment Report (FAR) of the Intergovernmental Panel on Climate Change (IPCC), published in 2007, provided a comprehensive summary of the state of the art of scientific knowledge.[3] It is expected that many trends that the FAR documented will be confirmed in the Fifth Assessment Report in 2014.[4] The 2007 Report gathered much evidence for the existence of climate change. The most striking observations were (IPCC 2007a: 2–4):

- A rise in average global air temperature by around 0.75°C and an increase in the heat content of the world's oceans during the century from 1906 to 2005.

---

[3] The IPCC regularly synthesises the state of the art of climate science and is widely considered as an authoritative and reliable source, even after controversies about its functioning following the discovery of several mistakes in the FAR in 2009. The stir that "Climategate" caused was settled quickly when independent advisory committees found no major flaws in the science reported by the IPCC (Ball/Johnson 2010).

[4] First releases of parts of the Fifth Report, which is work in progress at the time of writing, clearly point into this direction, see http://www.ipcc.ch/report/ar5/wg1/.

- A widespread melting of ice both in the Arctic sea and on mountain glaciers all over the planet, and of snow in mountainous areas in the Northern and Southern hemispheres.
- A rise in sea levels by an average of 1.8 mm per year since 1961 and by 3.1 mm per year since 1993.

The root causes of these trends have been attributed in large part to the "*enhanced* greenhouse effect" (IPCC 2007c: 946). Resulting from the fact that greenhouse gases (GHG) – essentially carbon dioxide (CO_2) and methane (CH_4) – absorb parts of the solar energy that the earth deflects back into space, the *natural* greenhouse effect is beneficial to human living conditions on Earth (Wigley 1999: 4, 44–45).[5] An amplification of this effect, however, enhances GHG concentrations in the atmosphere. It is such an increase in GHG emissions and concentrations that climate scientists have been witnessing in the recent past (IPCC 2007b: 36–37). The atmospheric concentration of the main GHG, carbon dioxide, has risen by about 30% above pre-industrial levels, from 270–280 parts per million (ppm) before 1750 to over 360 ppm in 2006 (Wigley 1999: 5; IPCC 2007a: 5). When it comes to the key driver for this increase in GHG concentrations, the FAR singles out the combustion of fossil fuels, noting that there was "*very high confidence*" that the global average net effect of human activities since 1750 has been one of warming" (IPCC 2007b: 37, 2007a: 5).[6] The identified warming effect of human activity appears to have already – with varying degrees of certainty – numerous negative repercussions for the planet (altered weather patterns, degradation of water quality and arable lands, shifts in eco-systems and adverse effects on human health such as heat-related deaths and the spread of tropical diseases) (IPCC 2007b: 31–33).

This study parts from these natural scientific parameters of the debate about climate change. It also assumes that dealing with or preventing its negative consequences necessitates urgent political action involving mitigation, i.e. the tackling of various anthropogenic sources of GHG emissions, and adaptation, which involves changes in practices and/or structures "to moderate or offset potential damages or to take advantages of opportunities related to climate change" (Toth 2008).

Since the planet's atmosphere represents a "global common", i.e. a constitutive element of a single ecosystem which is simultaneously used and shared by everyone and escapes anyone's exclusive sovereignty or jurisdiction, global climate politics amounts essentially to the intricate challenge of solving a highly complex collective action problem (Held *et al.*

---

[5] Without the natural greenhouse effect, the average surface temperature on earth could lie at –18°C (Pidwirny 2006).

[6] "Very high confidence" means that the chance that a finding is correct is 9 out of 10 (IPCC 2007b: 27).

1999: 384). Since the 1980s, policy-makers have actively attempted to provide a response to this challenge. A formal negotiation process was initiated in December 1990 when the UN General Assembly (UNGA) endorsed the creation of an "Intergovernmental Negotiating Committee for a Framework Convention on Climate Change" (INC), which was charged with delivering an international climate treaty by mid-1992 (UNGA 1990). Negotiated between February 1991 and May 1992, the resulting UNFCCC was opened to signature at the Earth Summit in Rio de Janeiro in June 1992 (Betsill 2005: 108). It entered into force on 21 March 1994 and by 2011 had been ratified by 195 parties (194 states and the EU). Between 1995 and 1997, parties negotiated a protocol to the Convention, introducing novelties like quantified emission reductions targets for industrialized, so-called "Annex I", parties and "flexible mechanisms" into the process. This treaty entered into force in February 2005. Ever since, attempts to reform the regime have been undertaken, but with limited success. By tracing EU activities through the five different time periods of regime evolution introduced above, the study provides a detailed account of these developments.

The Structure of the Study

Apart from this introduction, the study is divided into seven chapters and a conclusion. *Chapter 1* provides the *analytical framework* that serves as the basis for conducting the study. To design a comprehensive case study, it parts from a discussion of (i) the central concepts of influence and influence attempts, (ii) theoretical insights on the EU's foreign policy and the climate regime and (iii) methodological considerations on influence analysis. The longitudinal study relies on a combination of foreign policy analysis and influence analysis techniques. This allows a link to be made between the thick description of EU activities to a determination and subsequent explanation of its influence in the global climate regime. The description is facilitated by a theory-based selection for two embedded units of analysis: EU influence is traced with regard to (i) the emissions targets as the key norm of the regime and (ii) common but differentiated responsibilities (CBDR) as its main principle.

Chapters 2 to 6 comprise the *empirical study* of the EU's activity in and impact on the development of the global climate regime from a longitudinal perspective, offering analyses of five periods in the evolution of the regime, with a particular focus on the periods that were marked by major regime reform negotiations (1995–1997, 2007–2009).

Chapter 2 covers the period 1980s to 1995. It begins with a brief discussion of the historical foundations of the global climate regime, which predate the onset of negotiations on a Framework Convention in 1991. Subsequently, it examines the foreign policy behaviour and impact of the

European Union and its member states during these negotiations, which led to the adoption of the UNFCCC. It concludes that it was not so much the EU as such, but several of its more active member states that engaged in substantial climate diplomacy *vis-à-vis* the major player in these talks, the US. In the end, the strong engagement of these states guaranteed the Union on the whole some leverage over key provisions of the Convention. In the period between the adoption of the treaty and its entry into force, the need for a more binding approach emerged.

Chapter 3 engages in a detailed analysis of the EU's behaviour and influence during the talks leading from the first COP and the adoption of the "Berlin Mandate" in 1995 to the last-minute deal brokered essentially between the EU, the US and Japan on the Kyoto Protocol, adopted at COP 3 in late 1997. It shows how the Union attempted to lead the negotiation process through proactive proposals centred on its new narrative, adopted in 1996, of keeping global temperature rise below 2°C. This allowed it to gain some leverage over the magnitude of the emission reduction targets for industrialized countries enshrined in the Protocol. In return, it had to give in to US demands regarding provisions on the use of flexible mechanisms for reaching these targets. Attributing the EU a medium degree of influence during this period, the chapter explains the limits of its leverage essentially with a discrepancy between its external ambitions and internal disagreements.

The period from 1998 to 2007 marked, in many ways, a transition phase in the global climate regime, which is traced in *Chapter 4*. In the immediate aftermath of COP 3, efforts in this regime were concentrated on operationalizing key provisions of the Kyoto Protocol to prepare for its ratification. This process was not completed before late 2001, when COP 7 concluded the Marrakech Accords. In the run-up to these Accords, the EU had first been obliged to step up its diplomatic efforts, after the 2001 withdrawal of the US from the Kyoto Protocol ratification process. Later, it had to give in to the flexibility demands of Japan, Australia and Russia. It took until 2005, then, to finalize the ratification of the treaty. Convincing Russia proved particularly difficult, and the Union had to promise the country support for its WTO membership bid in return for the Duma's ratification. As soon as this was accomplished, the EU pushed for a renewed reform of the entire regime, in view of the expiration of the first commitment period of the Kyoto Protocol in late 2012. Several instances of successful influence-wielding during this period demonstrate the Union's capacity to mobilize other parties whenever its members act in unison and choose foreign policy tools suited for the context. Yet, the limits of leadership by example also crop up, as illustrated by COP 7, where it had to give up on its "environmental integrity" concerns when accepting watered down provisions operationalizing the flexible mechanisms of the Kyoto Protocol.

Chapter 5 analyses in detail how the EU attempted to influence the post-2012 negotiations which kicked off in December 2007 at COP 13 in Bali, Indonesia, and provisionally ended in December 2009 at COP 15 in Copenhagen. To begin with, special consideration is given to the context for global climate politics in the late 2000s. The IPCC's 2007 Fourth Assessment Report served to heighten the urgency with which the problem was perceived, while the rise of GHG emissions in China and other emerging countries made regime reform ever more necessary. In this context, the EU again tried to position itself as a global front-runner, issuing detailed proposals even before the official start of negotiations. The chapter traces the stream of EU foreign policy attempts to set the agenda and to determine the provisions discussed in the regime. At key moments in the negotiations, however, and especially during the final deal-making phase at the Copenhagen summit, it failed to convey its messages, being even physically excluded from some of the decisive meetings. The Union's comparatively low influence during this period is explained with reference to the altered geopolitical context and its own incapacity to strategically adapt to this evolving environment.

Chapter 6 addresses the EU's struggle to find its place in the regime during the three years after COP 15 (2010–2012). It traces the EU's efforts to (re)gain leverage over the global climate negotiations after a year (2010) marked by a more pragmatic diplomatic strategy. Following the incorporation of the Copenhagen Accord into the UN framework through the 2010 "Cancun Agreements", the Union stepped up its efforts again and was arguably instrumental to the adoption, at COP 17 in Durban, South Africa, of a new roadmap toward the conclusion of a climate agreement by 2015. Obstacles on the road toward this agreement appeared, however, as early as 2012 in the run-up to COP 17 in Doha. The chapter concludes with an assessment and explanation of the Union's influence on the resumed post-2012 talks.

Chapter 7 summarizes key findings and extracts patterns of EU influence on the global climate regime across time as a prerequisite for an *explanation* of this influence. It specifies a number of determinants of EU influence related to its actor capacity, foreign policy behaviour and to the external context in which it operates. It links these explanatory factors through several propositions formulated in the form of conjunctive causality statements, which distinguish between EU influence through bargaining and EU influence through arguing.

The *concluding part* of the book explicitly answers its three research questions and sets the findings into a broader academic and policy context, notably by considering the future of the EU's participation in global climate politics and exploring ways scholars can make sense of it.

CHAPTER 1

Analytical Framework

Studying the European Union's Influence on the Global Climate Regime

This chapter develops the analytical framework that guides the study. Three components of this framework are successively introduced: (i) the key concepts of influence attempts and influence; (ii) theoretical considerations derived from EU foreign policy studies and regime theory; (iii) the influence analysis methodology applied to the case study.

Building the Key Concepts: Influence Attempts and Influence

The central objective of this study is to investigate the EU's impact on global climate politics. To do so, a concept has to be designed that is capable of linking an actor-centric perspective, focussed on the EU, to the analysis of global politics. By referring to EU activities in the global climate regime as *foreign* policy, i.e. an area of politics directed at the external environment with the objective of *influencing* that environment, influence was identified as suitable to serve as this hinge. The concept captures the relationship between a purposive actor and its surrounding at the global level, at which this actor potentially causes change.

Employing a sufficiently specified concept of influence has several advantages for the type of study envisaged. First, as "all politics is the exercise of influence" (Dahl/Stinebrickner 2003: 34), manifold definitions and conceptualizations exist. This makes it not only possible but even necessary to build the concept of influence in a manner "appropriate to the substance of the phenomenon" that is studied (Goertz 2006: 16). Second, several influence analysis methods have already been applied to complex decision-making arrangements at the global level and can serve as sources of inspiration for designing an analytical framework capable of determining EU impact (Betsill/Correll 2008; Arts/Verschuren 1999). Finally, influence can be regarded as a "continuous concept" (Goertz 2006: 34): it allows for assessments about the EU's performance in global affairs in terms of gradations. The Union could, for instance, have abundant, substantial, little or no influence on specific decisions taken in the global climate regime. Assessing the EU's performance in terms of gradations makes it possible to specify its impact and effectiveness much more precisely than through a concept like leadership (Gupta/Grubb 2000). The latter allows, in essence, only for general, binary yes-no assessments: either the EU is a leader or not. At the same time, while the attribution of a leadership role demands the fulfilment of high normative criteria, an actor can be influential in global policy-making without necessarily (i) having to be a leader and (ii) having to be benign (Sjöstedt 1999: 228).

When it comes to defining and analysing influence, leading public policy analysts like Dahl and Stinebrickner lament the "absence of standard terminology" (2003: 12). Their own definition then also incorporates elements of various conceptualizations advanced by other scholars (e.g. Braam 1975; Nagel 1975). To them, influence is "a relation among human actors such that wants, desires, preferences, or intentions of one (…) actor (…) affect the actions, or predispositions to act of one or more actors in a direction consistent with (…) the wants, preferences or intentions of the influence-wielder" (Dahl/Stinebrickner 2003: 17). The definition highlights the existence of an influence-wielder and one or more influenced, which stand in some form of relationship to each other, while specifying the changes that the influenced is or are undergoing. Coming close to the definition of relational power as "getting another actor to do what it would otherwise not do" (Dahl 1957), this depiction of influence stresses a key challenge for any study of influence, which consists in distinguishing it from the closely related concept of power. Cox and Jacobson (1973: 3) made this distinction in the most convincing manner, defining influence as "the modification of one actor's behaviour by that of another" for the purpose of reaching the latter actor's aims, specifying that "[p]ower means capability (…). Power may be converted into influence, but it is not necessarily so converted at all or to its full extent".

This distinction between power and influence becomes central when the concept of influence is applied in and to foreign policy analyses. Foreign policy has regularly been defined with reference to influence in international relations (Hudson/Vore 1995: 215), as the "attempts by governments to influence or manage events outside the state's boundaries" (Manners/Whitman 2000: 2), as "those actions which, expressed in the form of explicitly stated goals, commitments and/or directives, and pursued by governmental representatives (…), are directed toward objectives, conditions and actors (…) *which they want to affect* and which lie beyond their territorial legitimacy" (Carlsnaes 2002: 333, emphasis added), or, in the formula that is also adopted here, as an "area of politics which is directed at the external environment with the objective of influencing that environment and the behaviour of other actors within it, in order to pursue interests, values and goals" (Keukeleire/MacNaughtan 2008: 19).

Consequently, classical foreign policy analysis distinguishes between "foreign policy making", the study of how the objectives of foreign policy are formulated internally, and "foreign policy implementation", which refers in essence to how decisions taken internally by foreign policy actors are expressed in concrete actions aimed at influencing others when these "actors confront their environment and (...) the environment confronts them" (Brighi/Hill 2008: 118; Webber/Smith 2002: 79–104). Against this backdrop of widely used definitions and understandings of influence in public policy and foreign policy analysis,[1] it becomes possible to undertake the crucial step of building the concept in a manner "appropriate to the substance of the phenomenon" that is studied here, i.e. the EU's influence as a foreign policy actor on the global climate regime. In this process, a distinction is made between two closely related concepts that are successively employed in the study: influence attempts and influence.

To enhance the analytical sharpness of the concept of influence, and to take account of the notion of foreign policy implementation as a set of actions undertaken with the *intention* of impacting an external context, it is first necessary to introduce the concept of *influence attempts*. These can be defined as *acts by an actor exerted with the purpose of bringing about change in the behaviour, preferences or beliefs of other actors in order to attain its aims.*[2] Influence attempts are analytically distinct from, but conceptually complement influence. An actor's influence is, in fact, the product of its successful exercise of an influence attempt.

Once defined as such, this concept can be dissected into various "constitutive dimensions", which implies that the "preliminary idea (...) formed" through the definition is expanded via the identification of necessary and sufficient conditions and/or causal mechanisms that need to be fulfilled to analytically ascertain the presence of an influence attempt (Goertz 2006: 6). Two core components can be detected:

1. **INTERACTION**: Influence attempts require some form of direct or indirect relation between a potential influence-wielder and one or more influence targets (Dahl/Stinebrickner 2003; Cox/Jacobson 1973).
2. **PURPOSIVE BEHAVIOUR**: The influence-wielder acts because it "wants to affect" the influence target (Carlsnaes 2002: 333).

Both components must be regarded as necessary conditions. Together, they are sufficient to determine that a foreign policy act qualifies as influence attempt.

---

[1] For a more detailed discussion and critique of common conceptualizations of influence, see Schunz 2010: chap. 2.

[2] The definition is inspired by Cox/Jacobson (1973: 3). Influence-wielders and influence targets can be states, but also non-state entities such as NGOs.

Various techniques of influencing have been identified in the IR and foreign policy literature. In International Relations, a set of six "acts of influencing" has regularly been evoked (Holsti 1995: 125–126): persuasion, offering rewards, granting rewards, threatening punishment, inflicting punishment and using force. In a foreign policy analysis context, influence attempts come above all in the form of foreign policy acts/tools. Analysts quite regularly evoke two types of foreign policy tools: diplomatic and economic instruments (Brighi/Hill 2008: 131–132; Webber/Smith 2002: 87–90). As such classifications tend to remain abstract, operationalization attempts like the World Event Interaction Coding Scheme (WEIS) have led to catalogues of foreign policy acts "as an aid to productive analysis" of foreign policy (Wilkenfeld *et al.* 1980: 117, 19). The WEIS catalogue identifies 22 verbs such as "to promise", "to grant" or "to reward" in order to pinpoint concretely what a foreign policy actor does when trying to exert influence. These classifications are integrated into a broader overview when it comes to specifying *EU* influence attempts in a subsequent section (see Table 2).

Based on its delimitation from the concept of influence attempt, influence is re-defined in this study as *the modification of one or several actors' behaviour, preferences or beliefs by acts of another actor exerted for the purpose of reaching the latter actor's aims*.[3] Four core components of the concept are identified, two of which overlap with the constitutive dimensions of influence attempts.

1. **INTERACTION**: Influence attempts require some form of direct or indirection relation between a potential influence-wielder and one or more influence targets.
2. **PURPOSIVE BEHAVIOUR**: The influence-wielder acts because it "wants to affect" the influence target.
3. **TEMPORAL SEQUENCE**: Actions by the influence-wielder precede any type of behavioural or mind change in the influenced (Braam 1975; Cox/Jacobson 1973).
4. **GOAL ATTAINMENT**: The behavioural or mind change in the influenced must go "in a direction consistent with (…) the wants, preferences or intentions of the influence-wielder" (Dahl/Stinebrickner 2003: 17). In other words, the influence-wielder's purposive behaviour is successful: its goal is attained. On this point, it has to be borne in mind that influence is a continuous concept. Partial goal attainment or shared influence-wielding with others do not rule out influence.

---

[3] It has to be noted that influence can also be aimed at avoiding change. In this case, an influence-wielder would try – and ultimately succeed in – altering the behaviour, beliefs or preferences of those who desire change.

With this, the "positive pole" of influence is determined (Goertz 2006). Considering also its negative pole (when is what we observe no longer influence), a fifth dimension comes into play:

5. **ABSENCE OF AUTO-CAUSATION**: Logically, observed behavioural or mind changes qualify only as influence if they can be – at least in part – attributed to the activity of the influence-wielder, and not exclusively to some other reason that may be inherent[4] to the influence target (Braam 1975; Huberts 1994).

All five components are necessary conditions. Together, they are sufficient to determine influence via conditional causal analysis.

While conditional causal analysis (Mackie 1974) tells us *that* certain conditions lead to the identified outcomes, it is unable to specify *how* and, above all, *why* precisely this happens (Mahoney 2003). Going a step further to – tentatively – explain why an actor has had influence amounts thus to accounting for an already established causal relationship. To "fully explain particular outcomes" (Mahoney 2003: 1), it has been suggested to resort to a different form of causal analysis by employing "causal mechanisms", i.e. "frequently occurring and easily recognizable causal patterns" in social reality (Elster 1998: 45). To do so, two causal mechanisms are integrated at the level of the constitutive dimensions of influence (and influence attempts) by specifying the first necessary condition ("interaction"). Since "influence is possible only when communication occurs" (Knocke 1990: 3), the relations between an influence-wielder and the influenced can take essentially two forms of communicative action: they can follow either a *bargaining* or an *arguing* logic of social interaction (Risse 2004). Arguing as a form of communication can be defined as "non-manipulative reason-giving" (Keohane 2001: 10). It is based on speech acts that can be described with verbs like to claim, to ask, to inform and to justify (Holzinger 2004).[5] Arguing typically aims at and involves a reasoned consensus among actors, who change their beliefs or preferences in the direction of what they perceive as the best argument (Kleine/Risse 2005: 9).[6] Bargaining as communicative action, by

---

[4] Obviously, a change in an influenced actor can also be the result of a third actor's influence. Influence of the analysed actor would in this case already be excluded through any of the first four conditions.

[5] It has been remarked that arguing as a mode of communication can also be used strategically for bargaining purposes. Yet, according to Risse (2002: 601), in a real negotiation process, this use of arguing will, once challenged by others, either quickly be unmasked as bargaining in disguise or it will transform into genuine arguing as quest for a reasoned consensus when a true exchange of arguments sets in.

[6] This is a basic, empirically useful definition of arguing. Far-reaching assumptions on the preconditions for arguing to set in between actors are made in Habermas' original theory of communicative action (1981). His normative criteria for consensus-oriented

contrast, depicts negotiations between two or more parties that are characterized by strategic interaction and the exchange of promises, concessions and threats (Holzinger 2004). It is targeted at positions and behavioural changes and typically results in a compromise in which actors' preferences and beliefs remain unaltered (Saretzki 1996). To further distinguish arguing from bargaining, Risse proposes to focus on the outcome of a negotiation process: a consensus based on arguing has been reached when the result is surprising, transcends the level of the lowest common denominator, and actors indicate similar reasons for it (2004: 302). Whenever these conditions do not hold, the result must be regarded as a compromise achieved via bargaining.[7] Each of the two interaction modes corresponds to a specific set of foreign policy tools, as further elaborated on below specifically for the EU case. Both influence and influence attempts can be argumentation- or bargaining-based.

As influence is a truly multi-dimensional concept, several additional clarifications need to be made to fully operationalize it. First, while it may be true that influence *attempts* can be isolated, incidental actions, both the intentions behind these acts and the effects they produce (the *actual influence*) need not be restricted to short-term behavioural changes. Influence attempts can also be exerted with a long-term strategy and have sustainable effects (e.g. when another actor's beliefs are permanently altered). Second, influence can affect agenda-setting or outcomes of political processes. Third, one can distinguish between different objects of influence, which can be one/more actor(s) or structures, understood as the formal and informal frameworks of interaction for actors in a given context (Giddens 1984). Such structures can become the ultimate aims of influence attempts via the intermittent display of "the modification of one or several actors' behaviour, preferences or beliefs". The operationalization of influence and influence attempts is summarized in Table 1, which serves as a typology for the study.

Another important clarification with regard to influence can be made by referring to the notion of "continuous concept" introduced above (Goertz 2006: 34). The *degree* of influence can be established on the

---

activity ("verständigungsorientiertes Handeln") are extremely demanding, including the necessity for actors to accept each other as equals and to share a common lifeworld ("gemeinsame Lebenswelt"). In the context of this work, no such normative assumptions are made. Rather, emphasis is placed on the application of influence acts that can be interpreted as argumentative action in concrete empirical contexts, and on the ex post identification of the conditions under which influence was exerted.

[7] Both bargaining and arguing are ideal-types, which can overlap and mix in social reality (Risse 2002: 601; Ulbert *et al.* 2004). Both are necessary to study reality, especially in a context of regime negotiations about international legal rules, which can hardly be understood by exclusively focussing on the analysis of strategic behaviour (Steffek 2005).

basis of an assessment that sets into relation (i) the significance of the influence-wielder's input *vis-à-vis* the final outcome of a negotiation process and (ii) the importance of the output of this negotiation process. As in the case of determining influence, establishing its degree requires interpretation in context. Several assumptions can nonetheless be made. For one, an actor's influence can be considered as highest when it attains its goals to the largest possible extent (*extent of goal attainment*). Focusing on the outcome, this actor's influence is highest if it attains its aims *and* the agreement has a high level of durability (e.g. through the creation of a durable structure) and/or of legal bindingness (e.g. an international treaty) (*durability of the outcome*). A tentative classification of degrees of influence could thus be to refer to influence as very high (high extent of goal attainment, high degree of durability), high (certain degree of goal attainment, some degree of durability), low (low to medium goal attainment and durability) or inexistent (no goal attainment) (for a more complex formula: Arts/Verschuren 1999: 419–420). Such classification represents a heuristic device to enable cross-case, cross-time comparisons.

Table 1: Toward a Typology of Influence

Dimensions / Perspectives	Objects	Phases in decision-making process	Time horizon
OUTSET-INFLUENCE ATTEMPTS	Actors (behaviour, preferences or beliefs) or structures targeted	Agenda-setting or outcome targeted	Short-term vs. long-term aims targeted
OUTCOME-INFLUENCE (EFFECTS)	Actors (behaviour, preferences or beliefs) or structures affected	Agenda-setting or outcome affected	Short-term vs. long-term effects

Setting the Theoretical Scene: Insights from EU Foreign Policy Analysis and Regime Theory

To specify the context in which the concepts of influence and influence attempts are applied in this study, this section embeds them into (pre-)theoretical considerations on (i) the EU, derived from EU foreign policy analysis, and (ii) on global climate policy, derived essentially from regime theory. Theories serve two key purposes here: first, enabling the transition from description to explanation by aiding in the selection of key explanatory variables of EU influence; second, helping to reduce the scope of the study via the selection of core units of analysis in the chosen case.

Influence and Influence Attempts Seen Through the Lens of EU Foreign Policy Analysis

Two closely linked debates from the literature on EU foreign policy analysis are of interest for the purposes of this study. A look into the debate on whether and under what conditions the EU *can be* a global actor helps to settle the important question of what EU foreign policy is, while yielding insights into the internal preconditions for EU external activity and impact. A second debate on the EU's capacities *as* a global actor helps to pre-specify its tools for exerting influence.

Bearing the definition of EU *foreign policy* as area of politics that aims at influencing its external environment in mind, and since the EU is a composite actor involving supranational institutions and member states, it is equally important to be clear about what *European Union* foreign policy exactly entails. It is here considered as the sum of foreign policies conducted by genuine EU actors (the European Commission, the Council Presidency, the High Representative – HR) or EU member states if they act explicitly on behalf of the EU or in line with its values and interests. If the Union can thus possess a foreign policy in its own right, it does not always appear as the same type of actor to its interlocutors. In some arenas it is represented by the Commission, in other contexts by the Presidency of the Council (or, since the Lisbon Treaty, the HR or the President of the European Council), and in still other fora the representation may change over time or depending on the issue.

Scholars have repeatedly attempted to specify the EU's capacity to act in its own right on the global scene (Carlsnaes 2007: 549). The concept of "actorness", as elaborated by Caporaso/Jupille (1998; drawing on Sjöstedt 1977) can be regarded as the most sophisticated of these attempts (Ginsberg 2001: 45–46). It conceives of the EU's capacity to act globally in terms of four categories (Caporaso/Jupille 1998): recognition (the EU's acceptance by others), authority (the EU's legal competence to act externally), autonomy (the EU's institutional distinctiveness/independence from its members) and cohesion (the degree to which the EU is able to formulate internally consistent policy preferences). Recognition excepted, all categories point to significant internal conditions the EU has to fulfil to be capable of exerting influence. Parting from an interdisciplinary (legal/political science) critique of "actorness", Schunz *et al.* (2012) developed an analytical framework that employs the more comprehensive concept of "actor capacity" to discuss the legal and foreign policy components of the EU's capacity to act externally. For them, EU actor capacity depends essentially on (i) the existence of legal competence to act in

EU primary law,[8] (ii) the de jure and de facto external representation arrangements (who speaks on behalf of the EU?; to what extent does it act independently from the member states?) as well as (iii) efficient internal decision-making and coordination processes[9] among member states and between these and the EU institutions. Moreover, actor capacity relies on the existence of (iv) relevant treaty objectives and common policy goals. Finally, the availability of (v) foreign policy tools, which form the basis for EU influence attempts, is an often neglected, significant component of the Union's capacity to act.

When it comes to these EU foreign policy tools, a small group of authors has attempted to come up with a classification (Smith 2003: 52–68; White 2001: 53–57; Ginsberg 2001: 49–50). The arguably most detailed elaboration stems from Karen Smith, who identifies both economic and diplomatic instruments (2003: 52–68). In the *economic* sphere, she distinguishes between positive (carrots) and negative (sticks) measures (Smith 2003: 60; Ginsberg 2001: 50). The EU can exert influence positively by, inter alia, concluding trade, cooperation or association agreements, reducing tariffs or providing aid. In negative terms, the EU can impose embargos or boycotts, delay or suspend agreements, increase tariffs, reduce aid etc. Smith refers to these latter tools also as "coercion" (2003: 22). In the *diplomatic* sphere, she identifies a range of EU instruments like issuing démarches or declarations, visiting other countries, imposing diplomatic sanctions, offering EU membership etc. (Smith 2003: 61). Smith's catalogue provides a useful starting point for specifying EU influence attempts, and can also be applied to EU foreign climate policy. Table 2 gives an overview of possible EU influence attempts, linking conceptual considerations on influence attempts made earlier to the concrete EU foreign policy literature. The table essentially distinguishes between influencing through the causal mechanisms of arguing (persuasion) and bargaining. While it is highly unlikely that the EU will resort to coercive means in a global environmental policy context, coercion is listed as a third possible causal mechanism. Taking these categories into account, Smith's catalogue needs to be adapted in two respects. First, her discussion of negative economic measures omits that such instruments cannot only be employed coercively, i.e. actually used to the detriment of the EU's interlocutors (thus "inflicting punishment" in Holsti's terms (1995)), but can

---

[8] The EU's participation in a multilateral forum regularly depends, moreover, on an overture created by international law. Via such a provision, the Union can be granted a legal status (e.g. full member, full participant) in UN bodies, endowing it with speaking and voting rights. This aspect of the legal preconditions for EU external actions is addressed in the analysis, where this is relevant.

[9] Decision-making refers to the definition of the EU's negotiation positions, while internal coordination depicts the processes of consultation among EU actors during international negotiations on the basis of such positions.

also have the status of threats ("threatening punishment", Holsti 1995). In the latter case, if sanctions are only invoked as a possibility, this activity falls under the broader category of bargaining. Second, diplomatic tools cannot only be used in a bargaining context, but also as means of persuasion. For data collection and analysis purposes, use is above all made of the second to last column (EU foreign policy instruments) of Table 2. How this can be done in a climate policy context is illustrated in the last column.

The EU's capacity *to act* and its capacities *as* a global actor constitute potentially important preconditions for its exercise of influence in global affairs, which are duly analysed in this study.

Table 2: How the EU Can Exert Influence – EU Foreign Policy Acts

Acts of influencing (Holsti 1995)	Foreign policy acts (expanded from Wilkenfeld *et al.* 1980)	EU foreign policy instruments (adapted from Smith 2003)	Possible examples of EU FP instruments in the field of climate change
CAUSAL MECHANISM: ARGUING			
Persuasion	Yield, comment, consult, approve, propose, demand, agree, request, reject, protest	Issue démarches, declarations, statements Visit Make proposals Initiate political dialogue Send envoys, experts Sponsor conferences Support action	Propose a negotiation text Initiate and sponsor a conference or a media campaign on the science of climate change in a third country Openly support climate change activities in a third country
CAUSAL MECHANISM: BARGAINING			
Offer rewards	Promise	Offer diplomatic recognition Offer membership Offer trade, cooperation or association agreement Reduce tariffs Increase quota Grant inclusion in the general system of preferential treatment Provide aid Extend loans Threaten with embargo (ban on exports) or boycott (ban on imports)	Offer access to technology Offer financial aid/loans Offer partnership

Acts of influencing (Holsti 1995)	Foreign policy acts (expanded from Wilkenfeld et al. 1980)	EU foreign policy instruments (adapted from Smith 2003)	Possible examples of EU FP instruments in the field of climate change
Grant rewards	Grant, reward	Grant diplomatic recognition Grant membership Conclude trade, cooperation or association agreement Reduce tariffs Increase quota Grant inclusion in the GSP Provide aid Extend loans	Grant access to technology Grant financial aid/loans Conclude partnership on climate change
Threaten punishment	Accuse, warn, threaten	Threaten diplomatic sanction Threaten to refuse recognition Threaten with embargo (ban on exports) or boycott (ban on imports) Threaten to… … increase tariffs, … decrease quota … withdraw GSP … reduce or suspend aid … delay conclusion of agreements … suspend or denounce agreements	Threaten tariffs on CO_2-intensive imported goods from countries without ambitious binding climate targets Threaten reduction or suspension of aid
CAUSAL MECHANISM: <u>COERCION</u>			
Inflict punishment	Reduce relationship, deny	Impose diplomatic sanction Deny recognition Increase tariffs Decrease quota Withdraw GSP Reduce or suspend aid Delay conclusion of agreements Suspend or denounce agreements	Impose tariffs on CO_2-intensive imported goods from countries without ambitious binding climate targets Reduce or suspend aid Delay conclusion of agreements

EU Influence and Influence Attempts in the Global Climate Regime Context

From the perspective of the global level of analysis, regime theory can help to single out, first, possible explanatory factors of EU influence and, second, the study's key units of analysis regarding major constitutive features of the global climate regime.

Regime theory has long been applied to the analysis of international cooperation, notably in the field of global environmental politics, in which several regimes – such as the UN climate regime – co-exist (Young 1994). Three main strands of regime theory can be distinguished (see Hasenclever et al. 1996 for an overview): (i) a neo-realist power strand that draws either on hegemonic stability theory or emphasizes the relational power (capacities) of various actors as important explanatory factor for the creation of regimes (Krasner 1991; Keohane 1984); (ii) a constructivist knowledge/ ideas strand that focuses on the role that ideas, identities or communicative action play in the formation and maintenance of regimes (Kratochwil 1989; Müller 1994); and (iii) a – mainstream – neo-liberal institutionalist interest strand, which operates on the assumption that regimes allow states – as rational agents seeking gains – to materialize common interests (Keohane 1989; Young 1989a). While the majority of institutionalist scholars holds that regimes are negotiated by fully rational state actors behaving strategically on the basis of fixed preferences, and that regimes are formed and persist when and as long as they "increase the welfare of their creators" (Keohane 1984: 80), Young (1989b; 1993) has contested this rigid view of states as pure utility maximizers and of bargaining as only distributive. He argues that rationality is always bounded, which makes it necessary for states to transcend pure self-interest and cooperate with the aim of expanding the overall collective benefits of all negotiating parties (Young 1994: 126–127). His approach has therefore been interpreted as a basis for building bridges to constructivist approaches (Risse 2002: 614).

While some of the potential determinants of EU influence – related to its own actor capacity – have been identified above, theories of regime formation and change can be crucial sources of additional explanatory factors for its influence on a global regime. By definition, EU influence on the climate regime will involve modifications in this regime, which is why insights about regime change can inform the design of the analytical framework for this study.

The tendency in much of the literature has been to rely on singular explanations: either power or ideas or interests have been regarded as the main driving factors behind regimes and their change, where all three may actually play a part (Young 1989b: 353; 1993: 435). Young suggests that regime transformation can result from any or a combination of changes in the political, economic or social contexts, in the inner dynamics of a regime, or in the issue(s) a regime deals with (1989a: 95–96). The study therefore relies on basic premises of his institutionalist approach with regard to bounded rationality, but enhances these further with insights of the constructivist strand, notably concerning communicative action in processes of regime formation and change. Further, the study also shares assumptions with Krasner's view that power, understood in material and non-material terms, determines who is involved in inter-state cooperation and by which rules this cooperation functions (Krasner 1991: 340). Purely rational choice and hegemonic stability accounts of regime creation/change are not retained.

Concretely, several potential explanatory factors of regime change and, by extension, of an actor's influence in an evolving regime context are identified. *Exogeneous* factors can be very diverse and related to, for instance, alterations in the socio-economic (e.g. a sudden global financial crisis) or political (e.g. an armed conflict in an important strategic theatre) environment that create a demand for or put pressure on a specific regime. Changes in the issue can come in the form of new scientific knowledge about the natural environment that, for instance, increases the urgency with which an environmental problem has to be treated through international cooperation. Factors related to the *inner dynamics of a regime* concern above all transformed patterns of interaction between actors. Following Krasner (1991), power asymmetries can play a role, as changes in the capacities of actors may also alter the way they interact. Further, domestically motivated preference changes that lead to altered strategic behaviour of actors can equally be of importance (Keohane 1989). Finally, changing perceptions and beliefs can determine any actor's activities and change the interaction between actors within a regime (Goldstein/Keohane 1993). By consequence, all these modifications in the internal dynamics of a regime can lead to changes in its set-up.

Bringing the EU-related and regime-related potential explanatory factors together, the causes of EU influence in the case of the global climate regime must thus above all be searched in (i) the context, (ii) the issue(s) under negotiation, and (iii) the internal dynamics of this regime (the ideas, interests, capacities (power) of the – state – actors and how they interact in the regime, including the EU and its actor capacity and activities). In other words, when assessing the EU's influence on the outcome of a regime process, all these items come into consideration as possible enabling or restraining factors of this influence.

Selecting possible explanatory factors of EU influence in a global regime context is one thing, reducing the scope of the analysis of the specific instance of EU foreign policy in the global climate regime another. In this respect, Krasner's (1983: 2) often employed "consensus definition" of a regime can be of use (Hasenclever *et al.* 1996: 179):

> implicit or explicit principles, norms, rules, and decision-making procedures around which actors' expectations converge in a given area (…). Principles are beliefs of fact, causation, and rectitude. Norms are standards of behavior defined in terms of rights and obligations. Rules are specific prescriptions or proscriptions for action. Decision-making procedures are prevailing practices for making and implementing collective action.

This broad definition is helpful "as a guide for empirical studies" because it identifies various constitutive dimensions of a regime (Hasenclever *et al.* 1996: 180). Following Aggarwal's suggestion for a refined terminology, which distinguishes between the rules and decision-making procedures on the one hand and "the principles and norms underlying the development of a regime [which] can be termed a 'meta-regime'" on the other hand (1985: 18), one can argue that the principles and norms constitute the core of any regime and stand hierarchically above rules and decision-making procedures on implementation. For an analysis of the ever more complex negotiations within the global climate regime, this implies that emphasis can be placed not on the many issues concerning the operationalization of the regime (the rules in Krasner's definition), but on the essential political discussions on issues related to its key principles and norms. Decisions on these elements will determine the course that regime development will take. Climate politics can thus essentially be regarded as a struggle over the content of what – in line with Aggarwal's definition – can be referred to as the climate meta-regime. This climate meta-regime consists, for the time being, of several key principles and a number of norms, i.e. rights and obligations for certain groups of parties. The most significant *principles* are the principle of common but differentiated responsibilities, the precautionary principle and the principle of equity, formally enshrined in Article 3 of the UNFCCC (Betsill 2005; Oberthür/Ott 1999). The equity principle implies that all actors in the regime should be treated equally. It is subject to different interpretations depending on whether it is applied to mitigation, adaptation, decision-making or participation in the regime (Metz 2000). The precautionary principle says, according to the Convention itself, that "where there are threats of serious or irreversible damage, lack of full scientific certainty should not be used as a reason for postponing such measures" (Art. 3.3 UNFCCC). Finally, the principle of common but differentiated responsibilities comprises the notion of a *common* responsibility of all states for the protection of the environment, but takes account of *differences* in both past and present contributions to environmental degradation and in the capacities to combat environmental problems (Rajamani 2000; Harris 2000: 226–228). The *core norm* is at the same time the aim of the entire regime: "The ultimate objective of this Convention and any related legal instruments (…) is to achieve (…) stabilization of greenhouse gas concentrations in the atmosphere at a level that would prevent dangerous anthropogenic interference with the climate system" (Art. 2 UNFCCC). It can be considered as a norm – a "standard of behavior" (Krasner 1983: 2) – because it obliges, if not in a legally binding then at least in a moral sense, those who ratify the Convention to work towards "achieving" a stabilization of GHG concentrations in the atmosphere. This requires a specific form of behaviour aimed at reducing emissions.

The exact denotation of both the principles and the rights/obligations is subject to regular (re-)negotiation. Different interpretations of the principles compete, and the determination of the "level that would prevent dangerous anthropogenic interference with the climate system" as well as the necessary actions to achieve this are also constantly contested. Collective decision-making is therefore needed to determine what they mean in a particular context. If the constitutive dimensions of the climate meta-regime represent thus the essence of global climate politics, this implies that impact on the global climate regime by and large passes by influence on the meta-regime. Following this logic, two embedded units of analysis for the case study can be selected:

1. The *CBDR principle*: it can be seen as a true "belief of rectitude" in the Krasnerian sense, as it incorporates a vision of the historical repartition of responsibility for environmental degradation in the world. The core question that it raises is a deeply political one, namely "who will do what?" in the regime. The principle has to be repeatedly filled with concrete meaning and becomes thus a key object of influence attempts.

2. The *core norm* (in the sense of obligation) of the regime embodied in the stabilisation objective (Art. 2 UNFCCC): filling this norm with meaning necessitates a definition of the emissions limit needed to avoid "dangerous anthropogenic interference" with the climate. This central choice of global climate policy is, in turn, deeply linked to a political decision on the nature (binding?/voluntary?) and calculation (per country or per capita emissions?/base year? etc.) of emissions reductions. While the targets and timetables for individual parties to the Convention and its Protocol could be interpreted as rules of a regime (Downie 2005: 66–67), the broader choice on whether obligations are binding or not (and for whom) and the concrete question of the aggregate amplitude they take, based on the individual repartition of the burden among parties, are more than what Krasner calls "pre-/proscriptions". They have to be considered as the fundamental norms of the climate meta-regime.

Given the centrality and highly political nature of the two items, influence on decisions on the inclusiveness and on the scope of the regime will enable actors to obtain leverage over other parties' emissions. Such influence is thus crucial for any actor that desires to impact climate regime development. It is therefore assumed that the results of a study of the EU's influence concerning these two issues allows for statements about its influence on the entire regime.[10]

Methodological Bases: Analysing and Determining Influence

On the basis of these conceptual-theoretical parameters of the study, the final step in the composition of the analytical framework consists in outlining the contours of the methodology applied to analyse and determine EU influence in a global regime context. "Measuring" influence essentially requires determining the causal relationship between an influence-wielder A's activities and an observed change in the behaviour, beliefs or preferences of actors B, C, D... in complex political contexts. Three "classical approaches" to influence analysis have regularly been identified: positional, reputation and process analysis (Betsill/Correll 2008; Arts/Verschuren 1999: 414). The arguably most developed tool for the assessment of political influence in complex global negotiation contexts combines process and reputation-based methods by effectively triangulating three perspectives (Arts/Verschuren 1999: 416–419): (i) the **E**go-perspective, i.e. the self-perception of the influence-wielder E about its impact; (ii) the **A**lter-perspective, which assesses the view that other players have of E's performance; and (iii) the **R**esearcher's analysis, which allows through the study of E's goal achievement for correcting potential misperceptions ("EAR" method).

The approach to influence analysis employed in this work reverses the logic of the EAR instrument by holding the researcher's process analysis central and by using the perceptions of the object being studied (EU) and of others (non-EU negotiators, observers) to corroborate findings.[11] Process-tracing generally involves the attempt "to identify the intervening causal process – the causal chain and causal mechanism – between an independent variable (or variables) and the outcome of the dependent variable" (Bennett/George 2005: 206–207). In this work, process analysis

---

[10] It should be noted that the two issues are not clear-cut, but, in practice, often linked to others, such as questions on the precise modalities of emissions reduction and, especially, finance and technology transfer, which may become central to reaching agreement on the core pillars of the regime. While focussing on the two issues helps to limit the scope of the analysis, other issues are, if necessary, touched upon.

[11] See Schunz (2010: chapt. 4) for a detailed explanation of the method.

can be considered as narration in search for patterns (Gysen *et al.* 2006). Narratives not only provide a concise account of social events as they unfold, but are also deeply causal in nature because "any explanation resides in its accounting for temporality and sequence" (Somers 1998: 771). Employing such a narrative approach to process analysis allows for a contextualized analysis of a foreign policy actor's influence attempts and their effects over time by identifying whether the necessary conditions for ascertaining influence are fulfilled. The method yields plausibility rather than probability statements about influence (Huberts 1994: 39; Gysen *et al.* 2006: 108; Arts/Verschuren 1999: 422). These statements are solidified through a reputation-based variant of influence analysis involving perceptions of EU foreign policy-makers and of non-EU climate negotiators and observers.

To allow also for data collection triangulation, a combination of three techniques is employed in this study: document analysis, semi-structured interviews and indirect or direct observation. First, the study relies on 32 qualitative interviews with EU, EU member states, third states and civil society representatives.[12] Second, it draws heavily on document analysis of primary sources including UN,[13] EU, EU member state and third country official and unofficial negotiation documents (position papers, negotiation syntheses, press releases etc.). Further sources, mainly used to interpret these documents, are news media, NGO, research institute and think tank coverages of the climate negotiations. A major tool for keeping track of these negotiations are the Earth Negotiation Bulletins (ENB), which provide in-depth accounts of the UN climate talks.[14] Finally, the study benefits from indirect (via webcast) and direct, non-participatory observation of UN climate talks.[15]

In practical terms, the process analysis follows a *five-step approach* that is applied to each of the time periods analysed in Chapters 2 (1991–1995), 3 (1995–1997), 4 (1998–2007), 5 (2007–2009) and 6 (2010–2012), with a special emphasis on the periods leading to the adoption of the Kyoto Protocol (1997) and of Copenhagen Accord (2009).[16]

---

[12] See Annex I for an overview. To guarantee anonymity, interviewees' names remain confidential. Moreover, each interview was attributed a random number so as to render any identifiable link between a specific interview(ee) and a given observation impossible. The transcriptions of the interviews are on file with the author.

[13] Archival research was carried out at the UNFCCC secretariat in Bonn between 9 and 11 February 2009.

[14] See http://www.iisd.ca/process/climate_atm.htm.

[15] The author attended, as party delegate, the UN negotiation sessions in Bonn (8–12 June 2009) and Barcelona (2–6 Nov. 2009) as well as COP 15/MOP 5 in Copenhagen (8–19 Dec. 2009).

[16] Chapter 2 (Historical Foundations) relies to a larger extent on secondary literature.

Step 1 covers an analysis of the global context involving a brief screening for (i) major scientific advances on climate change and (ii) important events in global climate politics outside the UN arena and other, per se unrelated major events that may impact on climate policies.

Step 2 serves to identify the EU's and other key actors' negotiation positions and strategies regarding the two issues of emissions reduction targets and responsibilities prior to the analysed time period. Instead of analysing how the EU attempts to influence all actors within the regime, the study focuses on key players from the three main coalitions that have – besides the EU – been identified as significant throughout the history of the climate regime (Betsill 2005: 108; Yamin/Depledge 2004: 30–59): (i) from JUS(S)CAN(N)Z (Japan, United States, Canada, Australia, New Zealand, with Switzerland and Norway joining during the mid-1990s, and Iceland), which became the "Umbrella Group" later, it focuses on the two major emitters *US* and *Japan*; (ii) from the more heterogeneous G-77/China bloc, the analysis focuses on *China* and *India* and two significant sub-groupings: the *Organization of Petroleum Exporting Countries (OPEC)* and the *Alliance of (42) Small Island States (AOSIS)*; (iii) *Russia* and other countries in economic transition are also considered.[17] The enquiry starts out with examining EU activities *vis-à-vis* these key actors and countries,[18] taking into account fluctuations within and across coalitions.

Step 3 provides for the crucial narrative of the global regime negotiations, with a focus on the EU's influence attempts. Emphasis is placed on tracing the negotiations conducted in the framework of COPs/MOPs and of preparatory sessions such as meetings of the Ad hoc Working Group on Long-term Cooperative Action under the Convention (AWG-LCA). The analysis transcends the UN framework to look into other global climate fora, where relevant. Following a discussion of the outcome of a given negotiation period, patterns are extracted to allow for a classification of EU influence attempts and for answering the first research question of *how* the EU tried to influence the climate regime.

With *step 4*, the study determines *whether* the EU actually exerted influence on the two issues under examination, thus answering the second research question. This necessitates a closer look at both the outcome and the story of the climate talks for the purpose of identifying turning

---

[17] Russia and the Ukraine over time joined the Umbrella Group. At the same time, Switzerland has left this negotiation bloc and formed a coalition with Mexico, South Korea and Liechtenstein, referred to as the Environmental Integrity Group.

[18] To further strengthen the choice of these actors, it has to be noted that they have all figured among the six greatest emitters of greenhouse gases in absolute terms in recent years. For 2007, China led this ranking with 24% of global GHG emissions, followed by the US (22%), the EU (12%), India (8%), Russia (6%) and Japan (PBL 2008).

points, i.e. points in time at which a relevant number of actors collectively decides to move from one negotiation phase to the next or to take a final decision on an item (Chasek 2001: 44–49; 150). A turning point can be observed when (i) several actors change their behaviour, converging at least to some extent into one direction, and/or (ii) proposals are eliminated from a negotiation process so that only a small number of options is maintained. Typically, such turning points can be observed at transitions between phases during a negotiation process, e.g. when actors move from the initial positioning phase to negotiations on precursor texts (Depledge 2005; Chasek 2001).

Table 3: Establishing EU Influence – Constitutive Dimensions

	Constitutive dimensions of the concept of influence	TURNING POINT X: condition fulfilled?
Interaction	Can an interaction between the EU and the other parties on this item be observed?	YES or NO
Purposive Behaviour	Did the EU want to change other actors' behaviour, preferences or beliefs on this item?	YES or NO
Temporal sequence	Did the EU approach the other parties first?	YES or NO
Goal attainment	Did other parties change their behaviour, preferences or beliefs in the direction of the EU? Does the outcome of this item under negotiation reflect (at least in part) EU aims?	YES or NO
Absence of auto-causation	Can the change in other actors' behaviour, preferences or beliefs be attributed (at least in part) to the EU, i.e. was it not the result of auto-causation in the other actors or of another factor that has to be considered more important than the Union's intervention?	YES or NO

Note: The dichotomy suggested by the YES/NO assessment per criterion is not so clear-cut, but requires interpretation. Its credibility relies on the empirical evidence provided.

At these points, in-depth narrative and conditional causal analyses become possible, allowing for establishing the share of EU influence. For one, it becomes possible to eliminate actors that can logically not have been influential: if their position was not in line with the outcome at a turning point t_1 and only changed afterwards, they cannot have been influential; if their position was, however, completely or partially in concordance with the decision taken at a turning point t_1, they may have been (at least partially) influential (Huberts 1994: 41–43; 57–59). If the EU remains among those who were potentially influential, it can be determined whether it actually exerted influence or not by checking whether the constitutive dimensions (necessary conditions) of influence are fulfilled (see

Table 3 for a visualization of this analysis). Concretely, this can be done by answering the following questions: has there been an interaction with other(s) in which the EU approached these other(s) first (*interaction, temporal sequence*)? With its (inter)actions, did the EU want to alter other actors' behaviour, preferences of beliefs on the analysed subject matter (*purposive behaviour*)? Has the EU *attained* its *goals* (at least partially and more than it would have without its actions) (Huberts 1994), i.e. have others changed behaviour, preferences or beliefs in the direction of the Union and/or does the overall outcome reflect EU aims? Can the change be attributed to the EU, i.e. was it *not the result of auto-causation* in the other actors or of another factor that has to be considered more important than the Union's intervention (counterfactual analysis)?

Besides establishing influence, the causal narrative also provides input into *step 5*, which serves to answer the third research question by plausibly *explaining* the EU's influence. To strengthen and extend the explanations already inherent in the narrative, two pathways are taken: pattern-matching and explanation-building (Yin 2003). Pattern-matching relies on the two sets of assumptions made in this chapter. On the one hand, two causal mechanisms may be at play in an instance of EU influence-wielding (bargaining or arguing) and the process-trace will show whether the EU's influence attempts and influence are based on the one or the other. On the other hand, explanatory factors that might account for the EU's influence on the climate regime (context, issue, regime dynamics, EU actor capacity and activities) can be taken as a basis for identifying the scope conditions under which these causal mechanisms are triggered (and EU influence becomes thus possible). Further explanatory factors may emerge from the empirical analysis and can serve to compose a denser account. By bringing these into the picture, the analysis transitions from the moderately deductive logic of pattern-matching to the inductive logic of explanation-building. While this type of explanation will round off the analysis of each time period, Chapter 7 composes a more general account on the basis of a cross-time comparison enabled by the longitudinal character of the study.

CHAPTER 2

Historical Foundations (1980s–1995)

EU Influence on the Set-up of the Global Climate Regime

This chapter provides an overview of the early evolution of the climate regime. After a brief historical excursus, covering the period until late 1990, it analyses EU influence during the talks that ended with the adoption of the UNFCCC at the Earth Summit in Rio in June 1992 (1991–1992). Finally, it discusses a third phase, during which further meetings of the negotiation body that had drafted the Convention prepared for the first COP (1992–1995) (for an overview of all UN climate negotiation sessions during this period and thereafter, see Annex II).

The Pre-negotiation Phase: From Scientific Circles to First Political Negotiations

A brief discussion of the most relevant milestones during the initial international debates on climate change demonstrates, first, that agenda-setting on climate change "had functioned as a learning process with scientists as 'teachers' and policy-makers as 'pupils'" (Sjöstedt 1999: 237). Second, many early discussions forcefully shape the debates until the present day (for exhaustive discussions of the historical evolution of the regime: Bodansky 1993, 1994, 2001; Pallemaerts 2004). Finally, the story of the early years shows that the EU on the whole, but especially some of its more active member states, were at the forefront of activities on climate change, with varying degrees of success.

Focusing on the scientific discussions first, although initial hypotheses on the link between atmospherical CO_2 and the warming of the planet had already been advanced at the end of the 19th century, the study of climate abnormalities was only accelerated from the middle of the 20th century on, primarily in the US (Pallemaerts 2004). After systematic measurements of CO_2 concentrations had been undertaken in the 1950s, US President Johnson was informed as early as 1965 by his Science Advisory Committee that the combustion of fossil fuels would "modify the heat balance of the atmosphere to such an extent that marked changes in climate, not controllable through local or national efforts, could occur" (White House 1965: 9; Bodansky 1993).

During the **1970s**, several steps were taken to further improve the scientific knowledge about long-term climatic alterations. The 1972 United Nations Conference on the Human Environment (UNCHE) in Stockholm called for the installation of centres for measuring air pollution (Pallemaerts 2004: 6). In 1979, the US National Research Council concluded from first computer models that the further increase of CO_2 emissions would result in climate change, and that there was "no reason to believe that these changes will be negligible" (cited in Bodansky 2001: 24). Organised by the World Meteorological Organisation in Geneva in that same year, the first World Climate Conference began to globalise knowledge of the issue by initiating a global climate research programme (Gupta 1998: 181). Started in 1980, the programme was coordinated by the WMO and the International Council for Science, later joined by the UN Environmental Programme.

In the **early to mid-1980s**, then, several major scientific conferences set the agenda for global political debates about how to tackle climate change (Bodansky 2001: 26–27). In 1985, a major international scientific conference in Villach, Austria, involving the WMO, UNEP and the ICSU, reunited experts from 29 – mostly European – countries, providing an overview of the state of the art of climate science and proposing the conclusion of a global treaty to counter climatic changes (Pallemaerts 2004: 7). This and similar events led to a gradual diffusion of knowledge from the scientific community to political arenas (Bodansky 1993). This trend was most visible in the US, where climatologists like James Hansen testified before Congress committees in 1988 (Bodansky 1993). At the same time, first systematic efforts to politically assess the findings of climate science were also undertaken in Europe, e.g. through an Enquête Commission of the German Parliament (Bodansky 2001: 27). Adding to a heightened political and public interest in the topic was the growing concern about the consequences of man-made damages to the ozone layer (Bodansky 2001: 27).

Three major political events would ultimately transform climate change into a political issue in this "watershed" **year 1988** (Bodansky 2001: 27). In early June, governments requested the WMO and UNEP to jointly set up the Intergovernmental Panel on Climate Change (IPCC), thus taking a major step in creating a body that would systematize and centralize research on climate change under intergovernmental oversight (Bodansky 1993: 464, 2001: 28). Second, in the wake of the Montreal Protocol and the Brundtland report, both adopted in 1987, a "Conference on the Changing Atmosphere" was organised in Toronto in June 1988 (Bodansky 1993: 461). Assessing ozone and climate change policy options alike, the over 400 participants from 48 (Western) countries issued a statement in which they called for global reductions of atmospheric concentrations of CO_2 of 20% by 2005 and the drafting of a comprehensive global convention to protect the atmosphere (Toronto Conference 1988: para. 22 and 30). Further, the Toronto Conference Statement made – for the first time in the context of atmospheric politics – reference to the "main responsibility" principle, which suggested that the "countries of the industrialized world are the main source of greenhouse gases and therefore bear the main responsibility to the world community" for acting against climatic changes (Toronto Conference 1988: para. 13; Bodansky 1993: 462). Third, climate change made its debut on the agenda of the UN General Assembly. In the autumn of 1988, a debate was held on Malta's initiative of treating climate as "common heritage of mankind". It resulted in a resolution referring to climate change as "common concern of mankind" necessitating a globally concerted response (UNGA 1989a: para. 1; Pallemaerts 2004: 9).

1989 saw an acceleration of the politicization trend of the previous years. In addition to discussions of the issue in the newly created IPCC and a call by the Paris G-7 summit for a framework convention on climate change (Bodansky 1993: 466), two significant conferences on the topic were held in the Netherlands, one of the most active European countries. In March 1989, a conference on the protection of the atmosphere in The Hague, jointly sponsored by the Dutch Prime Minister Lubbers, his Norwegian colleague Brundtland and French President Mitterrand, was attended by high-level representatives of 24 invited industrialized and developing countries (Pallemaerts 2004: 10). In their brief final declaration, the participants called for the creation of a "new institutional authority" designed to protect the atmosphere (Hague Conference 1989: principle a). More importantly, a second high-level conference was organised in Noordwijk in November 1989. Delegates from 66 countries and EC Commission President Delors met to discuss, for the first time, exclusively the issue of climate change in the framework of an intergovernmental forum (Bodansky 1993: 467; Pallemaerts 2004: 12). The final declaration included some important concepts: firstly, it defined an objective, namely that emissions should be reduced to "a level consistent with the natural capacity of the planet", which should be reached "within a time frame sufficient to allow ecosystems to adapt naturally to climate change" (Noordwijk Conference 1989: para. 8). Based on the idea that climate change was a "common concern of mankind", it also called for the adoption of national action plans by all countries, yet "according to their capabilities and the means at their disposal", which also meant that industrialized countries should financially help developing countries (Noordwijk Conference 1989: para. 7, 13). Observers of the process have interpreted the latter paragraphs as a clear sign that the gap between industrialized and developing countries was beginning to open in global discussions on the issue of climate change. This was combined with a growing politicization of the positions of the principal actors, with the developing world – due to numerical superiority – increasingly capable of affirming its position (Bodansky 1993: 467, 2001: 28; Pallemaerts 2004: 13). Concluding the year 1989, the UNGA adopted resolution A/44/862 on "The protection of global climate for present and future generations of mankind", which reiterated what conferences of the previous years had stated, i.e. the need for designing a framework convention (UNGA 1989b: para. 12). Further, it clarified that parties considered the UN to be the "appropriate forum" for negotiations on such a convention (UNGA 1989b: para. 5). With this, the decisive step in transforming climate change from a scientific concern into a topic framed as needing intergovernmental cooperation was seemingly taken (Pallemaerts 2004: 14).

1990 was marked by two major advances in terms of scientific knowledge about the climate regime. At its very end, the UN ultimately took the issue from the hands of scientists (Chasek 2001: 124–125). In August, the IPCC presented its first report including some alarming findings about unprecedented temperature rise on the assumption that a business-as-usual scenario of continued fossil fuel combustion was followed by a majority of countries (IPCC 1990). These findings were discussed at the Second World Climate Conference, held in Geneva in November 1990. Where its predecessor had been a purely scientific reunion, this conference comprised a high-level political segment, attended by representatives of 137 countries and the EU. Several parties, including the EU, Australia, Canada, the Scandinavian countries, Japan and Switzerland offered to stabilize CO_2 emissions at 1990 levels by 2000 (Pallemaerts 2004: 16). Although these expressions of intent were welcomed, the final Ministerial Declaration included a more general formula, urging industrialized countries simply "to establish targets and/ or feasible national programmes" (SWCC 1990: para. 12). Further, the Declaration re-iterated the differentiation between developed countries – who "must show the way" – and the developing world that was to act according to its capacities (SWCC 1990: para. 5). The whole build-up on the topic of climate change culminated in a December 1990 UNGA resolution, which launched a "unique" intergovernmental negotiation process on the adoption of a framework convention (UNGA 1990: para. 1). The Intergovernmental Negotiating Committee (INC) created to this effect was placed under the authority of the UN Secretary General.[1] It was to deliver its final result by June 1992, in time for the UN Earth Summit in Rio (UNGA 1990: para. 7).

---

[1] The set-up of the INC under UN auspices has been interpreted as a success of the developing world. Many industrialized countries had preferred negotiating in technical bodies such as the WMO or UNEP (Bodansky 1993: 473–474).

Historical Foundations (1980s-1995)

Turning to the **EU's involvement** in these early talks on climate change, it is first important to note that when the topic appeared on the agendas of policy-makers in 1988, the EU(-12)'s actor capacity was fairly limited – and this despite the fact that the "Single European Act had inserted environmental protection policy into the EEC, and from the start provided for external action" (Eeckhout 2011: 141).[2] Following a 1986 European Parliament resolution and report on the topic, the Commission issued a first climate-related communication in 1988, which arguably marked "the commencement of formal climate policy making" in the EU (Jordan/Rayner 2010: 53–55). In December of that same year, conclusions by the European Council then stated explicitly that "the Community and the Member States are determined to play a leading role in the action needed to protect the world's environment", especially regarding the "greenhouse effect" (European Council 1988: Annex I). In 1989, an Environment Council resolution confirmed this desire and demanded that the EU be effectively implicated in the international negotiations on climate change (Council 1989: point 2). A March 1990 Environment Council called for a common position on climate change to underpin this ambition (Lescher 2000: 49). Member states diverged on the core of this position, namely the adoption of an emissions reduction target. Countries like Germany and the Netherlands had already adopted quantified targets unilaterally and advocated a common European target, while the UK, despite its national target of stabilization at 1990 levels by 2005, was against taking a decision on this issue within the EC before the start of international negotiations (Jordan/Rayner 2010: 56; Lescher 2000: 50). An agreement was finally reached under the impulsion of the Environment Commissioner and the 1990 Dublin European Council conclusion. Under the heading "the environmental imperative", the heads of state and government called on the EU to accept "a wider responsibility (…) to play a leading role in promoting concerted and effective action at global level" (European Council 1990: Annex II). This broad call for greater responsibility was translated into a specific commitment when the Environment and Energy Council adopted the position that the EU should stabilize its emissions at 1990 levels by 2000, following a recommendation by the European Commission (Pallemaerts 2004: 42; Brambilla 2004: 247). Derogations were granted to the cohesion countries (Spain, Portugal, Greece, Ireland), but also to the UK (Lescher 2000: 50). While the EU had thus managed to forge a basic common position, its representation in international fora was, in the absence of clear rules, mostly guaranteed

---

[2] Article 130r, para. 5 SEA read: "Within their respective spheres of competence, the Community and the Member States shall co-operate with third countries and with the relevant international organizations (…) without prejudice to Member States' competence to negotiate in international bodies and to conclude international agreements."

through the activities of a limited number of member states, but also of the Commission, which was now regularly invited to international meetings and conferences.

The EU's Influence on the UN Framework Convention on Climate Change (1991–1992)

The Intergovernmental Negotiating Committee met five times between February 1991 and June 1992 to conclude a deal on the UNFCCC. The five-step analysis focuses on the negotiation process regarding the two key issues, major turning points in this process as well as its outcome, highlighting EU influence attempts and their effects.

The Context: Major Developments in Global Politics and Climate Science

The negotiations were kicked off in a unique context. Scientifically, the 1990 First IPCC report highlighted that the existence of the greenhouse effect and its anthropogenic causes were now undisputed, and that unprecedented global temperature rise by about 0.3°C per decade could be expected for the 21st century, while pointing also to existing uncertainties due to an incomplete understanding of sources, sinks, clouds and oceans (IPCC 1990). Politically, the fall of the Iron Curtain had altered the global balance of power and made cooperation within the UN on this truly global issue appear more likely. Beyond the UN arena, the 1989 Paris G-7 meeting had already advocated an "umbrella convention on climate change" (cited in Bodansky 1993: 466), an aim that was reiterated a year later in Houston (Sebenius 1991: 111). Finally, a momentum for action on climate change had built over several years, with public pressure mounting especially in industrialized countries due to a growth in extreme weather events (Bodansky 2001: 27).

Key Actors in the Global Climate Regime and their Positions

The negotiations would essentially oppose two major negotiating blocs:[3] the developing world, gathered under the G-77 and China umbrella, and the members of the Organisation for Economic Cooperation and Development (OECD). A brief overview of the main positions on the two studied issues (reduction targets and CBDR principle) is followed by a discussion of the EU's actor capacity.

---

[3] Another group was composed of the ex-Warsaw Pact states of Eastern Europe and the Commonwealth of Independent States, including Russia. This group arguably did not play a big role in these talks: while Russia often sided with the US, the other countries tended to support EU positions (Paterson/Grubb 1992: 304).

By 1991, with the exception of the US and Turkey, all *OECD countries* had not only formulated an emissions stabilization target, but also clarified their more general approach to these negotiations (Gupta 1998: 182; Sebenius 1991: 111). Generally, the industrialized countries accepted to lead the way in the talks and on possible commitments (Bodansky 1993: 478). Canada, Australia, and New Zealand, but also most non-EU European countries were in support of the approach chosen for the ozone regime, i.e. the adoption of quantified emission reduction targets and timetables. They had mostly declared that they would be prepared to stabilize their emissions at 1988 or 1990 levels by the year 2000 (for an overview: Paterson/Grubb 1992: 301; Dasgupta 1994: 134–135). A smaller group of countries, including the US and the Soviet Union/Russia, opposed such a rigid approach. Sceptical of international regulation, the US argued against the precautionary principle and for further research and national approaches to climate change (Bodansky 2001: 28–29; Paterson/Grubb 1992: 302, 304). Finally, Japan opted for an ambiguous "best efforts" approach (Paterson/Grubb 1992: 303).

At the same time, *developing countries* were collectively arguing for the differentiation of commitments in line with the "main responsibility" principle evoked in earlier international climate conferences, and for technological and financial aid by the industrialized countries to help the poor in their fight against climate change (Bodansky 2001: 30, 1993: 479–480; Paterson/Grubb 1992: 300). Generally, countries of the G-77/China bloc were thus not willing to consider emissions reduction efforts for themselves. Big differences existed, however, regarding the concerns about climate change: while the group of small island states (AOSIS) called for immediate action, OPEC countries questioned the science and the need for reducing emissions altogether (Paterson/Grubb 1992: 299–300; Bodansky 1993: 480–481).

When examining the *European Union*'s actor capacity and position before the onset of the negotiations leading to the UNFCCC, a first observation is that it had addressed the issue of climate change fairly swiftly, and had, despite the internal divergences described above, come to a common position as early as 1990 (Wettestad 2000: 28; Skjaerseth 1994: 27). Its general position was characterized by the desire for a legally binding, comprehensive international agreement which would enshrine the obligation of stabilizing emissions at 1990 levels by 2000. On the question of responsibilities, the EU acknowledged the responsibility of industrialized countries, without specifying what, if anything, it expected from developing countries.

While refining its position for the international negotiations, the EU also began sketching out the contours of an internal climate policy (Jordan/Rayner 2010: 57–59). In 1991, the Commission made a proposal for a climate package relying on four core pillars: measures to improve energy efficiency and promote the development of renewable energy sources (programmes SAVE and ALTERNER), a combined carbon and energy tax, and the set-up of a monitoring programme for greenhouse gas emissions (Wettestad 2000). It took until 1993 to finalize talks on this package: while the tax never materialized, due to opposition notably by the UK, the other proposals were gradually adopted (Jordan/Rayner 2010; Haigh 1996). The lengthy talks on these measures also implied that the Union had to try to defend, during the global negotiations, an external position not underpinned by common internal policies.

Shortly before the first INC session, on 4 February 1991, the Council decided unanimously to give the Commission the mandate to negotiate a Convention on climate change (Brambilla 2004: 165). As the European Community did not possess a legal status in the INC, it would, however, de facto be represented through the Council Presidency,[4] which also made efforts to coordinate activities among all member states, arguably achieving a certain degree of uniform representation of the EU's positions (Brambilla 2004: 165). Commission representatives did however reportedly also actively participate in the INC process, and – according to some analysts – equally played a significant role in promoting the unity and consistent representation over the course of the various presidencies (Jachtenfuchs 1996: 114–116). Despite these coordination efforts, each EU member also had its own representation (see Barrett 1991: 187, footnote 18). In the absence of a clear common negotiation strategy, the EU's joint action would therefore often be of an ad hoc nature. It was only at the Rio summit itself that the EC was then granted the status of "full participant" through a UNGA decision, which gave it the right to attend and speak, but no rights to vote or be elected (Schumer 1996). For the summit, the Council mandated the Commission to represent the EC's interests in all areas of exclusive competences, while it kept issues of shared competence to itself (Brambilla 2004: 165–167).

The Negotiation Process and the EU's Influence Attempts

The **first two sessions of the INC** (4–14 February 1991, Chantilly, Virginia; 19–28 June 1991, Geneva) were mostly dedicated to procedural matters, such as the appointment of Jean Ripert (France) as INC Chair, and the creation of two working groups (WG). WG 1 was to deal with commitments both in terms of concrete emissions reductions and in terms of finance and technology transfer, whereas WG 2 had the task of preparing the legal and institutional mechanisms that would form the backbone of the future

---

[4] In 1991, Luxembourg and the Netherlands held the Council Presidency, followed by Portugal in the first half of 1992.

convention (INC 1991a: 23; Paterson 1996: 51–56). Further, parties began to state initial positions. Draft proposals were circulated by the US, Australia, Germany and the UK (Bodansky 1993: footnotes 205 and 206). The US proposal, an "inaction plan" according to critics, underscored its opposition to quantified targets and timetables, and proposed instead a range of national policies covering all GHGs (Paterson 1996: 54). Most other proposals included clearly defined targets and timetables. On the issue of differentiation, while China and India insisted, on behalf of the G-77/China, on industrialized countries' responsibilities, the US called for contributions from developing countries in accordance with the means at their disposal and their capabilities (Dasgupta 1994: 133). By contrast, other OECD countries and the EU were more willing to accept the "main responsibility" principle.

Substantial talks began only in the **second week of INC 2**. WG 1, which – due to its mandate on commitments – is of most interest here, discussed inter alia a Japanese proposal on "pledge and review", a concept that foresaw voluntary national mitigation actions to be reviewed internationally and received support from the UK and France (Bodansky 1993: 486). Further, a compromise formula on a phased "comprehensive approach" was proposed by the UK and the US. Bridging the gap between those favouring a hard target and the US "no target" position, it introduced the ideas of using credits for GHG cuts (beyond solely carbon dioxide) and sinks, i.e. processes or activities which remove GHGs from the atmosphere such as forest or land management (Brenton 1994: 188; Bodansky 1993: 486). In both cases, the majority of EU members and the Commission were opposed to the proposals (Dasgupta 1994: 136–137).

INC 3 (9–20 September 1991, Nairobi) saw no substantial advances in either of the WGs. Only at the very end of the session a mandate was given to the co-chairs of the two groups for preparing coherent negotiating texts (INC 1991b). At **INC 4** (9–20 December 1991, Geneva), "states tended to reiterate their previously enunciated positions, reintroducing proposals and language that had been omitted from the co-chairs' drafts" (Bodansky 1993: 488). The main event of this session was a break-up of the G-77/China bloc over the definition of a common position on commitments (Bodansky 1993: 488–489). While a group of 44 countries led by China and India reiterated the stance already taken beforehand (leaving the question of industrialized countries' emissions targets open), AOSIS proposed that developed countries stabilize their carbon dioxide emissions at 1990 levels by 1995 and reduce them thereafter (Paterson 1996: 58). At the end of the session, delegates decided to combine the negotiating texts of both WGs into one "Consolidated Working Document".

INC 5 (18–28 February 1992, New York) was to negotiate on this, heavily bracketed document in order to reach a final agreement. Despite a noticeable acceleration of the negotiations, it could, however, not deliver on this aim. While the G-77 had come to grips with a coherent position, the industrialized countries were still split on the question of commitments. Almost the entire session was thus spent with intra-OECD talks (Kjellen 1994: 160–161). The US continued to oppose targets and timetables, refusing all language about "stabilization", while the other industrialized players generally agreed on the necessity of targets, but could not settle on their precise shape (Bodansky 1993: 490). Up to this point, the strategy of the EU and many OECD countries had been to attempt convincing the US with arguments, with European negotiators believing that it would be "possible to persuade the US of the political feasibility of committing itself" to the EU's stabilization target (interview with Dutch negotiator Vellinga in ECO, 20 December 1991: 4–5, cited in Paterson 1996: 59). This approach changed under growing time pressure, when it became evident that a deal could only be made if a compromise with the US would be struck. No agreement could be reached, however, at this session, which forced the Chair to convene another, originally not scheduled meeting.

In the run-up to this ultimate INC session, Chair Ripert gathered the **Extended Bureau**, a smaller circle of around 24 delegates, the presiding officers of the two WGs and representatives of key countries including the US, China, Japan, Russia, Brazil, and – from the EU – France, Germany and the Netherlands (Bodansky 1993: footnote 242). At this meeting in Paris (15–17 April 1992), delegates unanimously convinced the Chair to produce a compromise draft for a convention (Borione/Ripert 1994: 88–89). He agreed under the condition that he would not be forced to propose text on the highly sensitive issue of targets and timetables. Arguably, this Extended Bureau meeting marked therefore a first, albeit late, major turning point in these negotiations. The Chair's agreement to shoulder the indispensable job of eliminating brackets from the text left parties to focus on the politically most sensitive issue: how much GHG reductions would be agreed to (Chasek 2001: 130).

The two weeks **in between the Extended Bureau meeting and the reconvened fifth INC meeting** would prove decisive for the adoption of the Convention. Talks were held at the highest political level, for instance between the US President and both Commission President Delors and the German Chancellor Kohl (Lescher 2000: 60–61). A deal regarding the issue of commitments was finally made between the UK, supported by some EU members, and the US. In late April, UK Environment Secretary Howard visited Washington to negotiate with US State Department officials on text concerning the objective of the convention. These textual proposals were introduced at the resumed INC 5, and made it virtually unchanged into Art. 4.2(a), (b) of the UNFCCC, discussed below (Bodansky 1993: 491). In the transatlantic negotiations, Howard gave up on the EU targets and timetables approach and agreed with the US on much softer language around the issue of emission reductions objectives (Cass 2007: 76). The meeting must therefore be regarded as the second major turning point in this process.

The **second part of INC 5** (30 April–8 May 1992, New York) marked then the final turning point in the negotiations (Chasek 2001: 130–131). On this occasion, the INC split into three WGs: one on commitments and finance led by the INC Chair, one on the objective and principles, and one on institutions, dispute settlement and final clauses. The INC discussed all elements of the Chair's draft, and lifted final brackets by brokering compromises on the most sensitive issues, notably regarding targets and timetables (Bodansky 1993: 491). While most sections, including the one on principles (Art. 3 UNFCCC), were quickly accepted by the parties, the UK/US compromise proposal on targets initially came under heavy criticism by the developing countries and EU representatives like Environment Commissioner Ripa di Meana, who called the text "completely unacceptable" (Bodansky 1993: footnote 251). Following intense internal consultations in the G-77/China and in the EU, however, both blocs finally agreed to the text to ensure the overall deal. This latter explicitly excluded commitments for developing countries.

The Outcome: the United Nations Framework Convention on Climate Change

The final product of one and a half years of negotiations was the United Nations Convention on Climate Change, an international treaty signed at the Earth Summit in Rio in June 1992. Its entry into force, on 21 March 1994, would require the ratification of 50 parties.

Explicitly called "framework convention" because of its general character, the UNFCCC set the climate regime an ambiguous objective: "to achieve, in accordance with the relevant provisions of the Convention, stabilization of greenhouse gas concentrations in the atmosphere at a level that would prevent dangerous anthropogenic interference with the climate system. Such a level should be achieved within a time frame sufficient to allow ecosystems to adapt naturally to climate change, to ensure that food production is not threatened and to enable economic development to proceed in a sustainable manner" (Art. 2 UNFCCC). This "declarative goal" did not impose any legally binding obligations on the parties, but would nevertheless acquire significance as a general point of reference in climate talks (Bodansky 1993: 451). During the period after the adoption of the Convention, the objective was interpreted very differently: some parties (like AOSIS) considered that Article 2 stipulated immediate mitigation action, whereas others believed that it indicated the need for gathering further scientific evidence of man-made climate change (Rowbotham 1996: 34; Brambilla 2004: 41).

Central to the treaty and to the climate regime as such are the five principles listed in its Article 3: the principle of common but differentiated responsibilities, the principle of equity, the precautionary principle, the principle of sustainable development, and the principle of free trade. These principles were perceived as "interpretation aides" for the treaty by some, whereas others regarded them as "subjective rights" which may grant parties certain privileges (Ott 1996: 65; Bodansky 1993: 501). The relative uncertainty about their status caused controversies which would shape negotiations for decades to come after the Rio summit.

The Convention further defined various obligations, distinguishing between general and specific duties, applicable to different groups of parties. General obligations for all parties were stipulated in Articles 4.1 (information and data collection requirements), 5 (research and systematic observation), 6 (education, training and public awareness) and 12.1 (reporting). A major distinction between groups of countries was made through Annex I, which lists the OECD countries. According to Articles 4.2(a) and (b), negotiated during the UK/US meeting prior to the decisive INC session, each Annex I party "shall adopt national policies and take corresponding measures on the mitigation of climate change, by limiting its anthropogenic emissions of greenhouse gases and protecting and enhancing its greenhouse gas sinks and reservoirs. These policies and measures will demonstrate that developed countries are taking the lead in modifying longer-term trends in anthropogenic emissions consistent with the objective of the Convention" (Art. 4.2(a) UNFCCC). Further, in "order to promote progress to this end, each of these Parties shall communicate, within six months of the entry into force of the Convention (...) detailed information on its policies and measures (...), as well as on its resulting projected anthropogenic emissions by sources and removals by sinks of greenhouse gases (...) *with the aim of returning individually or jointly to their 1990 levels these anthropogenic emissions of carbon dioxide and other greenhouse gases*" (Art. 4.2(b) UNFCCC, emphasis added). Interpreted as the "heart of the UNFCCC" at the time of the entry into force of the treaty (Mintzer/Leonard 1994), this article never really acquired any substantial significance thereafter due to both the soft formulation of the target and its lack of legal bindingness. The Convention itself called for a review of the adequacy of the substantial provisions of these paragraphs (Art. 4.2(d) UNFCCC). Finally, the treaty included further specific obligations for Annex I parties in terms of more wide-reaching reporting duties (Art. 12.2 UNFCCC) and for Annex II parties (Annex I minus economies in transition) regarding the financial assistance and technology transfer to developing countries (Art. 4.2 UNFCCC) (Depledge 2005: 21).

For the purpose of this study, the discussed obligations for different groups of parties and the principles of Article 3 represent certainly the most significant provisions of the treaty, as they embody the core pillars of the climate regime. Other aspects of the UNFCCC nevertheless need to be highlighted. Its Article 7 created the conference of the parties as the main decision-making body in the regime. The COP adopts its own rules of procedure. It meets once yearly. Moreover, a secretariat (Art. 8 UNFCCC) and two Subsidiary Bodies were created (Art. 9 and 10 UNFCCC). In terms of procedures, the UNFCCC foresaw consensus or, as a last resort, a "three-fourths majority vote of the Parties present and voting" for the adoption of amendments (Art. 15.3 UNFCCC) and consensus on the adoption of a Protocol (Art. 17) (Pallemaerts/Williams 2006: 36–37). Further, the Convention included provisions on various key issues like the financial mechanism (Art. 11) and dispute settlement (Art. 14), whose significance has been discussed elsewhere (Yamin/Depledge 2004: chaps. 10, 12). Finally, it is important to mention that the UNFCCC provided for flexible ways of implementation, as it allowed inter alia for a joint implementation of policies by different parties (Art. 4.1 (a) and (d) UNFCCC).

Assessing and Explaining the EU's Influence Attempts and Influence during the Negotiations on the UNFCCC

The UNFCCC laid the formal foundations for the global climate regime whose development is analysed in this work. As the deliberations leading up to its adoption provide a first insight into the dynamics of global climate talks, they convey significant background information for the study of subsequent negotiation rounds. Further, their analysis allows for tracing the EU's influence attempts and influence over time.[5]

To establish the Union's influence on the two issues selected for in-depth inspection in this work, a closer look needs to be taken at the three turning points identified for the INC negotiation process. The first turning point – the Extended Bureau Meeting in April 1992 – facilitated bargaining among parties and thus ultimately decision-making on the UNFCCC. It was thus crucial for the overall outcome of the talks, but did not lead to any concrete decisions on the key items analysed here, and can thus be neglected. By contrast, turning point 2 (on targets), i.e. the preparatory meeting between the US and the UK, as well as the third identified turning point at INC 5 (on both issues) marked central milestones regarding the key pillars of the regime. At these points in time, the EU's influence can be established by zooming in on the negotiation processes

---

[5] The analysis of this period is primarily based on document research and secondary literature as well as the statements of negotiators and observers of the talks (see, e.g., the contributions in Mintzer/Leonhard 1994).

and carrying out the conditional causal analysis based on the five constitutive dimensions of the concept of influence.

Before assessing the EU's influence, a word should be said about its actor capacity. The story of the negotiation process illustrated that it may be difficult to treat the EU as a uniform actor in the context of the INC talks. Assessments of its capacity have been contradictory: while some observers of the process have gone as far as remarking that "the EC (as opposed to its Member States) in fact only played a limited role in the negotiations leading to the Climate Change Convention" (Haigh 1996: 181), others noted that the "member states and the Community were intertwined in such a way that the EC could be seen as a unitary actor using multilateral diplomatic channels" (Sbragia 1998: 298–299). In final analysis, the evaluation of the Union's capacity is not as much a matter of opinion than of the definition of EU foreign policy one employs. In this work, EU foreign policy is considered the sum of the activities of EU institutions and member states, if the latter contribute to an overall EU objective. And even if one cannot speak of "the EU" without a certain degree of caution for this period, one can certainly speak of a common European position. Even if member states acted thus seemingly inside *and* outside an EU context – and some, notably the UK and France, did so more than once in this process –, their activities must be interpreted as a contribution to European influence *if* they advanced the Union's overall aims. While this is not to imply that the EU possessed full actor capacity during the entire period, its loose coordination ensured a degree of coherent representation through the Presidency and other member states at key moments in the talks.

This minimum degree of coherence, but also the individual foreign policy activities of EU member states ensured that it managed to exert some influence on the creation of the climate regime to partially reach its aims in these negotiations. In the words of (then non-EU country) Sweden's negotiator Kjellen, "without doubt, Washington and various EC capitals were the central actors in the final phase of negotiations" (Kjellen 1994: 163). A closer look at the two issues of interest in this study, the emission reduction objective and the CBDR principle, brings greater clarity to this general assessment. First, the *emissions reduction targets* embodied in the overall objective of Arts. 2 and 4.2 (b) resulted from an obvious compromise reached at the first relevant turning point between the US opposition to targets and the EU's desire to stabilize emissions at 1990 levels by the year 2000 (Bodansky 1993: 491). Throughout the entire negotiation process, the EU had, via the Council Presidency and most of its member states, promoted its stabilization goal for the year 2000, particularly by attempting to "convince the US to change its position" on this item (Sbragia 1998: 298–299; Paterson 1996: 59). At early stages in the process, the Union had tried to persuade the Americans through arguments only. A close inspection of the activities surrounding the decisive turning point that ultimately led to the agreement on the targets demonstrates that it was only when the EU shifted into a bargaining mode that a final agreement became possible. According to numerous commentators, not least the INC Chair himself, the UK's role – especially through the meeting between Environment Secretary Howard and the US State Department negotiators prior to INC 5, part 2 – was "pivotal to reaching [this] agreement", which was then introduced and formally accepted by the totality of parties at that session (Borione/Ripert 1994: 83; Cass 2007: 76). When visiting the US, Howard had the support from some of his European colleagues, but "whether this can be regarded an EC contribution is a matter of opinion" (Lescher 2000: 61; Haigh 1996: 181). Formally, he had certainly no EU mandate to compromise on the Union's targets and timetables approach, but it appeared as clear at the time that a bargain was in any case inevitable, if the Union was to reach its overarching goal of getting to an agreement that included the US as the biggest emitter at all. According to Haigh's counterfactual interpretation, "without the machinery provided by the EC for discussion between ministers it [Howard's deal with the US] may not have happened" (1996: 181–182). The UK on its own would certainly have lacked the clout to negotiate a deal with the US and subsequently defend this against the entire rest of the world (OECD and G-77/China). In that sense, the UK's activities may be regarded – ex post – as a form of implicit task-sharing within the EU.[6] Through the activities of several member states which contributed to its overall aims, the EU was thus able to exert influence on the negotiations and final outcome on this issue. All the necessary conditions for establishing influence are indeed fulfilled: the Union stated its position early in the process, and clearly was, together with some other OECD countries (e.g. Canada), among the agenda-setters for this item (*temporal sequence*); it interacted closely with other major players, notably the US (*interaction*), and with a clear intention of impacting on these actors to alter the final outcome of negotiations (*purposive behaviour*); moreover, it attained its minimum objective of having some, albeit vague target mentioned in the treaty (partial *goal attainment*). Finally, in counterfactual perspective, the slight US change of position during the final negotiation stages must be regarded as the result of other parties' pressure, first and foremost the EU's, rather than of internal developments in Washington, as further explored below (*absence of auto-causation*). Second, concerning the issue of *responsibilities*, i.e. "who

---

[6] That some member state representatives and Commissioner Ripa di Meana first rejected the US/UK compromise might not have been a sign of EU unity, but can also be interpreted as personal frustration about an outcome that was much less ambitious than what these personalities had hoped for.

should do what" when it comes to reducing emissions, the inclusion of the CBDR principle and, moreover, the distinction between groups of countries (Annex I and the rest) was a rather uncontroversial matter throughout the negotiation process, formally agreed to by all parties at INC 5, part 2. Both developing and developed countries supported it, but for different reasons. For the developing world, the industrialized countries had the "main historic responsibility", while for the industrialized countries, notably the US, the emphasis placed on their greater financial and technical capabilities was of key importance (Bodansky 1993: 503; Steffek 2005). The inclusion of this provision can thus be interpreted largely as a consensual general agreement, and a significant success of the developing world, enabled through a deliberately ambiguous formulation allowing for different interpretations at the time of adoption – and afterwards (Steffek 2005: 239). The EU had remained very timid on this issue throughout the negotiation process, respecting the concerns of the developing world without actively lobbying for them. As necessary conditions for attributing influence (*purposive behaviour*, *interaction*) are thus unfulfilled, it can be concluded that the Union did not exert any influence on this item.

While it is thus possible to establish *that* the EU exerted influence on the negotiations pertaining to the creation of the UN climate regime with regard to the issue of targets, the determination of its share of influence demands further counterfactual argumentation. First, when it comes to assessing the EU's share of influence *vis-à-vis* other OECD countries (besides the US), even if it was certainly not the only actor demanding a change in position from the US, it was the most fervent defender of the 2000 stabilization target within the OECD. It must thus be considered as very plausible that it was above all the EU that ensured "the adoption of a convention with a soft stabilisation target for all industrialised countries" (Yamin 2000: 49). With this, it achieved its minimum objective, pushing the US – against its will – towards some form of a target short of its red line. Second, when assessing the relative influence of the EU and the US on the final deal reflected in the UNFCCC, Haigh credits the EU with having determined the overall approach of stating a target in the treaty at all: "Despite its non-binding character, Article 4 (2) of the Convention would certainly have been much weaker without the EC's prior position" (Haigh 1996: 162). Yet, the US strategy of holding out had equally been successful in fighting off binding targets in the short term (Paterson 1996: 62; see also Nitze 1994 who speaks of a "success for US diplomacy"): "had the U.S. not taken such a hard line on commitments, the Convention would no doubt have been stronger" (Hunter *et al.* 2002: 618). Even so, the US did concede a partial defeat regarding its aim of completely avoiding any mention of targets (Paterson 1996: 62). The EU and US shares of influence on the final outcome regarding this item were thus of comparable magnitude.

Historical Foundations (1980s–1995)

In synthesis, the overall degree of EU influence must be assessed as medium. The *extent of the EU's goal attainment* was clearly medium: the Union ensured its minimal aims of concluding a treaty comprising the biggest emitter and some form of a target for industrialized countries, but did not attain its ultimate objective of including legally binding targets and timetables in the Convention. The *durability of the outcome* must equally be assessed as medium: the UNFCCC is an international treaty that is still in force and provides the central point of reference for the global climate regime, but the pillars of this treaty, the targets and principles, are of a soft nature, characterized by a limited degree of bindingness. Further in line with the typology of influence (see Chapter 1), the EU's influence on the objective of the climate regime (Art. 2 UNFCCC) can be characterized as more enduring (Bodansky 1993: footnote 296) than that of the US, albeit only in default of stronger bargaining leverage: in retrospect, the Union helped to ensure the adoption of a treaty that would determine all global climate negotiations that followed (Schröder 2001: 36). It not only contributed to the fact that a target was mentioned in the treaty, but also to the necessity to review the adequacy of this target (Art. 4.2 (a), (b), (d) UNFCCC). US influence on the treaty negotiations, by contrast, was more short-term, bargaining-based and instantaneous, allowing it to reach immediate aims in line with preferences that were essentially less related to climate change than to domestic debates about potential losses of competitiveness and jobs under a strict emission reductions regime.

Tentative explanations of the EU's influence may be found in the fact that it had a fairly coherent, basic position on key items under negotiation and that some member states were willing and able to advance this position actively and through a variety of channels in the negotiations. The limits to this influence regarding the adoption of a more ambitious target, notably *vis-à-vis* the US, can be explained by internal factors and one significant external variable.

Concerning internal factors, without an established foreign policy system and the necessary instruments, the EU did not possess the tool-kit to devise a common strategy that would have built enough momentum among OECD countries to convince the US of the necessity to integrate legally binding targets into the treaty. Arguing as prime negotiation strategy was clearly insufficient in the negotiation context and proved therefore also unsuccessful. Moreover, internal divisions – the UK acting partially outside the EU framework with Howard's visit to the US, but also on issues such as Japan's "pledge and review" proposal – hampered a better strategic performance. The UK's attitude may be explained by preferences that slightly diverged from those of France, Germany or the Netherlands, notably in its desire to use all its leverage and its "relationship with the US to constrain European foreign policy" and, most importantly, the Commission (Cass

2007: 65, 76). Intra-EU foreign policy objectives of specific countries thus limited the effectiveness of the Union's external activity as a collective actor, preventing it from a potentially greater influence on the global scene. Given these internal constraints, the deal eventually struck between the UK and the US was maybe the best the EU on the whole could have achieved.

A major set of external factors limiting EU influence in these negotiations is to be seen in the overall importance, power and difficult bargaining position of the US. Its importance was not so much due to its economic and technological state of advancement, but simply to the nature of the problem: a treaty without the US – as the largest emitter – would have left a large share of global emissions uncovered by any global agreement. So it was ultimately the US lifestyle and domestic political debates circling around competitiveness concerns that guaranteed its de facto "veto power" in the negotiations, and allowed the George H.W. Bush administration to hold out until the end with a position that "represented a triumph of election year politics over environmental science" (Harrison 2000: 107; Paterson 1996: 99–100). Against such resistance, the individual foreign policies of the other parties to the negotiation process were obviously not sufficiently compelling. By consequence, all players made the strategic choice to accommodate US concerns, as they were unwilling "to sign an agreement without the participation of the US. These countries determined for themselves that an otherwise well-structured convention with non-binding language on short-term targets that could be signed by the US was preferable to a similar convention with binding language that was not signed by the US" (Nitze 1994: 188; also Dasgupta 1994: 149).

The Road to COP 1 (1992–1995)

During the post-Rio period, and while awaiting the ratification and entry into force of the treaty, the notion of "prompt start" had gained ground: between 1992 and the first conference of the parties in March 1995, six further meetings of the INC were held in order to deal as quickly as possible with issues related to the operationalization of the regime and to prepare for the review of the "adequacy of commitments" stipulated in Art. 4.2(d) UNFCCC (Bodansky 1994: 34). Although discussions during this period did not per se focus on the two key issues of targets and responsibilities, the further development of the treaty-based regime was evoked at several INC meetings, albeit without any substantial results. This section therefore provides only a brief trace of these taks in search of EU influence (attempts).

In the EU, the ratification process was used to hold discussions on how the stabilization target of Art. 4(2) UNFCCC could be fulfilled. Talks soon stalled over the question of whether the Union should adopt a carbon dioxide and energy "eco-tax" (Lescher 2000: 69; Jordan/Rayner 2010: 60–61).

Proponents of this measure (above all the Commission, Denmark, Germany, the Netherlands, Belgium) tried to make the joint ratification of the UNFCCC conditional on the adoption of such a tax, but did not succeed against other EU members, especially the UK, which was most strongly opposed to the idea of common European taxation (Lescher 2000: 69, 72). While the ratification of the UNFCCC was finally achieved in February 1994 (Council 1994a), the tax never came (Pallemaerts 2004: 43; Haigh 1996). Instead, the EU adopted an "essentially symbolic" package of measures aimed at the promotion of energy efficiency (SAVE programme) and renewable energies (ALTENER programme) as well as a decision to create a monitoring mechanism for GHG emissions in the EU (Pallemaerts 2004: 43–44). At the same time, no substantial advances were made regarding the definition of a common EU position on further GHG reductions. To better deal with internal struggles about this issue, the Environment Council created, in October 1994, an ad hoc working group of national climate experts. The group was charged with mediating between EU members in order to prepare the Union's negotiation position for COP 1, effectively taking, at least to some extent, the right of initiative out of the hands of the Commission (Lescher 2000: 72).[7] Although the creation of this group did not immediately result in a substantial common position for COP 1 (Schumer 1996), it did represent a major institutional innovation: the permanent institutionalization of a forum for exchanges between member states on issues related to global climate policy. In the global arena, although the UNFCCC was not yet in force, the EC was already treated as a party to the Convention. A regional economic integration organisation (REIO) clause inserted in Art. 22 UNFCCC, endowed it with all the rights of a full member to the INC (Lescher 2000: 73). In practice, the Council Presidency would exercise these rights on behalf of the Union, representing it in all fora.

Turning to the global debates, **sessions 6 to 8 of the INC** (7–10 December 1992, Geneva; 15–20 March 1993, New York; 16–27 August 1993, Geneva) started slowly, dealing with practical issues related to the operationalization of the regime. Much time was spent on consultations regarding the modalities of technology transfer and the financial mechanisms, without any significant advances (INC 1992, 1993a, 1993b). Among the most interesting decisions taken during these sessions was the election of a new President and Bureau. At INC 7, the Argentinean Raul Estrada-Oyuela, who would later play a decisive role during the Kyoto Protocol talks, took over from Jean Ripert. Further, the landscape of negotiation blocs changed when the OECD countries split

---

[7] The novelties of the Treaty of Maastricht, in force since late 1993, and their consequences for the Union's foreign climate policy are discussed in the relevant section of Chapter 3 on the EU's role in the Kyoto Protocol negotiations.

into two groups: the EU-12(/15 from 1995 on) and the newly formed JUSCANZ (Japan, US, Canada, Australia, New Zealand) coalition.

It was only during the **final three sessions of the Committee** (7–18 February 1994, Geneva; 22 August–2 September 1994, Geneva; 6–17 February 1995, New York) that Working Group 1 on "commitments" actually took up topics related to a further development of the climate regime, when it started discussing the "adequacy of commitments", as stipulated in the Convention (Oberthür 1994: 299). At **INC 10**, Germany, the designated host of the first conference of the parties, introduced a paper with "elements of a protocol for [consideration at] COP 1", which stated, inter alia, that Annex I parties should reduce their CO_2 emissions "by the year (x) (...) by (y) %"; the paper was, however, not discussed at this session (Oberthür 1994: 299). Instead, parties could only generally agree on the necessity of pursuing further emissions reduction efforts, without specifying in what framework such efforts should be carried out, leaving commentators to compare the slow pace of the talks to the first INC sessions in 1991 (Oberthür 1994: 299, 302–303).

Shortly after INC 10, the time window for substantive proposals for a new protocol or agreement to be discussed at COP 1 was effectively closed. Although the Convention was not yet ratified, parties operated under the assumption that proposals for any new legal agreement would have to be submitted six months before its (potential) adoption, as stipulated by the so-called "six-month rule" of Art. 17.2 UNFCCC. Before this deadline in late September 1994, only two parties provided written submissions that fulfilled the rule. The first comprehensive proposal had been introduced by AOSIS. It suggested the adoption of a protocol to the Convention, whose Article 3 would introduce "targets for greenhouse gas reductions" requiring that "each Annex I country shall reduce its 1990 level of anthropogenic emissions of carbon dioxide by at least 20 percent by the year 2005" (INC 1994a). Non-Annex I parties could make commitments on a voluntary basis (INC 1994a: Art. 3.3). A couple of days later, referencing the AOSIS proposal and target, Germany presented a more formal version of its "elements paper", sketching out some key components for a future framework convention. The German submission re-stated the stabilization target for carbon dioxide emissions by the year 2000 (at 1990 levels) and called for a reduction thereafter (INC 1994b: I.1). The precise scope of reductions as well as the target year were, however, not specified (INC 1994b: I.1). Some parties, like the Netherlands or Denmark, supported the Toronto target of 20% CO_2 reductions by the year 2005, linking thus the EU's to the AOSIS approach, while such numbers were completely unacceptable for other OECD countries, notably the US (Victor/Salt 1994: 28; Lescher 2000: 75). **INC 11** continued talks on this agenda item and held, as the Chair's report notes, "fruitful and constructive, but not fully conclusive discussions", which ultimately ended without decisions (INC 1995: 50).

CHAPTER 3

From the Berlin Mandate to the Kyoto Protocol (1995–1997)

EU Influence on the First Development of the Global Climate Regime

This chapter provides an in-depth analysis of the EU's influence attempts and their effects during a period that led to the substantial development of the global climate regime. It analyses the negotiations that were kicked off at COP 1 in March 1995, developed over eight meetings of the Ad Hoc Working Group on the Berlin Mandate (AGBM) and COP 2 in Geneva, and were concluded with the adoption of the Kyoto Protocol at COP 3 in late 1997 (for an overview, see Annex II).

The Context: Major Developments in Global Politics and Climate Science

The years 1992 to 1995 had already been dominated by the political and socioeconomic adjustments to the fall of the iron curtain. These themes continued to dominate agendas also during the following years.

The end of the Cold War had left the United States as the "sole superpower" in an international system otherwise characterized by a crumbling of multi-ethnic states both in what had been the Soviet Union and in Central and Eastern Europe. This unique "unipolar moment" as well as the profound political and economic transformations in the two regions were bound to have implications for the functioning of global multilateral institutions. At the same time, the breakdown of the economic systems in the former Warsaw Pact states also had immediate environmental consequences. Highly pollutant and energy-intensive factories had to close, resulting in an immediate reduction of air pollution and, notably, greenhouse gas emissions. Further, the economic downturn experienced in many Western countries, growing "unemployment and excessive national public expenditure deficits beleaguered most governments", decreasing their interest in environmental, including climate, policies (Schröder 2001: 61). This tendency was reinforced by a decline in public interest and media attention on the topic of climate change after the Rio summit (O'Riordan/Jäger 1996: 27).

In contrast to these potentially adverse political and socioeconomic background factors, significant advances in knowledge about climate change were bound to have a positive effect on the negotiations. In late 1995, the IPCC published its comprehensive Second Assessment Report (SAR), which "set the [scientific] context for the negotiation of the Kyoto Protocol" (Grubb *et al*. 1999: 7). Compared to the First Report of 1990, which had solely confirmed the existence of the greenhouse effect, emphasized its unnatural intensification through human activity and provided loose estimates as to future temperature rise, the SAR was more affirmative in its statements (IPCC 1995a). Its centre-piece, the report of the IPCC Working Group on Science, made clear that "greenhouse gas concentrations have continued to increase" due to discernible human influence (IPCC 1995b: points 1 and 4), and that this trend would lead to an average temperature rise of 2°C over the course of the 21st century (IPCC 1995b: point 5). The report generally had the potential to increase the pressure in the regime, while also providing ample examples of policy solutions to tackle the problem (IPCC 1995c; Grubb *et al*. 1999: 14–17). Nonetheless, it also pointed to the continued existence of uncertainties (IPCC 1995b: point 6).

Key Actors in the Global Climate Regime and their Positions

Key Actors Other than the EU

Besides the EU, three core blocs were involved in the negotiations on a protocol to the UNFCCC: developing countries traditionally cooperated under the G-77/China umbrella; JUSCANZ (Japan, United States, Canada, Australia, New Zealand), which later became JUSSCANNZ (with Switzerland, Norway and Iceland), had formed a new negotiating group of non-EU industrialized countries after the UNFCCC negotiations; finally, the countries in economic transition, including Russia, cooperated to defend their special interests in the talks.

JUS(S)CA(N)NZ,[1] an "unnatural alliance" for some, emerged gradually as a counterpart to the EU during these negotiations. It united countries for whom emissions reductions were, for a variety of reasons, less politically desirable and arguably less easily achievable than for the EU (Grubb *et al*. 1999: 34). Central to this group's behaviour throughout the talks was the position of the US and, in the final stages, of the Japanese hosts of the decisive COP 3. These two countries also represented the largest GHG emitters in absolute terms within the coalition.

---

[1] For reasons of coherence, the acronym JUSSCANNZ, although this may not always be entirely accurate, will be used throughout this chapter.

From the Berlin Mandate to the Kyoto Protocol (1995-1997)

In the early 1990s, the *United States* was by far the most important carbon dioxide emitter in the world. Its share of the total CO_2 emissions of all industrialized countries in 1990 lay at 36% (UNFCCC 1997e: 60).[2] As the world's largest coal producer and second largest producer of oil and gas, the US had become used to a "low-price energy culture", making it highly dependent on the use of fossil fuels (Grubb *et al.* 1999: 31). This dependence outweighed the threat perception about climate change, which was expected – at the time – to manifest itself above all through extreme weather events in some regions of the country (Oberthür/Ott 1999: 18–19). Consequently, climate change was thus predominantly framed in economic terms (Damro/Mendez 2003; Baker 2006). The political system of the US facilitates an economic framing of the subject, since Congress plays a significant, indirect role in the definition of negotiation positions for international treaty-making processes. The US Constitution states that the President "shall have Power, by and with the Advice and Consent of the Senate, to make Treaties, provided two-thirds of the Senators present concur" (Art. II, section 2). As a result, "it is Congress, not the executive branch or the President, that has the final say over U.S. environmental policy, both at home and abroad" (Paarlburg 1997: 149). By consequence, the US State Department, formally in charge of the negotiations, acts de facto under manifold restrictions from the legislator (Harris 2001: 19–22). The US Congress, in turn, is traditionally open for input from various lobby groupings. Both members of the House of Representatives and Senators defended very diverse interests regarding climate change during the period analysed in this chapter, most of which have to do with the short-term economic well-being of their constituencies. While party affiliation at times determined their stance on climate policies – with Republicans generally more hostile towards environmental legislation, and Democrats more inclined to take action on green issues – no clear-cut division between the different camps can be discerned (Harris 2001: 20). Quite a number of Democratic Senators from coal-producing states tended to prioritize the economic concerns of their voters over measures to reduce GHG emissions in the 1990s. Further, big business arguably played an important part in shaping positions in this period (Harrison 2000: 92–93). After mid-term elections in 1994, the Clinton and Gore administration that had originally been positively inclined toward climate action saw itself confronted with a Congress dominated by Republicans – and sceptics among fellow Democrats – who

---

[2] The numbers cited here are 1990 figures, taken from the first national GHG reports. They have been chosen to allow for comparability and because they provided the basis on which the various parties formulated their preferences and negotiation positions at that time (UNFCCC 1997e: 60).

believed that climate policies represented a threat to economic growth and "American jobs" (Bryner 2000: 124). This would have a considerable impact on the US negotiation position. Prior to the Berlin COP, this position was outlined in the October 1993 Climate Change Action Plan, which called for a stabilization of GHG emissions at 1990 levels by the year 2000, covering four gases (CO_2, methane, N_2O, HFC) (Clinton/ Gore 1993: figure 2). This stance was further specified in the 1994 Climate Action Report: for the US, the negotiations were to determine "a new 'aim' that would provide specific guidance for international commitments beyond the year 2000" (Department of State 1994: 193). Further, the climate regime was to display a number of characteristics, including being flexible (which meant, in essence, cost-effective), comprehensive and equitable, i.e. "engaging all countries in the global effort while recognizing differences in national circumstances and capabilities" (Department of State 1994: 193). This position would evolve only slightly over the course of the negotiations.

Similar to the US, *Japan*, with 8.5% of the total emissions of all industrialized countries, did not at the time expect to be subject to many negative consequences of climate change (UNFCCC 1997e: 60). The country's rationale for undertaking significant efforts to conserve energy as early as the 1960s had been based on economic rather than environmental considerations: both the Japanese government and its business community feared the increased economic competition from other Asian countries, and thought it necessary to embark on economic modernization towards greater energy efficiency (Schröder 2001: 40–41). The general attitude towards the issue of climate change in Japan can therefore be characterized as a mixture of "impact-skepticism" and "techno-optimism" (Fermann 1993: 292). This general stance on climate change was translated into a negotiation position through lengthy and conflict-ridden discussions between the Environmental Agency, responsible for environmental policies and more ambitious with regard to a reduction target, and the Ministry of International Trade and Industry (MITI), in charge of energy policies and protector of Japanese economic interests (Matsumura 2000: 11–16; Kameyama 2004: 72–73). In the early phase of the Kyoto Protocol negotiations, Japan would thus further defend the previously adopted position of promoting a stabilization of CO_2 emissions at 1990 levels by 2000 (AGBM 1995b: 35). By contrast, it did not have a clear stance on the differentiation of commitments and potential efforts to be undertaken by developing countries. This position evolved considerably when the country embraced the issue of climate change to raise its foreign policy profile by proposing to host COP 3 (Kameyama 2004; Schröder 2001: 43; Matsumura 2000).

For the *other JUSSCANNZ members*, the domestic circumstances were quite diverse. While Canada and Australia, with 3.3% and 2.1% of the total CO_2 emissions of Annex I countries in 1990 respectively, had conditions regarding energy production and consumption that were similar to those of the US, New Zealand (0.2%), Norway (0.3%) and Switzerland (0.3% of total industrialized countries' emissions in 1990) all possessed a fairly high share of renewable energy sources in their domestic energy production (UNFCCC 1997e: 60; Grubb *et al.* 1999: 33–34). Despite the resulting interest differences and diverging perceptions of the problem, the five countries advocated fairly similar positions on the issue of targets, calling for indicative emission reduction objectives (Andresen/ Butenschon 2001: 339–340; ENB 1995c: 1, 2, 1995f: 2). Their stances on the issue of differentiation were less clear.

Although the heterogeneity of the *G-77/China* – regarding attitudes, preferences and negotiation positions – had further increased over the years, the group chose to continue to cooperate for a variety of reasons (Grubb *et al.* 1999: 35; Oberthür/Ott 1999: 24). First and foremost, a major unifying factor for this coalition was its members' high vulnerability to the manifold facets of climatic alterations, intensified through an often striking incapacity to adequately react to these problems. Out of "a tradition of Third World solidarity", they therefore decided to cooperate (Rajan 1997: 265). Moreover, many of the smaller countries of this group faced the challenge of lacking the administrative and diplomatic capacities to prepare and follow up on the climate negotiations by themselves (Gupta 2000). To gain more diplomatic weight, these countries depended thus on an alliance with resourceful parties such as India, China or Brazil. For the latter, being able to speak on behalf of such a large group of countries also implied a strategic advantage. Finally, the vast majority of countries in this group shared the same vision regarding questions of equity and their right to develop (Rajan 1997). Despite these reasons for cooperation, differences within the G-77/ China concerned, firstly, the cleavage between the emerging and the least developed countries. The key actors in this group were generally the largest countries, *China* and *India*. Altogether, they had a fairly limited share of global emissions at that time, which were, however, projected to rise steeply in the medium-term future (IPCC 1995a: 37; ENB 1995h: 2). Both countries were also beginning to be affected by climate change in the form of extreme weather events (droughts, floods). This explains why each of them had started to take actions to reduce their energy intensity (Oberthür/Ott 1999: 28). Their key positions in the regime discussions were clear: encouraging developed countries to adopt more adequate GHG reduction commitments under a follow-up agreement to the Convention, while refusing any obligations for themselves.

Together with other major actors in this bloc (e.g. Mexico, Brazil), they thus advocated above all equity and development concerns (Grubb *et al.* 1999: 36). In spite of this point of convergence, *least developed countries* had more basic demands. Often already victims of climate change, a problem to which they had contributed little, they argued not only for developed countries to take up their responsibilities to mitigate climate change, but also requested help in their own fight against the consequences of climatic variations (IPCC 1995a: 37). Second, the G-77/China group was split along another line: the coalition of small island states (*AOSIS*), directly threatened by climate change and thus among those who advocated most urgently the swift adoption of ambitious measures, stood in stark opposition to the oil-producing countries (*OPEC*) (Grubb *et al.* 1999: 36). OPEC feared that strict carbon reduction measures would threaten their major source of income and therefore tried to slow talks down. Not surprisingly, the negotiation positions of the two groups prior to COP 1 could thus hardly have been more different. In the protocol proposal that AOSIS had presented in 1994, it had called for reductions by developed countries in the range of 20% by 2005 and opposed obligatory commitments for developing countries (INC 1994a). By contrast, OPEC, led by Saudi Arabia, basically advocated that climate science was too inconclusive to be discussing further measures at all (Grubb *et al.* 1999: 35–36).

Turning to a third group of players in the negotiations, the ex-Soviet Union and its former sphere of influence in Eastern Europe had undergone dramatic changes since the very first talks on climate change. The economies in this region had been extremely energy-intensive, and their collapse left countries like Russia or Ukraine with a much-improved record regarding GHG emissions. As a country with enormous fossil fuel reserves and with a largely indifferent attitude toward potential impacts of climate change – with some voices in the internal debate even arguing that global warming might have positive impacts on parts of the country – *Russia* was not necessarily in favour of negotiating a protocol with targets for itself. Although it first did not have a clear position on the new negotiation process (Oberthür/Ott 1999: 23), this would change when it realized that it could use its "hot air" (the emissions saved since 1990, when it had 17% of the global total CO_2 emissions) and sell it to other countries (UNFCCC 1997e: 60). Both Russia and the *Ukraine* would then begin to join the JUSSCANNZ group in its call for targets coupled to greater flexibility (Grubb *et al.* 1999: 34–35). By contrast, *countries from Eastern Europe* that hoped to join the EU were gradually moving towards the Union's stance on climate policies, which they would eventually have to take over as part of the latter's environmental acquis (Oberthür/Ott: 23–24; Grubb *et al.* 1999: 34–35).

The European Union: Actor Capacity, Negotiation Positions and their Foundations

The Union's positions on the issue of climate change have to be understood against the background of its unique character as a multi-state entity. In 1995, Sweden, Finland and Austria joined the EU, further augmenting the diversity of approaches to climate policies among its then 15 members (Skjaerseth 1994). While all member states, with the exception of the UK, were net energy importers at the time, not all of them shared the same feeling of responsibility towards the global environment (Grubb et al. 1999: 30). Countries of the North, including the three new member states, Denmark, the Netherlands, Germany, and the UK, were generally more concerned about climate change as an environmental problem, while the "cohesion countries" (Spain, Portugal, Greece, Ireland) showed greater interest in their economic development (Lacasta 2008: 5; Schreurs/Tiberghien 2007: 36–40). This divergence was intensified through the different emissions (and energy production) patterns across Europe: Germany and the UK had a fairly high share of the EU-15 total CO_2 emissions (roughly 29% and 17% respectively in 1994), while Italy's (13%) and, notably France's (11%) emissions were comparatively low. Differences existed also between smaller countries of comparable population size, like, for instance, Belgium (3.8% in 1994, i.e. 11.6 t of CO_2 per capita) and Portugal (1.5% in 1994, i.e. 4.5 t of CO_2 per capita) (EEA 1997: 25–26).

Despite the resulting low degree of interest and preference homogeneity between the North and South of Europe at that time (Delreux 2008: 147–148; Skjaerseth 1994), the EU had since the late 1980s managed to establish a common approach which perceived climate change not only as an environmental, but also as an economic issue, framing it as an opportunity for modernization of economic structures in line with the precautionary principle and sustainable development considerations (Scheipers/Sicurelli 2007: 445–450; Baker 2006; Damro/Mendez 2003: 79). While expected climate change impacts in Europe may have played only a minor role in the definition of this approach, the general idea that the EU's 24% share of the total CO_2 emissions of industrialized countries in 1990 could be reduced, had a more notable impact (UNFCCC 1997e: 60). In addition to its adherence to the precautionary principle, the EU had also identified a "vested interest" in reducing its energy consumption to become more independent from its suppliers and enhance its competitiveness (Oberthür/Ott 1999: 15). Finally, the EU's belief in and support for multilateralism and international law played a role in determining its willingness of pursuing globally concerted solutions to the problem (Scheipers/Sicurelli 2007: 448).

EU actor capacity is a function of the existence of legal competences for climate change activities, of decision-making and coordination procedures for such activities as well as of external representation arrangements, treaty objectives and tools.[3] Throughout the studied period, the EU-15 functioned under the provisions of the Treaty of Maastricht and in accordance with established practice regarding internal coordination and external representation (Brambilla 2004: 160). In legal terms, the European Community was endowed with the legal personality that allowed it to enter into international treaties (Art. 281 TEC). The EC had then also ratified the UNFCCC in 1994 (Council 1994a) and was – through a specific regional economic integration organisation (REIO) clause inserted into the Convention (Art. 22 UNFCCC) – allowed to participate as a full member in the negotiations on the Kyoto Protocol. This implied that it enjoyed the same rights as the other parties regarding such important matters as tabling, speaking and voting. If the EC had to vote, it would do so on behalf of all its members, preventing them from exercising individual voting rights (and vice-versa) (Art. 18.2 UNFCCC). From the perspective of international law, its actor capacity was thus undisputed. From the perspective of European law, its foreign policy activity depended above all on the existence of substantial and procedural competences in primary law, permitting it to exercise its rights within the international climate regime. Substantially, European Community activity on climate change had its treaty basis in Art. 130r TEC, which was a slightly adapted version of the same article in the SEA (Maastricht): "Community policy on the environment shall contribute to the pursuit of the following objectives: – preserving, protecting and improving the quality of the environment, (…) promoting measures at international level to deal with regional or world-wide environmental problems." Paragraph 4 of this Article underscored the fact that environmental protection was not an exclusive EC competence:

> Within their respective spheres of competence, the Community and the Member States shall co-operate with third countries and with the competent international organizations. The arrangements for Community co-operation may be the subject of agreements between the Community and the third parties concerned, which shall be negotiated and concluded in accordance with Article 228. The previous sub-paragraph shall be without prejudice to Member States' competence to negotiate in international bodies and to conclude international agreements.

The right of negotiating and concluding international treaties had therefore to be shared between the EC and its member states (Eeckhout

---

[3] EU foreign policy tools were generally discussed in Chapter 1 as part of the question how the EU can exert influence. They are further highlighted in the process analysis.

2011: 141). This provision had particular consequences for the procedures that needed to be followed when the EU defined a negotiation position internally and when it represented that position externally. If Article 130s TEC generally granted the EC member states the right to negotiate international treaties, it left an opportunity for them to authorize the Commission to conduct these negotiations following the provisions of Article 228 TEC-Maastricht, which represented a codification of pre-1993 practice (Brambilla 2004: 160). The decision-making rule that was to be applied depended on the decision-making mode used for internal legislation (Art. 228.1, 2 TEC-Maastricht). In practice, the member states kept, in line with Art. 130r, para. 4 TEC-Maastricht, the right to negotiate to themselves throughout the period analysed in this chapter (Oberthür/Ott 1999: 66). The EU's international negotiation position was defined by the Environment Council, deciding by unanimity (Delreux 2008: 151; Groenleer/van Schaik 2007: 985). Positions were prepared by member state representatives in the Ad Hoc Working Group on Climate Change created in the autumn of 1994, which was itself served by an Expert Group on Common and Coordinated Policies (Lescher 2000: 72; Oberthür/Ott 1999: 65–66). Input into the negotiation position came from the Commission, and, to a lesser extent, the European Parliament, which formally only had to be consulted on Commission proposals (Art. 228, para. 3 TEC-Maastricht; Pinholt 2004). Coordination during the process of position-building was assured by the Council Presidency. For instance, a major coordinating role in the definition of a numerical emissions reduction target was arguably played by the Dutch Presidency in the first half of 1997 (Kanie 2003). Negotiation positions resulting from this complex internal decision-/foreign policy-making process were regularly fairly rigid, leaving narrow margins for manoeuvre to the negotiators. Once such a position had been formulated, the Council Presidency, assisted by its predecessor and successor in the "EU Troika", was charged with the task of internal coordination, taking the form of regular exchanges of information and consultation during COPs and, especially, at the very end of the negotiations at COP 3 (Interview EU representative 5). Moreover, the Troika on the whole would represent the position of the EU *vis-à-vis* third parties in the international negotiations (Brambilla 2004).[4] All in all, the Union's negotiators were credited with having managed to ensure the EU's external coherence, making it appear as a unitary actor throughout most of the studied phase (Oberthür/Ott 1999: 17). Individual member states hardly took the floor unless when supporting EU positions (Delreux 2008: 143). All in all, the EU certainly possessed actor capacity throughout the period 1995 to 1997: on top of its

---

[4] After Germany in late 1994, France and Spain assumed the Presidency in 1995. In 1996, Italy and Ireland took over, followed by the Netherlands and Luxembourg in 1997. At COP 3, the UK (Council Presidency in 1998) completed the Troika.

overall stance, economic and political clout, it was able to act as a foreign policy player thanks to its legal competences, and the input and coordinating role of various presidencies. Yet, this capacity would vary over time, notably with regard to the coherence of its representation (Grubb *et al.* 1999: 112).

The EU's ultimate aim in the climate negotiations between 1995 and 1997 was to reach a comprehensive global agreement with clearer mid-term emissions reductions commitments than those enshrined in the UNFCCC. Prior to COP 1, its position was not as unified and detailed as it would become during this negotiation process, in parallel with a successive consolidation of its internal climate regime. The Union's overall approach to these negotiations was expressed in the Environment Council conclusions of December 1994, in which the Ministers reaffirmed their initial goal for the negotiations of reducing CO_2 emissions to 1990 levels by 2000, but also invited the Commission to propose a set of "policies and measures (…) aimed at progressive limitations and reductions", e.g., "by 2005 or 2010" (Council 1994b: 6–7). This position was further clarified in a submission of the German EU Presidency for the eleventh negotiation session of the INC, also in December 1994, which demanded the adoption of a protocol containing policies and measures as well as targets and timetables. In this paper, the EU called on other Annex I countries to commit themselves to "comparable efforts for the period after 2000" (UNFCCC 1994: 12–13). A quite indeterminate statement was made regarding the issue of responsibilities:

> Industrialized countries need to take the lead in limiting emissions (…) The European Union reiterates its view that a reasonable balance between industrialized and developing countries commitments should be maintained, for instance in the form of further requirements for non-Annex I Parties on reporting and limitation of emission growth for certain more advanced developing countries (UNFCCC 1994: point 9, page 14).

A submission by the French Presidency, early in the AGBM process – referring also to the 1994 Council conclusions – underscored this stance and named as the Union's ultimate objective the adoption of a "global protocol" (AGBM 1995b: 15–21).

The Negotiation Process and the EU's Influence Attempts

This section provides a focused narrative process trace of the negotiations as they unfolded from COP 1 in March/April 1995 until the adoption of the Kyoto Protocol at COP 3 in December 1997. It pays specific attention to the two embedded analytical units (the core norm of reduction obligations and the CBDR principle), each of which provides for a sub-plot of sorts in the narrative. A third embedded sub-plot deals with the

EU's foreign policy activities and their internal preparation. If relevant, its activities outside the UN arena are mentioned.[5] Following the story, the Kyoto Protocol, which was not only the principal outcome of this negotiation process, but also set the framework for the negotiations that ensued, is briefly analysed. Moreover, patterns of the Union's influence attempts are systematically extracted from the story to allow for a subsequent establishing of its influence.

The period between the summer of 1992 and spring 1995 had been marked by an absence of substantial advancements regarding the development of the climate regime. The report of the last INC session, which had explicitly focused on the review of Article 4.2 UNFCCC, had elegantly noted "fruitful and constructive but not fully conclusive discussions" on the adequacy of commitments (INC 1995: 50–51). This diplomatic language hardly concealed the fact that no agreement had been reached beyond the general notion that the commitments under the Framework Convention represented only a first step toward meeting the objective embodied in its Art. 2 (Oberthür 1994).

The precise shape of the outcome of the first conference of the parties, to be held in Berlin from 28 March to 7 April 1995, was therefore completely open. Only two proposals were on the table: an AOSIS draft protocol and a proposal from the German COP presidency (see Chapter 2). The German paper had called for a stabilization of emissions at 1990 levels by 2000, and for reduction efforts by Annex I countries following a targets and timetables approach after that (INC 1994b: 3–4). Further, it had emphasized:

> "We should continue to work towards balanced commitments on the part of industrialized and developing countries, for example by means of further reporting commitments for non-Annex I Parties and commitments to limit the rise in emissions in the case of certain more advanced developing countries."

This formulation would later become part of the EU's position.

On these grounds, the discussions held at **COP 1 in Berlin** took a slow start. Essentially, the COP was to deal with four issues: reviewing both the adequacy of commitments under Art. 4.2 UNFCCC and the first national communications, deciding on the institutional as well as procedural frameworks of the climate regime (ENB 1995a: 2).

The latter two, purely organizational matters were dealt with during the opening plenary (ENB 1995b: 1). Two decisions are worthy of mention, as they would gain importance during the further negotiation process. First, no consensus was reached on the adoption of the rules of

---

[5] To visualize the different sub-plots, the level of analysis discussed in a paragraph is highlighted in bold.

procedure, which implied that the draft rules of procedure would be applied during the COP. This, in turn, meant that decisions generally had to be taken by consensus (UNFCCC 1995d: 9).[6] Further, the parties elected INC Chair Raul Estrada-Oyuela as chairman of the Committee of the Whole (COW), the body that was going to deal with the leftovers of the INC process (ENB 1995b: 2). The Argentinean would later also be elected Chair of the Ad Hoc Working Group on the Berlin Mandate (AGBM) and would remain in this function until the end of 1997.

During the first meeting of the COW, debates circled around the most important agenda item, the adequacy of commitments and the specific question of what outcome this COP should produce. The G-77/China highlighted the responsibilities of developed countries, stating that the negotiations should above all lead to the implementation of existing commitments (ENB 1995c: 1). AOSIS, supported by Norway, called for the adoption of its own protocol proposal (ENB 1995c: 1). France, the EU Council Presidency at the time, expressed the Union's preference for the conclusion of a protocol mandate and the creation of an ad-hoc working group to negotiate such a protocol (ENB 1995c: 1). It further reiterated the EU's positions clarified at INC 11 (see Chapter 2). The US, after prior submissions in which it had clarified that "we do not come (...) with specific proposals" (UNFCCC 1995b: 76), made reference to the Clinton Climate Action Plan and highlighted the need to come to an agreement with broad international participation covering the period beyond the year 2000 (ENB 1995c: 2). Other countries from the JUSSCANNZ group were more explicit. Australia called for guidelines for the negotiation of a protocol that would include emission reduction efforts for all countries (ENB 1995c: 2). Similarly, New Zealand wanted a clear mandate for a protocol, legally binding targets and the inclusion of major developing countries in future emissions reduction efforts (ENB 1995c: 2). Finally, a group of countries, including Russia and Saudi Arabia, argued that climate science was not yet fully reliable and that talks on a new agreement should be postponed until the second IPCC report, expected for late 1995 (ENB 1995c).

As the first week of the conference evolved, it was especially the position of this latter group that slowed talks down and would, ultimately, lead to a split of the G-77/China bloc. When it had become patently clear that growing tensions between the progressive stance of the AOSIS coalition and the advocates of a status quo, mostly OPEC, made it impossible for this bloc to define coherent positions, Chair Estrada interrupted the negotiations to give the group time to consider its options (ENB 1995d: 2).

---

[6] This situation has remained unchanged ever since COP 1.

A smaller faction of 42 like-minded states, including AOSIS members, Argentina, South Korea, South Africa, China, and – as lead country – India, broke from the G-77/China coalition. This group managed to present a common approach, referred to as "green paper", on 2 April (ENB 1995e: 2). In its "Proposed elements of a mandate for consultations on commitments in Article 4, paragraph 2 (a) and (b)", it clearly outlined its stance on the issues of targets and timetables and inclusiveness. On the one hand, it called for a negotiation process to be based on the protocol proposal by AOSIS, whose "first priority" it was to set "specific and legally binding reduction targets (for example, Toronto Targets) within specific time-frames for emissions by Annex I Parties" (UNFCCC 1995c: points 7 (a) and (c)). On the other hand, it demanded that "the consultations will not introduce *any new commitments whatsoever* for developing country Parties" (UNFCCC 1995c: point 4, emphasis added). The proposal had been the fruit of discussions between the like-minded countries and environmental NGOs, who had purportedly helped to draft it (Oberthür/Ott 1999: 46).

Reactions to this paper from JUSSCANNZ as well as from the remaining G-77 members were quick and mostly skeptical (ENB 1995e: 2). By contrast, as the first major party to support the green group's proposal, the EU stated that it was prepared to exclude discussions of developing country targets from the future negotiation process. This decision had been enabled by a move of the German hosts, whose previous proposals had timidly indicated the contrary, namely the expectation for some form of effort from developing countries (Oberthür/Ott 1999: 46; INC 1994b; Steffek 2005: 243). It was supported by all EU members, who partially accepted the rationale of the developing countries' equity-based argumentation and certainly did not want to jeopardize the agreement on a mandate for a new negotiation process (Interviews EU, US representatives 5, 17). As JUSSCANNZ and OPEC continued to be more reluctant to granting such a wide-reaching guarantee to the developing countries, the negotiations focused very much on the question of developing country obligations in the final days, opposing a coalition of the majority of G-77/China, the EU and the NGOs to JUSSCANNZ, OPEC and business interest groups (Oberthür/Ott 1999: 46). In the end, given their members' interest in reaching an agreement to start a new negotiation process, the US and other JUSSCANNZ members gave in to the pressure from the green coalition and the EU, anticipating that the issue of developing country participation could still be re-discussed at later stages in the negotiation process (Interview US representative 17). This anticipation was made on grounds that the wording of the original demand in the "green paper" of the 42 like-minded countries, which spoke of no "new commitments whatsoever" (UNFCCC 1995c: point 4), had been slightly modified: still maintaining that "no new commitments" should be introduced, the passage in the Berlin Mandate called for a reaffirmation of commitments in Article 4.1 and the duty "to continue to advance the implementation of these commitments in order to achieve sustainable development" (UNFCCC 1995e: 5; Interview US representative 17). Generally, the ultimately adopted "Berlin Mandate: Review of the adequacy of Article 4, paragraph 2 (a) and (b), of the Convention" began a negotiation "process to enable it to take appropriate action for the period beyond 2000" (UNFCCC 1995e: 4). This process was to be completed "as early as possible in 1997" (UNFCCC 1995e: 6). The Mandate called, as a priority, for the adoption of quantified emissions reduction targets for industrialized countries (UNFCCC 1995e: 5). Further concessions to the JUSSCANNZ/OPEC bloc included a reference to additional analysis and assessment activities as basis for the talks, and the open formulation that this process could lead to a protocol or "another legal instrument" (UNFCCC 1995e: 4; Oberthür/Ott 1999: 47). An ad hoc group of parties, the AGBM, was created to conduct talks (UNFCCC 1995e: 6).

At the **first meeting of the AGBM** (21–25 August 1995, Geneva), discussions focused on the necessity and scope of the analysis and assessment foreseen in the Berlin Mandate (AGBM 1995b: 9). The EU did not play an active role in this round of talks, arguing solely, and together with most of the developing countries, for faster proceedings (ENB 1995g: 2–3). No substantive progress was reached, leaving the authors of the Earth Negotiations Bulletin to observe that "the sense of urgency was not readily apparent" (1995g: 7).

Moving beyond analysis and assessment, **AGBM 2** (30 October– 3 November 1995, Geneva) continued discussions on the basis of a list of issues identified by parties (AGBM 1995c) and other written proposals (AGBM 1995b). Debates were mostly restrained to reiterations of positions stated earlier. For instance, the US had already made it clear that it considered that "we are not – at this stage – negotiating", but rather called for further talks on the science (AGBM 1995b: 82–83). As part of this exercise, US representatives presented findings indicating that the GHG emissions of the most advanced developing countries would exceed those of the industrialized world by the middle of the 21st century, making it necessary to come to a truly global solution to the problem of climate change (ENB 1995h: 2). The session was further marked by one substantial and partially new proposal, introduced on 30 October 1995 by the Spanish Presidency on behalf of the European Union. This first major EU influence attempt in the AGBM process was a fairly formal one: the outline of a structure for a protocol supplementing the UNFCCC (AGBM 1995d: 37–53). The proposal comprised six brief articles and three annexes. Its first article suggested the introduction of "policies and measures" (PMs) to be adopted by Annex I Parties. Three types of PMs were distinguished in Annexes A, B and C: measures common to all parties, high priority measures, and measures "for inclusion in national programmes as appropriate to national circumstances" (AGBM 1995d: 42). The proposal further included the EU's previous position on quantified reduction objectives "to be set within specific timeframes" (AGBM 1995d: 39). It reflected the EU's preferred "command and control" approach, with the Spanish paper even stating that "The EU has always been committed to a combined approach" including PMs and quantified targets (AGBM 1995d: 38). The submission did not, however, contain specific proposals on targets or numbers (Lescher 2000: 81). Regarding the issue of responsibilities, it reiterated the second line of the respective Berlin Mandate passage: the EU sought provisions on "continuing to advance the implementation of existing commitments by all Parties" (AGBM 1995d: 40). In first reactions, OPEC and China called such a proposal "premature", and no substantial advances were then also made at the session (ENB 1995h: 8).

In the report on the second session, delegates expressed hope that the next meeting – **AGBM 3** (5–8 March 1996, Geneva) – would "present an initial opportunity [for] narrowing the range of options" (AGBM 1995e: point 34), but the session did not live up to this expectation. Following the more in-depth discussion of the EU's protocol proposal, which had partially set the agenda for these talks, it did, however, begin to see a conflict emerge between the EU, supported by Japan, and several other parties (US, OPEC): while Europeans and Japanese advocated PMs linked to quantified emissions targets, the latter were opposed to either the PMs (like the US) or both targets and PMs (like the OPEC members) (AGBM 1996b: points 41, 42). On the issue of targets, parties began to state more substantial proposals: apart from AOSIS' reintroduction of its protocol draft, Germany, as an individual party and outside the EU context, suggested an overall 10% reduction of CO_2 emissions by 2005 and of 15–20% by 2010 (compared to 1990 levels) (ENB 1996a: 6). Without mentioning any clear commitments, the Italian EU Presidency, referring to the second IPCC report adopted in mid-December 1995, indicated that a GHG concentration level of 550 ppm should guide the setting of a target (ENB 1996a: 5). Australia, the US and Canada supported the idea of targets, linked to flexibility, but did not table any numbers (ENB 1996a: 6). Differences surfaced on the more technical questions of whether to adopt a multi-gas, gas-by-gas or comprehensive approach (AGBM 1996b: point 44(e)). All in all, parties slowly entered into a more substance-oriented negotiation mode by clarifying their positions with regard to the Berlin Mandate.

AGBM 4 and the second conference of the parties coincided (8–19 July 1996, Geneva). Regarding the two issues of targets and inclusiveness of the regime, talks were mostly held under the AGBM track. The main foci of the debates were quantified emissions targets, and, linked to this, policies and measures. Positions, as far as expressed, had been gathered in several compilations of proposals (AGBM 1996a and c). They testified to the marked differences in approach between the EU, supported by most of the developing countries, favouring policies and measures and opposing differentiation on the one hand, and the US and most of JUSSCANNZ, demanding more flexibility, on the other hand (AGBM 1996a). The Japanese position had now moved to occupy the ground somewhere in between, advocating differentiation and flexibility, but equally seeing possibilities to combine policies and measures with targets (AGBM 1996c: 2). Despite these differences, the EU repeatedly pushed for an acceleration of the negotiations, arguing for the preparation of a draft protocol as soon as AGBM 6 (ENB 1996b: 7).

The high-level segment of COP 2 brought new impetus to the debates (Grubb *et al.* 1999: 54). The discussion of the Second IPCC Assessment Report was taken as an opportunity by the US to considerably clarify its position. In a speech on 17 July 1996, US Under-Secretary of State for Democracy and Global Affairs Timothy Wirth stated that – on the backdrop of the latest science – "we must do better", which is why "the US recommends that future negotiations focus on an agreement that sets a realistic, verifiable and *binding* medium-term emissions target", to be achieved through flexibility mechanisms such as "reliable activities implemented jointly, and trading mechanisms around the world" (Audio 1996, 17 July 1996, emphasis added). According to Wirth, PMs did not fulfil the criteria of being "realistic" or "achievable". Although it did not specify any reduction target, this statement opened up new possibilities for collaboration, as it "was in fact the first time that any major Party had specifically called for quantified commitments adopted under the negotiations to be made binding" (Grubb *et al.* 1999: 54). With a simple declaration, the US had thus unexpectedly joined the group of parties (above all AOSIS, the EU) who acknowledged the urgency with which wide-reaching measures against climate change had to be taken. At the same time, it had also positioned itself clearly *vis-à-vis* those who wanted to slow negotiations down, especially OPEC, but also Russia (ENB 1996b: 4). Yet, Wirth's forceful statement in favour of the concept of flexibility, introducing the idea of emissions trading, also effectively coupled the adoption of legally binding targets – the core of the whole regime and main purpose of the Berlin Mandate – to flexible trading mechanisms. This stance exacerbated the differences between the advocates of flexibility from the JUSSCANNZ group and the EU, supported by many developing countries (Jordan/Rayner 2010: 62).

The EU greeted the US change in position and, together with the majority of the G-77/China, took the Americans up on their promise to accelerate the negotiations (ENB 1996b: 10). In the final days of the COP, a "Friends of the Chair" working group drafted a non-binding declaration, which the majority of parties adopted in plenary against the opposition of 16 parties, including Russia and Saudi Arabia (ENB 1996b: 10, 13). The Geneva Declaration reiterated the Berlin Mandate, endorsed the SAR and reminded Annex I countries "to limit and reduce emissions" faster (UNFCCC 1996: point 2). Further, the Ministers "instruct[ed] their representatives to accelerate negotiations on the text of a legally-binding protocol or another legal instrument" including "policies and measures" and on "quantified legally-binding objectives for emissions limitations (…) within specified time-frames" (UNFCCC 1996: point 8). The latter emphasis on legally binding targets and timetables marked the true novelty for the AGBM talks.

The negotiations in the Working Group had to wait until their sixth session to see some form of acceleration. Until then, talks would stagnate in the positioning phase. For **AGBM 5** (9–13 December 1996, Geneva), the UNFCCC secretariat had prepared a synthesis of proposals by the parties as a basis for further discussions (AGBM 1996e). This document and the debates at this session demonstrated to what extent the talks had become more complex after the US had opened up to the idea of legally binding reduction targets. Setting such targets required the prior settling of numerous other issues regarding inter alia gas coverage, level and timing, distribution of commitments and degrees of flexibility (ENB 1996c: 4). Following an invitation from the Chair to submit new proposals, the EU made its second major influence attempt in these negotiations when it submitted an "elaboration" of its draft protocol structure (AGBM 1996d: 19–24). The proposal contained different types of commitments and introduced a new category of parties, "Annex X parties", which were to adopt quantified emission targets. The submission suggested that Annex X should include not only developed parties, but also other, not further specified ones. This indeterminacy caused some confusion in the talks, and the EU had to defend itself against criticism from the G-77/China bloc by clarifying that Annex X covered Annex I and new OECD members (ENB 1996c: 4).[7] After its strong position-taking at COP 2, the US also made two written submissions in 1996. In its "Elements of a new legal instrument" paper of 21 October, it called for "a shift away from unrealistic, near-term targets" to "legally-binding, medium-term targets that are both realistic and achievable" (AGBM 1996d: 50–54, 51). In its more extensive "non-paper" of December, it underscored the need for a longer-term goal, and "for all nations, including developing nations, to take actions to limit greenhouse gas emissions" (AGBM 1996a: 26–37, 26).

---

[7] The EU provided a list of Annex X countries in July 1997 (AGBM 1997e).

On the issue of targets, it made several proposals regarding, above all, a multi-year instead of a single-year target, banking and borrowing of emissions, a flat-rate approach for all Annex I countries and comprehensive gas coverage. Further, the paper contained elaborations on the flexibility mechanisms championed by the US (emissions trading, joint implementation). During the debates at AGBM 5, the US "hard commitments to soft targets" and long-term approach came into apparent conflict with the EU's, AOSIS and Norway's preferences for legally binding early Annex I actions (ENB 1996c: 4). On differentiation, however, the US and the EU were united in their argument for a flat-rate approach – against other JUSSCANNZ members (Canada, Japan, Australia and Russia), who supported different targets for Annex I countries (ENB 1996c: 4). Finally, regarding flexibility, which had become inextricably linked to the targets, both JUSSCANNZ and the EU supported joint implementation. The Union was, however, much more cautious about emissions trading, which it did not believe to be able to replace PMs (ENB 1996c: 5). The session concluded with parties authorizing the Chair to provide a compilation of these proposals (AGBM 1996g: 8).

Already for AGBM 5, many EU member states, including the UK, Germany and the Netherlands, would have liked to see **the European Union** produce a concrete proposal for reduction targets (Kanie 2003: 349–350). Under the impression of the SAR, the Union had, in fact, geared up its efforts, calling for global average temperatures not to exceed "2 degrees above pre-industrial levels" (Council 1996: para. 3). This, in turn, "should guide global limitation and reduction efforts" (Council 1996: para. 6). The idea of proposing concrete emission reduction targets that would meet this overarching objective had been tabled under the Italian Presidency in the first half of 1996, and discussed in depth under the Irish Presidency in the second semester of that year (Kanie 2003: 349–350). Prior to AGBM 5, the Environment Ministers had, however, been unable to define a common position due to difficulties related to sharing the burden of emissions reductions (Kanie 2003: 350; Agence Europe 1996). The incoming Dutch Presidency of early 1997 made the definition of such a position one of its key priorities. Consequently, the topic was taken up again in discussions of the Council Ad Hoc Working Group on Climate Change in February 1997, and earlier proposals of a 15% reduction by 2010 were re-discussed. In these debates, agreement on burden-sharing among the member states could, however, only be reached for a 9.2% reduction (Agence Europe 1997a; Vogler 2008). Nonetheless, on 3 March – the first day of AGBM 6 – the Environment Council took the decision that the EU would propose to reduce the emissions of three gases (CO_2, N_2O, CH_4) by 15% until 2010 (compared to 1990 levels) and that the burden-sharing of roughly "10% would be enough until Kyoto", as the remaining 5% could be distributed in accordance with the outcome of the negotiations at COP 3 (Council 1997; Agence Europe 1997a; for an in-depth discussion: Kanie 2003; Jordan/Rayner 2010: 63). This proposal would become the EU's third major influence attempt during this negotiation process.

Discussions at the **sixth AGBM session** (3–7 March 1997, Bonn) were thus continued on the basis not only of a framework compilation of positions, but also of the updated EU position and a comprehensive US submission (AGBM 1997a, b, c). The reception of the Union's target proposal on the second day of the meeting was decidedly mixed. Some parties, mostly from the G-77/China bloc, praised the ambitious numbers the EU had put forward, while others, notably JUSSCANNZ members like the US or Australia, criticized the insufficient burden-sharing agreement, calling the EU's proposal "unrealistic" (ENB 1997a: 1). The US had itself, in its submission of 17 January 1997, abstained from any kind of numerical specification, but had provided a quite detailed "Draft Protocol Framework" instead (AGBM 1997b: 78–87). In this document, it introduced the idea of "emissions budgets", underscored its position on flexibility, and provided for the voluntary inclusion of new countries into Annex I, while demanding measures for the "advancement of the implementation of Article 4.1" (obligations) for *all* countries (AGBM 1997b: 79, 82). Further, it suggested that the "Parties shall adopt, by [2005], binding provisions so that all Parties have quantitative greenhouse gas emissions obligations and so that there is a mechanism for automatic application of progressive greenhouse gas emissions obligations to Parties, based upon agreed criteria" (AGBM 1997b: 87). Similar proposals with regard to flexibility and non-Annex I commitments were introduced by Australia and New Zealand, inter alia in a joint submission with the US (AGBM 1997b: 65; ENB 1997c). The fact that both the EU and the US had now made fairly detailed and different proposals led to an intensified transatlantic polarization of the debates (ENB 1997c: 8–10). While most developing countries supported the EU approach, JUSSCANNZ was not completely behind the US position. Notable differences existed regarding the issue of differentiation among developed countries. Contrary to the US, Australia and Japan were strongly in favour of differentiating between Annex I parties. In a roundtable discussion on this issue, a Japanese representative argued that differentiation was an integral part of the Berlin Mandate, which had emphasized the "different starting points" of countries (ENB 1997b: 2). He further explained that three options existed for differentiation: a formula-based approach working with one specific indicator (e.g. per capita emissions), a selective approach (based on GDP) or a purely political negotiation approach (ENB 1997b: 2). During the same discussion, Australia used the EU's burden-sharing approach as example for differentiation (ENB 1997b: 2). At the end of the session, all main proposals – certainly regarding the two topics of interest here – with the exception of concrete (numerical) positioning on targets by key industrialized countries, were thus on the table for the remainder of the negotiations. For the following meeting, the UNFCCC secretariat was asked to prepare a negotiating text, following the "six-month deadline", which stipulated that half a year before COP 3 no fundamentally new ideas should be injected into the debates (Art. 17.2 UNFCCC; ENB 1997c: 1).

The fact that the end of the proposal phase was drawing to a close might have led several parties to clarify their stances also **outside the UN arena**. One forum for this was the G-7+1. At a summit held in Denver on 20–22 June 1997, the four EU members (France, Germany, Italy, the UK) advocated the Union's 15% reduction target, but only obtained a weak statement in favour of an agreement with some form of a target for 2010 (G-8 1997: point 16). Remarkably, given the clarity of the Berlin Mandate on this issue, the richest countries of the world also declared: "Action by developed countries alone will not be sufficient to meet this goal. Developing countries must also take measurable steps, recognizing that their obligations will increase as their economies grow" (G-8 1997: point 17). Some days later, at the UNGA Special Session (23–27 June), which was to assess the implementation of the 1992 Earth Summit decisions, several developing countries demonstrated their commitment to concluding a strong agreement in Kyoto, while US President Clinton explained his incapacity to exercise leadership given the continued resistance from Congress (ENB 1997d).

Talks at **AGBM 7** (28 July–7 August 1997, Bonn) were, for the first time, held on the basis of a 128-page synthesis of proposals that had been prepared by the Chair in late April 1997. The "negotiating text" presented alternative suggestions – in legal language – on all major issues, but no longer explicitly attributed specific submissions to particular countries (AGBM 1997d; Depledge 2005: 166). In several submissions prior to the negotiation session, the EU had given extensive, in-depth input into the negotiation process, commenting on the Chair's proposal and completing its own position (e.g. AGBM 1997f: 35–51; 1997q). It had also further clarified its emissions reduction proposal through the definition of an intermediate target of –7.5% by 2005 (compared to 1990 levels) in support of its overarching aim of reducing emissions to a level that would allow limiting global temperature increase to 2° Celsius (AGBM 1997g: 6). By contrast, none of the large countries of the JUSSCANNZ coalition had made a specific numerical proposal yet, with the US explaining that "it was not possible to decide what kind of numerical target might be undertaken without knowing what constraints would be imposed on such a target" (ENB 1997e: 2). Negotiations at AGBM 7 were conducted in four "non-groups": in addition to the ones previously established on commitments under Art. 4.1 as well as on institutions and on mechanisms, two new groups dealt with PMs and targets and timetables (ENB 1997e). Regarding the issue of *targets*, the "negotiating text by the chairman" listed 16 (!) alternative formulations (AGBM 1997d). During the debates on this topic, the non-group on quantified emissions limitation and reduction objectives (QELROs) took up the issues of a budget or multiple-year approach and of gas coverage. While the former remained unresolved, the comprehensive US proposal of covering six gases became generally accepted (ENB 1997f). Moreover, this group discussed – without reaching consensus – differentiation among Annex I countries (ENB 1997f). On the key conflict opposing supporters of hard targets and PMs (mainly the EU) to advocates of flexibility and soft targets (the US, Australia, New Zealand), the first possibilities for compromise emerged. It became evident that the EU was prepared to accept flexibility demands, notably trading, if adequate reduction commitments were made by all industrialized parties. In return, the US, despite its opposition to differentiation, demonstrated that it was prepared to accept the EU's internal burden-sharing if trading would become a prominent feature of the final agreement (ENB 1997g: 12–13). As for the issue of *responsibilities*, the non-group on Article 4.1 ("further commitments") originally discussed on the basis of two texts: a negotiating text and a text by the group's Chair (ENB 1997g: 5). The latter text contained two alternatives regarding the inclusiveness of the regime. One supported by the developing countries and the EU stressed the CBDR principle and, in line with the Berlin Mandate, refused the introduction of commitments for non-Annex I parties. The other stated that such parties should take measures contingent upon the implementation of Annex I country commitments, finance and technology transfer (ENB 1997g: 5). In a previous proposal, the EU had recognized that "in the long term emissions of greenhouse gases from countries not included in Annex I must also be regulated if the long term objective of the Convention is to be met (…) this should be considered as one element in the first review of the Protocol" (AGBM 1997g: 14). This soft wording contrasted with the approach inherent in the US framework protocol proposal of "automatic application of progressive (…) obligations" to all parties (AGBM 1997b: 87). This latter proposal had already been absent from the Chair's negotiating text and was effectively discarded by Estrada during the talks as not falling under the Berlin Mandate (ENB 1997g: 13; Oberthür/Ott 1999: 229). The EU's formulation was equally not considered further. While parties had thus managed to develop their positions during this session, reflected in the reports by the chairs of the informal groups (AGBM 1997h; Depledge 2005: 166), no agreement on any of the issues under debate could be reached. Recognising the importance of targets to the final agreement, the Chair demanded that "two Parties' target definition" (i.e. those of the US and Japan) should be unveiled (ENB 1997g: 3).

Before the final AGBM and COP, the pressure on the JUSSCANNZ members to advance clear positions on the reduction targets had thus become intense. Especially the host country of COP 3 was expected to demonstrate its willingness to reach an agreement: on 7 October 1997, after lengthy internal negotiations, Japan was then also first to yield to the pressure when it presented a 5% emissions reduction proposal over the period 2008–2012 for three gases (CO_2, N_2O and CH_4) (Matsumura 2000: 11–15; Schröder 2001: 44–45). The proposal represented a compromise between the Ministry of International Trade and Industry, in favour of stabilization, the Environment Agency of Japan, advocating effective mitigation policies and thus higher targets, and the Foreign Ministry, which was above all concerned with Japan's image as a global player (Kameyama 2004: 72–73). It can be interpreted as seeking middle ground between the EU and the US (Schröder 2001: 46; Matsumura 2000: 14). In the US, the debate about an international climate agreement had, in the meantime, become very complex. Whereas the Republican-dominated Congress remained hostile towards environmental regulation, public opinion did not display great concern about climate change (Harrison 2000: 104–105). Against this backdrop, the Senate had, a few days before AGBM 7, adopted a brief resolution sponsored by Senators Byrd and Hagel. Adopted by a 95-0 vote, it sent a very clear message to the US negotiating team: the Senate would ratify a newly negotiated international climate treaty only "if the protocol or other agreement also mandates new specific scheduled commitments to limit or reduce greenhouse gas emissions for Developing Country Parties within the same compliance period" (Senate 1997: point 1(A)). Under these circumstances, the Clinton administration had not yet decided on a target proposal before AGBM 8.

The **EU** used the period prior to AGBM 8 to reaffirm and further refine its negotiation position. On 1 October 1997, the Commission published "The EU Approach for Kyoto", which explained the logic behind the Union's internal strategy for emissions reductions and its international negotiation position (European Commission 1997). In this document, the Commission bluntly stated what the EU demanded from other parties in the global talks. Firstly, "all industrialized countries must be committed to comparable action", i.e. GHG reductions of 15% by 2010 (European Commission 1997: 18). Secondly, "it is important (…) that the more developed among the developing countries gradually assume bigger responsibilities when their level of development justifies it. There is no room for free riders on this issue" (European Commission 1997: 19). With this latter position, the Commission signalled, albeit ambiguously, that the EU was not going to demand any (binding) commitments from the developing countries in an agreement that would result from the Berlin Mandate negotiation process. After this process, however, the situation needed to be reassessed (Interview EU representative 5).

At the beginning of **AGBM 8** (20–31 October 1997, Bonn), talks were held on recent proposals, some of which had found their way into the new negotiating text (AGBM 1997i). On the issue of targets, the G-77/China had been able to reach a compromise:[8] it called for 7.5% reductions by 2005, 15% by 2010 and 35% by 2020 (compared to 1990 levels) for the same three gases as in the EU and Japanese proposals, and for a phasing out of other GHG gases, including the three additional ones proposed by the US (ENB 1997h: 1). Despite this slight difference with the EU's approach, the numerically largest negotiation bloc thus backed the Union's proposal, with many of its members giving up support for the more ambitious AOSIS proposal, and adding an additional target demand for the long term. The G-77/China's proposal was further strongly opposed to the notion of flexible mechanisms. Only hours after this proposal had been publicized, US President Clinton unveiled the last missing pieces in the US negotiation position. In a speech held at the National Geographic Society in Washington, he suggested stabilizing the emissions of six gases (the three also mentioned in all other proposals plus HFCs, PFCs, and SF_6) at 1990 levels during the budget period 2008–2012, and to reduce them over the period 2013 to 2018 (National Geographic Society 1997). Further elements of this proposal concerned the – at this point not surprising – introduction of flexibility mechanisms and, importantly, the meaningful participation of key developing nations, a demand that was de facto imposed on the US negotiators by the Byrd-Hagel resolution (National Geographic Society 1997). With regard to this latter point, the US would express its disappointment at AGBM 8 with the fact that the proposal by the Chair did not mention commitments by all parties in the medium term, arguing that the Kyoto agreement should constitute a starting point for also thinking about developing countries commitments (ENB 1997h: 2). With its lack of ambition regarding targets, the US proposal was, reportedly, met with disappointment and scepticism by other negotiating blocs such as AOSIS as well as by the environmental NGO community (ENB 1997i: 2). In the plenary of 27 October, the EU criticized the US proposal as even less ambitious than the Japanese, and altogether insufficient "to produce the outcome the world needs" (ENB 1997j: 2). With all elements now on the table, AGBM 8 concluded the negotiations regarding some technical

---

[8] This proposal was arguably the fruit of internal concessions by OPEC on the 2010 target and by AOSIS on the issue of a compensation fund for oil-producing countries.

points, while leaving the major political issues undecided, as parties remained unwilling to consider compromise formulas on issues like emissions trading, gas coverage or PMs, until an agreement on the targets was crafted. Chair Estrada suggested that a future negotiation text should best re-state all ideas that were on the table, but leave the numbers out for separate consideration (Audio 1997a, 29 Oct. 1997), and follow the approach that "nothing is agreed until everything is" (ENB 1997k: 6). He would prepare a new text that would be discussed at a reconvened AGBM 8 session the day before the Kyoto conference was to start (ENB 1997k: 11).

The **month of November 1997** marked the final opportunity for parties to informally explore options for agreement before the COP. Intensive bilateral talks were conducted, inter alia, between the US and the EU, Japan and the EU, and the US and Japan (ENB 1997b: 16–17, Interviews EU, US representatives 5, 17). On 8 and 9 November, Japan invited ministers from a selected number of players from JUSSCANNZ (among others the US, Canada, Australia, New Zealand), from the EU (especially the UK, Germany, the Netherlands, Luxembourg, France, Italy, the European Commission), and from the G-77/China (inter alia China, India, Brazil, Argentina, Indonesia, Samoa, Saudi Arabia) as well as Russia and Chair Estrada for a final preparatory meeting (Oberthür/Ott 1999: 78). Although positions were further clarified, no major disagreements were solved at this meeting.

The **resumed AGBM 8** of 30 November 1997 was as inconclusive as its predecessors (Oberthür/Ott 1999: 80; ENB 1997l). In the absence of clear results, almost all decisions needed to be taken at COP 3 (1–11 December 1997, Kyoto). Not many party proposals were actually off the negotiation table: the changes made in the Chairman's "revised text under negotiation" were predominantly linguistic and stylistic rather than content-specific, reducing the text to a manageable 32 pages, which still contained many brackets, however (UNFCCC 1997a). It thus reflected the entrenched positions at the outset of the conference. The analysis that follows focuses on tracing the evolution of the negotiations on the key topics analysed in this work. Concerning the issue of targets, four problems still had to be resolved: defining the assigned amounts of emissions reductions, deciding on a differentiation between parties, adopting a commitment period or one target year, and establishing the gas coverage. *Article 3* of the revised negotiation text listed the options (UNFCCC 1997a). Its paragraph 1 mentioned three alternatives with regard to the question of emissions budgets/commitment periods (alternatives A and B) or target years (alternative C, suggesting 2005, 2010 and 2020), joint or individual fulfilment (with only alternative A allowing for both), and two alternatives regarding the number of gases, listed in Annex A (three vs. six). The Article did not specify any reduction targets. Paragraph 2 took up the issue of differentiation of emission reduction commitments by industrialized countries, suggesting either a flat-rate approach (preferred by the US and the EU) or an indicator-based approach following a complex procedure listed in draft Annex B (Japan, other JUSSCANNZ members). Further, the bracketed paragraphs 7 and 8 suggested two consecutive emissions budget procedures. The proposed alternatives on all issues reflected above all the US, Japanese and EU proposals. On the issue of responsibilities, *Article 10* of the draft, reportedly inserted by Chair Estrada himself, allowed for emission reduction measures by non-Annex I countries on a voluntary basis (Oberthür/Ott 1999: 229–230). At the same time, *Article 12* of the draft agreement reaffirmed and restated "existing commitments in Article 4.1 of the Convention", "without introducing any new commitments for Parties not included in Annex I", reflecting formulas used in the Berlin Mandate (UNFCCC 1997a).

To conduct negotiations on these items, the COP split into numerous groups (inter alia on QELROs, on gas coverage and differentiation, on financial issues, on institutions and on commitments under Art. 4.1) **during the first week**. Regarding *targets*, the first major change of position at the conference came, to everyone's surprise, from the US. Its negotiators suddenly signalled openness for differentiation when calling for a working group on this issue to allow for the discussion of a proposal by Russia for a "big bubble" of all industrialized countries, inspired by the EU's burden-sharing construct (ENB 1997m: 1). In the first meeting of the negotiating group on QELROs on 2 December, chaired by Estrada himself, discussions were held on this issue (ENB 1997n: 1). Canada proposed, for example, to take on individual GHG reduction targets of 3% and 5% by 2010 and 2015 respectively, both referring to mid-term points of budget periods (ENB 1997n: 1). Further, the idea of a 0–5% range of reductions, with differentiation between Annex I countries, inspired by the Japanese proposal, was taken up (ENB 1997n: 2). In press conferences, the EU's negotiators indicated that a consensus on the issue of five-year budget periods was emerging, but that they were, at this stage, not prepared to accept a 0–5% differentiation range, as such a decision would require ministerial approval (ENB 1997n: 2). In the following days, various proposals with regard to targets were tabled, inter alia on the differentiation of Annex I countries into three groups (ENB 1997o: 2). To bring about a decision, Chair Estrada took the initiative to propose reduction objectives of 10% for the EU, 5% for the US and 2.5% for Japan (Schröder 2001: 79). For the EU, such a proposal seemed unacceptable, as it violated its continued preference for a flat-rate approach. In a first reaction, its negotiators therefore also defended again the idea of an overarching target and its own "bubble" approach, before suggesting that other countries try the same, e.g. the US under the North American Free Trade Agreement (NAFTA) (ENB 1997o: 2). Despite these exchanges, no major advancements were made on draft Article 3 by the end of the first week of the COP. As far as *responsibilities* were concerned, the only serious proposal to change draft articles 10 and 12 in the Chair's negotiating text during COP 3 was made by New Zealand in a plenary session of 5 December. The delegation called for "progressive engagement" of major developing countries in the form of limits of emission growth after 2014 (ENB 1997p: 2). This proposal was strongly supported by the US, and more moderately by Japan and the EU, who had made similar proposals before AGBM 7 and in the Commission's "The EU Approach for Kyoto" communication (ENB 1997p). It was met with fierce resistance from the G-77, notably India, China, and Brazil, whose delegate stated that the developed country attitude ("If you don't deliver, we won't deliver") would not be accepted. Rather, he made his group's stance quite clear: "until you deliver, we don't discuss" (ENB 1997p: 2). As a reaction, the EU made reference to the Berlin Mandate, which precluded any further commitments by non-Annex I countries, but also called for a review process to talk about these commitments *after* Kyoto (ENB 1997p: 2). The developing countries' unequivocal stance on this issue, however, made a change to the draft negotiation text impossible (Schröder 2001: 87–88; Interviews EU, US representatives 5, 17). The first week of negotiations ended with a Committee of the Whole meeting during which some pieces of text agreed in the various working groups were discussed: while several technical and institutional issues had been settled, no consensus was reached on most of draft Articles 3 and 10 (ENB 1997q: 1). Further hotly contested issues included emissions trading and the concept of sinks. This state of affairs was reflected in the Chair's heavily bracketed non-paper, circulated on Sunday, 7 December, before the start of the ministerial segment of the COP (UNFCCC 1997b). In this version of the negotiating text, Article 3 on commitments had become even more complex, whereas Articles 10 and 12 had remained – substantially – unchanged.

All came down to the **second week of the COP**. The arrival of US Vice-President Al Gore on 8 December 1997 arguably "marked the beginning of the final phase of the conference" (Oberthür/Ott 1999: 85). In his highly mediatised speech on that day, Gore not only reiterated US positions (legally binding targets, flexibility mechanisms, participation of developing countries), but also instructed his negotiators to show more flexibility in the talks, thus effectively clearing the way for final deal-making (ENB 1997r: 1). Such bargaining would also be necessary: prior to Gore's intervention, the EU had already underscored its position against differentiation and hazardous flexibility, arguing again for PMs. Further, in an effort to ensure the support of G-77/China, it had called debates about developing countries commitments at this stage unhelpful and not in line with the Berlin Mandate (ENB 1997r: 1).

In the **COW** that followed the high-level segment, negotiations concentrated on the Chair's non-paper, with countries remaining unwilling to reveal new offers regarding QELROs, on which all other items clearly hinged (ENB 1997r: 2). Aware of the importance of new numbers, the Chair set a deadline for 3 o'clock in the afternoon of 9 December to have countries state their positions on quantified reduction targets (ENB 1997r: 2). In the meantime, active "backroom diplomacy" was conducted, involving bilateral and trilateral talks between the three key delegations – the US, COP host Japan, and the EU (Schröder 2001: 25; Matsumura 2000: 18–20). In these talks, the Union had increasingly become represented by the UK's Deputy Prime Minister Prescott, who formed, together with the Netherlands and the Council Presidency from Luxembourg, the EU Troika (Oberthür/Ott 1999: 86; Delreux 2008: 147). In bilateral exchanges between the US and Japan, the former began, for the first time, to evoke the idea of "symbolic reductions", showing preparedness to make a "small percentage cut" (Schröder 2001: 79). The results of these talks found their way into a revised draft by the Chair, presented late on 9 December, for the first time including numbers agreed among the "big three" parties (UNFCCC 1997c; ENB 1997s). Article 3.1 of the new negotiation text contained the idea of an overall reduction target for all Annex I countries of 5% for three gases (CO_2, CH_4, N_2O) between 2006 and 2010 (compared to 1990 levels), and differentiation between them (UNFCCC 1997c: 3). Targets for the other three gases that the US had wanted to include were to be decided at COP 4 (Art. 3.3, UNFCCC 1997c: 3). Under this proposal, the EU would have to reduce its emissions by 8%, the US by 5% and Japan by 4.5% (Annex A, UNFCCC 1997c: 24). Article 10 on voluntary developing country commitments had remained unchanged. The new text represented the basis for discussions until early in the morning of 10 December when Chair Estrada interrupted the talks by observing that draft Article 3 was in need of further informal and then formal consultations on the final day, concerning mainly the issues of each party's individual commitment to the 5% overall Annex I target, the number of gases covered and the timing of the commitment period (ENB 1997s: 2).

Until the final COW, scheduled for the evening of 10 December, important efforts were made informally to come to an agreement among the key players on these issues. The diplomatic operations carried out in the background involved the highest political echelons, with telephone conversations between the Japanese Prime Minister Hashimoto and US Vice-President Gore and among leaders such as US President Clinton, UK Prime Minister Blair and the German Chancellor Kohl (Oberthür/Ott 1999: 88; Matsumura 2000: 20). While it is close to impossible to reconstruct these private exchanges, judging by their outcome, they must have involved further preparedness for compromise on the part of the US, Japan and the EU regarding the central issue of the amount of quantified emissions reductions. The negotiation skills of Chair Estrada were then, according to observers, especially crucial for striking the final balance between them (Interview EU representative 5; Schröder 2001: 79; Andresen/Agrawala 2002: 48). From the EU side, it was the British Deputy Prime Minister who communicated the Union's preparedness to deviate from the flat-rate approach and negotiate an outcome in which the EU and the US met in the middle of the spectrum delimited by their respective positions. The EU was to accept an 8% emissions reductions target by 2010, while the US settled for −7%. Japan, as the host, could not afford to adopt a much lower target and thus ended up with a more ambitious aim than it had originally envisaged (−6%) (Schröder 2001: 80). Other Annex I countries, like Canada or Australia, basically made voluntary pledges unrelated to the outcome of the talks among the trio EU-US-Japan. Altogether, these pledges added up to a 5.2% reduction by Annex I countries. This outcome came in a package with an agreement on the commitment period: the EU had given in to US demands for a five-year period and an agreement was reached on the period 2008–2012 (Schröder 2001: 81). Finally, the issue of gas coverage was also settled by a compromise: all six gases proposed by the US were included, but the base year was changed to 1995 instead of 1990 for SF_6, HFCs and PFCs (Grubb *et al.* 1999: 74–75). Going beyond Article 3, the overall deal included further concessions to the US in the form of enhanced flexibility, which the EU was now prepared to accept.

The outcomes of these talks, as well as those of the various working groups, were considered in the **concluding COW** on the basis of a final draft protocol text shortly after 1 o'clock am on **11 December 1997** (UNFCCC 1997d). Estrada imposed a deliberation on this draft article by article, starting with "the letters, not the numbers" of the central Article 3, then discussing all the other articles, before returning to the emissions reduction targets referenced to in this article (and detailed in Annex B) (Audio 1997b, 11 Dec. 1997). Regarding the issue of *targets*, Article 3 of the new negotiation text contained the main fruit of the compromise between the US, the EU and Japan. Its first paragraph included the obligation for industrialized countries to "individually or jointly, ensure that their aggregate anthropogenic carbon dioxide equivalent emissions of the greenhouse gases listed in Annex A do not exceed their assigned amounts, calculated pursuant to their quantified emission limitation and reduction commitments inscribed in Annex B", regarding six gases over the period 2008 to 2012 (UNFCCC 1997d: 4). Estrada's article-by-article approach functioned smoothly until discussions came to paragraph 10 of draft Article 3, which allowed for the possibility of emissions trading (Audio 1997b, 11 Dec. 1997). Strong opposition against this flexibility tool from the majority of members of the G-77/China, forcefully voiced by the Chinese delegation, stalled the negotiations for quite some time, jeopardizing the overall agreement (Audio 1997b, 11 Dec. 1997; ENB 1997t: 12). After a short break, Estrada proposed to adopt the idea of emissions trading, central for getting the US and Japanese approval of the agreement, in a different article (16 bis, now 17 KP), effectively shifting the decision to a later conference of the parties (ENB 1997t: 12).[9] Article 3 could thus be adopted, and later also the numbers contained in Annex B. In return, the majority of the G-77/China obtained that the draft Article on voluntary *commitments* for developing parties (Article 9 in this final negotiating text, Article 10 in previous versions), and thus a central demand of the US delegation, was deleted, despite support for this article by AOSIS (Oberthür/Ott 1999: 230). Commitments for all parties were, however, included in draft Article 11 (later to become Art. 10 KP; UNFCCC 1997d: 11–12), which essentially reiterated previously existing obligations under the UNFCCC (Yamin 1998: 123). While the US had thus obtained its desired linkage between targets and flexible mechanisms, it had failed on the issue of developing countries commitments. The EU, in return, did not achieve a fundamental commitment to PMs. Article 2 of the Protocol contained an exhaustive list of PMs, but made none of them obligatory (for the in-depth discussion of the EU's influence, see below). Discussions ended towards 10 am on 11 December 1997, paving the way for the ultimate approval of the Protocol in the final COP plenary later that day (ENB 1997t: 8).

The Outcome: the Kyoto Protocol

The outcome of this 30-month negotiation process was the Kyoto Protocol (KP), an international treaty complementing the soft-law approach of the UNFCCC with some "harder" provisions regarding the ultimate objective of the regime (Art. 2 UNFCCC) and the means for reaching this aim (Bodansky 2001). Its entry into force would require the ratification of 55 parties to the UNFCCC, including Annex I parties representing 55% of the total emissions of all Annex I parties (Art. 25.1 KP).[10] This section provides a brief discussion of the core features of the new treaty as far as they are relevant for the specific thematic focus of this study, for the appreciation of the EU's influence and the overall understanding of the further analysis (for legal analyses of the Protocol, see Depledge/Yamin 2004; Oberthür/Ott 1999; Yamin 1998).

---

[9] The new text in what became later Art. 17 KP was a compromise formula proposed by the UK, on behalf of the EU (Audio 1997b, 11 Dec. 1997; Yamin 1998: 122).

[10] This was achieved in late 2004. The Protocol entered into force on 16 February 2005.

At the heart of this treaty lay the *legally binding numerical emissions reductions obligations* for Annex I countries in its Article 3.1. They provided a further specification of the overall objective of the regime embodied in Art. 2 of the Convention, which remained applicable. Art. 3.1 KP reads:

> The Parties included in Annex I shall, individually or jointly, ensure that their aggregate anthropogenic carbon dioxide equivalent emissions of the greenhouse gases listed in Annex A do not exceed their assigned amounts, calculated pursuant to their quantified emission limitation and reduction commitments inscribed in Annex B (…) with a view to reducing their overall emissions of such gases by at least 5 per cent below 1990 levels in the commitment period 2008 to 2012.

Emissions reduction pledges covered a range of six gases (Annex A) and were differentiated among parties, with country percentages listed in Annex B of the Protocol: the 15 EU members committed collectively to a –8% target, the US to –7%, Japan and Canada to –6%, Russia and New Zealand to stabilization, and Australia was allowed an 8% increase. Together, the individual targets added up to a total of 5.2% GHG reductions for the period 2008–2012 (compared to 1990 levels). Article 3 KP further contained specific rules reflecting the complexity of the negotiations on the targets and covering such issues as sinks and calculation (Arts. 3.3, 3.4, 3.7) or banking (Art. 3.13) (for details, see Yamin 1998: 118–199; Oberthür/Ott 1999: 121–123).

Further obligations with regard to reporting duties for Annex I parties to the new treaty were specified in Art. 5 KP: each of these parties had to put into place a "national system for the estimation of anthropogenic emissions (…) of all greenhouse gases not controlled by the Montreal Protocol". No new *responsibilities* were introduced for non-Annex I parties. Art. 10 KP essentially only reaffirmed the existing obligations under Article 4.1 of the Convention (Oberthür/Ott 1999: 232–233; Yamin 1998: 123).

For the purpose of this study, these provisions represented the central results of the negotiations. Two other articles were of particular importance to the EU: Article 2 KP listed a set of policies and measures including, for instance, "the enhancement of energy efficiency" or the promotion of "research on renewable sources of energy". The "such as" provision of this article implied that the measures were non-binding, reducing the list to an arguably useful, but indicative compilation of potential ways of cutting GHG emissions (Yamin 1998: 116). This was obviously very far from what the EU had originally intended with its protocol proposal (Grubb *et al.* 1999: 126). A second provision of specific interest to the EU was Article 4 KP on the joint fulfilment of obligations under the Protocol. It opened the way for the Union to "share the burden" and collectively meet its obligation during the Kyoto commitment period (Pallemaerts/ Williams 2006: 39–40). Yet, on this point again, the EU did not quite manage to reach its aim: it had originally planned to reserve the concept of joint fulfilment to regional integration organisations only, but later had to accept it as general option for all parties (Oberthür/Ott 1999: 140–145).

The treaty remained fairly ambiguous when it came to the major additional novelties, namely the three flexible mechanisms: joint implementation (Art. 6 KP), the Clean Development Mechanism (CDM) (Art. 12 KP) and emissions trading (Art. 17 KP) (Yamin/Depledge 2004: 136–196; Yamin 1998: 121–122). Joint implementation as a concept allowed for the exchange of "emissions reductions units" between Annex I countries. Emissions trading, at the time not clearly defined, involved a system of tradable emissions permits between parties (Yamin/Depledge 2004: 156–159). Finally, the CDM, as a newcomer of the final days of the Kyoto talks, allowed for industrialized countries to reduce emissions by financing climate-friendly projects in developing countries in return for "certified emissions reduction units" (Werksman 1998). Much uncertainty prevailed with regard to how these mechanisms – particularly emissions trading, which had almost been the cause for a last-minute breakdown of the negotiations – would be implemented in practice. As Articles 6.2, 12.7 and 17 KP shifted the discussions on the concrete use of each of these mechanisms effectively to future conferences/meetings of the parties, they would become the centre of discussions and sources of discord during the negotiations after 1997 (Dessai et al. 2003).

Concerning the institutional set-up of the regime, the Protocol relied to a large extent on the institutions that already existed under the Convention, with the exception of introducing a separate "meeting of the parties" (abbreviated MOP in the past, now referred to as CMP) (Art. 13 KP; Yamin 1998: 124). This meeting, although held in parallel to the UNFCCC COP, is legally distinct from the latter, as the parties to the UNFCCC and the Kyoto Protocol did not necessarily have to – and indeed in practice do not – overlap (Yamin 1998: 124).

Finally, several articles foresaw procedures for periodical reviews of and amendments to the Protocol. A review should take place in the light of the "best available" science, and "at regular intervals and in a timely manner" (Art. 9 KP). Article 3.9 stipulated that, if it was found to be necessary to alter emissions reductions targets, "the Commitments for subsequent periods for Parties included in Annex I shall be established in amendments to Annex B to this Protocol." These provisions opened the way for continued negotiations about the sufficiency and functioning of the regime, and would therefore acquire a certain importance in the further evolution of the global climate regime.

The EU's Influence Attempts: Extracting Patterns

A first observation with regard to the Union's influence attempts throughout this process concerns the negotiation context itself: the discussion of the process illustrates that the foreign policy activities on climate change were, despite growing politicization of the issue, very much confined to the UN negotiation arena.[11] The decisive talks were clearly held during the AGBM sessions and COPs (Interviews EU, US representatives 5, 17). Outside events (and related influence attempts), like the G-7+1 summits and several bilateral or small-scale multilateral meetings between key players beyond the UN framework, may have been complementary to the wider multilateral efforts, but the crucial foreign policy acts were clearly exerted under the UN negotiation track, at least until the final days of COP 3.[12]

The EU's foreign policy acts regarding these negotiations can therefore be analytically divided into exchanges with key partners (above all the US, but also Japan) *beyond* and, above all, *within* the UN framework (Interviews EU, US representatives 5, 17). Outside the UN arena, constant exchanges between climate negotiators were assured, primarily, by formal talks at Troika level, with the Presidency, but also other Troika members (like the UK at COP 3), playing a key role (Delreux 2008; Kanie 2003; Cass 2007). Bilateral meetings between single EU member states and other parties, such as China, Japan or Russia as well as consultations between, for instance, the UK and Australia or New Zealand also served the purpose of building confidence, explaining common positions and exchanging information (Oberthür/Ott 1999: 63–64). Finally, the EU was engaged in all major fora discussing climate change at the time, including the G-7+1 and other summits (Oberthür/Ott 1999: 59–63). In the more restricted realm of the UN negotiations, the EU, represented by the Troika, met with all major delegations on a regular basis, but had particularly intense discussions with the US and Japan throughout the entire process (Interviews EU, US representatives 5, 17). While these formal and informal exchanges proved important to bolster the flow of information, the most significant EU foreign policy acts tended to be rather formal and take the form of written submissions to the UNFCCC secretariat and oral statements during the negotiations.

---

[11] In this sense, the negotiations were much more restricted than they would become in the period thereafter (Interviews EU, US representative 5, 17).

[12] Notable other fora that were not highlighted in the general story included the OECD, where an "Annex I Expert Group" provided an arena for discussing contentious issues (such as PMs or emissions trading), but "could not heal the divides" among countries (Oberthür/Ott 1999: 59–60).

Given their centrality to the EU's foreign policy strategy, a brief recapitulation of the Union's most striking influence attempts throughout the negotiations serves to unveil their key characteristics:

- **AGBM 2** (30 October 1995): The EU proposes the outline of a structure for the protocol, prominently featuring PMs.
- **AGBM 5** (December 1996): The EU submits an elaboration of its draft protocol structure including various types of commitments and an Annex X that would allow new parties to join the Annex I country group.
- **AGBM 6** (3 March 1997): The Environment Council announces the EU's GHG reductions target of −15% until 2010 compared to 1990 levels for three gases. This proposal is later completed through a mid-term target of −7.5% by 2005.
- **BEFORE AGBM 8** (autumn 1997): The Commission recalls, in a communication, "The EU Approach to Kyoto". Bilateral exchanges are intensified.
- **COP 3** (December 1997): After a rather defensive start, the EU becomes a central player in the final bargain on targets and all other interlinked elements of the Protocol.

From an analytical perspective, this overview demonstrates that the range of foreign policy tools the EU employed was not very extensive. In essence, the main instruments that its influence attempts were based on were diplomatic. In line with the WEIS code, they can be described as "to make proposals", but also – in the later stages – "to demand" (Wilkenfeld et al. 1980). These proposals would often be linked to dialogues, visits and conferences, which would be exploited to explain the proposals, exchange positions, and search for commonalities. They were based mainly on conclusions by the Environment Council, which provided the negotiation position for the global climate talks and formed the basis for the Union's written submissions (through the Presidency) to the UNFCCC secretariat. Occasionally, Commission communications or conclusions of the European Council would also be employed as influencing tools. The propositions were often not directly targeted at any actor, but remained fairly general. Hardly ever did the EU employ any other foreign policy instruments to supplement this diplomatic approach. Its proposals for technology transfer and financing of climate mitigation or adaptation activities in developing countries, for instance, did not contain any incentives for developing countries to join into mitigation efforts, even on a voluntary basis.

Several characteristic features of the Union's main diplomatic tool – the written submissions – can be distilled from its contributions. First, the EU's proposals were substantially quite detailed – unlike, at least in the first fifteen months of the negotiations, the proposals of other parties – clarifying on most issues what it had decided that it wanted the outcome of the negotiations to look like. In the first instance, the Union concentrated to a large extent on its preferred regulatory approach, with extensive proposals on policies and measures. During this stage, its aim was clearly the creation of durable structures: it wanted to impact on the design of the key features of the future climate regime in accordance with its own "command and control" environmental policy tools. This approach contained a good deal of pragmatism. Until the US u-turn at COP 2, the EU had apparently thought that the Americans would never accept a new treaty with legally binding targets (Grubb et al. 1999: 54–55). It wanted thus an agreement with clearly defined measures, which would ensure that even reductions that were not binding could realistically be achieved. After COP 2, its stated position became gradually broader and more actor-focused, explaining also explicitly what it expected from other groups of countries: comparable efforts from industrialized countries, and the preparedness for assuming obligations in the medium to long term from advanced developing countries (European Commission 1997). These proposals, as seen and further analysed below, would set the agenda for the later stages of the negotiations, especially regarding the discussion of targets. The degree of detail of the proposals implied that many of the EU's positions were not very flexible, and had seemingly not been developed with an idea of potential concessions to other parties (and the corresponding internal fall-back positions) in mind. The EU essentially posited its approach and argued for its usefulness. This indicates that it may have wanted (consciously or not) to engage other parties in a process of arguing during the early stages of the process, interpreting the negotiations as a sort of competition for the best, most convincing ideas. This pattern recalls certain findings for the UNFCCC negotiation process, during which the EU had, for a long time, attempted the same, believing that the US could be convinced of the necessity to stabilize emissions by the year 2000. However, when the other industrialized parties held back their proposals on crucial agenda items for such a long time, the EU was the first major party to come out with its 15% GHG reduction proposal. This proposal has been interpreted as "more apparent than real" (Jordan/Rayner 2010: 63), but it was actually both: on the one hand, it represented a sincere wish by some of the European Environment Ministers of being capable of substantially reducing emissions in this range; on the other hand, it constituted a complicated position aimed at bargaining (Matsumura 2000: 18; Cass 2007: 78; Interviews EU, US representatives 5, 17). The fact that only 10% reductions were covered by the EU's internal burden-sharing agreement strongly indicates the importance of the latter interpretation. For the first time, the other parties – and those more reluctant to substantial reductions were clearly targeted (the US, other JUSSCANNZ members) – thus received the signal that the Union was generally willing to compromise, entering subtly into a bargaining mode. In the final stages of the talks, the EU multiplied such signals on other issues (e.g. emissions trading) taking on a more pragmatic stance and – again a parallel to the UNFCCC negotiations – were led by the British negotiator, in this case Deputy Prime Minister Prescott, to agree to a compromise formula. Second, and linked to this, the timing of the EU's detailed proposals was striking. The Union was, besides AOSIS, the first major player to make wide-reaching proposals about the shape of a new climate agreement. This proactive approach to the negotiations became very visible when it exposed, as the first Annex I party, its position regarding the key issue of the Berlin Mandate, the setting of targets. This move, at a critical moment in the AGBM process, earned it "renewed profile and initiative" (Grubb et al. 1999: 58). Despite inflexibility, the strategy of "frontloading" very detailed proposals arguably ensured the EU an important agenda-setting role and can partially explain its influence on this process, as further elaborated below.

In synthesis, a clear pattern of EU influence attempts in the Kyoto Protocol negotiations emerges from the process trace: the Union relied heavily on the quality, appeal, persuasiveness and timing of formal diplomatic tools. Early in the negotiation process, it exposed a rather well-prepared and wide-reaching position, which it gradually refined. Throughout the talks, it argued and reached out on the basis of this position, tending to become rather defensive in the final stages of the process. The EU's approach to multilateral negotiations and international lawmaking, in this case through the development of an existing legal regime, appeared thus as a unique mixture of a *legal-formalistic form* (tools of UN conference diplomacy) with a *highly political substance* (with a position centred on targets, but otherwise little "give and take").

The EU's Influence in the Kyoto Protocol Negotiations

As could be observed from the phased process analysis, the talks on the Kyoto Protocol went through several phases. Transitions between these phases regularly coincided with the evolution of the textual bases on which the deliberations were conducted (Depledge 2005: 166). This allows for isolating turning points in the narrative, enabling a determination of EU influence on the basis of a conditional causal analysis covering the five constitutive dimensions of the concept (see Chapter 1).

The negotiations remained in a pre-negotiating and issue definition mode for the first half of the AGBM process. AGBM 4/COP 2 then marked a first turning point because it provided a different framing for the further talks, with the US joining the group of parties in favour of legally binding emissions reduction targets. Entering into the more concrete positioning phase, talks were marked by several key proposals, notably regarding targets, before or during AGBM 8. As these proposals considerably reduced the number of options on the table, AGBM 8 can be interpreted as the second turning point. Final decisions were, however, mostly taken towards the end of COP 3, which marks the third turning point. Where these turning points concerned primarily the emissions reductions target, one more turning point needs to be added.

This *first, major turning point* during the negotiations on the Kyoto Protocol could be observed very early in the process. Regarding the issue of responsibilities (*who* is to take on emissions reductions obligations?), the evolution of the negotiations at COP 1 in Berlin proved crucial for the final outcome. As seen, the proposal of the AOSIS coalition prior to the conference and the negotiation position issued by the like-minded group of G-77/China members during the COP had categorically opposed the start of protocol negotiations if these should lead to the adoption of (binding) obligations for developing countries. The "green paper" issued by this group had therefore demanded that "the consultations will not introduce any new commitments *whatsoever* for developing country Parties" (UNFCCC 1995c: point 4, emphasis added). This position was quickly supported by the EU, eager to begin a new negotiation process and fearful that the opposition of developing countries could prevent a decent outcome of this COP.[13] This broad coalition of the majority of G-77/China and the EU, supported by the environmental NGO community, would build sufficient pressure to impose its view on the reluctant JUSSCANNZ and OPEC countries. The Berlin Mandate that was ultimately adopted contained a range of safety clauses for developing countries, including several references to the CBDR principle (UNFCCC 1995e). It made it patently clear that the newly begun negotiation process should "not introduce any new commitments for Parties not included in Annex I" (UNFCCC 1995e: point 2 (b)). This provision would become the point of reference throughout the further negotiation process whenever attempts were made by the US and others – including the EU, with several submissions calling for voluntary developing countries' commitments, and New Zealand with its sharply contested "progressive commitments" proposal in the final week at COP 3 – to convince key developing countries to do their share in the emissions reduction efforts, either voluntarily or in a future commitment period. On this basis, the G-77/China, assisted at times by Chair Estrada, employed its numerical superiority to successfully prevent any mention of new commitments for its members in the Kyoto Protocol. De facto, COP 1 in Berlin thus pre-emptively eliminated all other potential paths

---

[13] And this was despite the fact that it had previously timidly alluded to the necessity of emissions reductions by the most advanced developing countries (see EU submission to INC discussed above – UNFCCC 1994: point 9).

on this very issue, locking it in, and demonstrating thus an interesting example of enduring influence of the G-77/China bloc. This influence was fundamentally grounded in the group's structural achievements of the past: its successful insistence on the prominent insertion of the CBDR principle in Art. 3 of the UNFCCC during the period 1991–1992 would allow the bloc to constantly refer to this provision. This influence based on the exploitation of previously created structures was rooted in the size of the coalition, traditional attempts at keeping the lines closed in support of shared interests, but also in apt coalition-building with the NGO community and key countries from the EU at crucial moments in the talks (Rajan 1997: 283). At this turning point, and thus on the issue of the sharing of responsibilities between developed and developing countries during the Kyoto Protocol negotiations in general, the EU clearly failed to exert any influence. With its at times ambiguous position – which acknowledged the rights to development of the non-Annex I countries in the immediate future, but at the same time the findings of climate science that called for the eventual inclusion in collective mitigation efforts of those developing countries whose emissions were projected to rise steeply – it did not succeed in changing any other party's preferences or behaviour on this issue. Quite on the contrary, it had itself to change its position on the possibility of advanced developing countries emission reductions at COP 1, and repeatedly failed to build any sort of momentum for even the slightest mentioning of such (voluntary) action in the new treaty afterwards (absence of *temporal sequence*, no *goal attainment*) (Yamin 2000: 63). This, in turn, meant that the EU was unable to gain any leverage over one of the core pillars of the reformed climate regime.

COP 2 in Geneva marked a *second* "decisive turning point in the negotiations" (Grubb *et al.* 1999: 54), particularly regarding the core norm of emission reduction targets. Up to that point, the negotiations had been fairly inconclusive, with parties working on the assumption that the final outcome would not be legally binding (Grubb *et al.* 1999: 54–55). US Under-Secretary of State Wirth's speech at COP 2, in which he called for negotiating legally binding targets linked to flexible mechanisms, changed this approach completely. The consequences of this US change in position were manifold: in the short run, it allowed for the adoption of the (non-binding) Geneva Declaration, in which the ministers "instruct their representatives to accelerate negotiations on the text of a legally-binding protocol or another legal instrument", thus setting the negotiations on a clear track towards legally binding, quantifiable emissions reductions (UNFCCC 1996). Proposals by countries that had initially refused legally binding targets (Russia, OPEC) were thus clearly off the negotiation table. Moreover, the core of the AOSIS proposal (high emissions reduction targets) was, as aptly noted by an observer quoted in the ENB, virtually "dead in the water". Once obligations were to become legally binding, they would necessarily tend to be less ambitious because the US and other JUSSCANNZ members would never accept – in their view – exceedingly high binding objectives (ENB 1996b: 13). Finally, reduction targets became inextricably linked to the notion of flexibility mechanisms, effectively opposing the US and its coalition partners' quest for cost-effectiveness to the EU's command-and-control approach of PMs. Through a simple declaration of intent, the US exerted undeniable influence on the further course of the talks. The US change in position has been interpreted as a result of its reception of the latest scientific advances in the Second IPCC report. Wirth's statement made several references to the science, criticizing those who still contested it, thus effectively targeting his speech also at a domestic audience (Grubb *et al.* 1999: 54; Harrison 2000: 104–105). The modification of the US stance was thus clearly not the result of any other party's influence attempts. This, in turn, means that the EU did not exert any influence at this turning point. It neither was at the origin of the US change of beliefs or preferences, nor had it itself been successful in demanding a legally binding outcome (absence of *purposive behaviour, temporal sequence, interaction and goal attainment*). By contrast, and in a similar way as on the issue of responsibilities at COP 1 (where the EU followed the G-77/China), it was fairly easy for the Union's negotiators to support the US proposal (Grubb *et al.* 1999: 54). As a by-product, the US science-based and domestically motivated mind change effectively also altered the EU's negotiation strategy: where it had concentrated on PMs before, it quickly shifted its attention to targets after COP 2.

The *third* major turning point in the Kyoto Protocol negotiations occurred arguably before and at AGBM 8 when some of the key players would finally make public statements about their negotiation position regarding quantified emission reduction targets. While the positions on the details of an emission reductions article in the future treaty (gas coverage, commitment year or period, differentiation among Annex I countries) had been known for a while, the numbers that several critical parties were prepared to commit to had been held back until October 1997. From the major emitters, only the EU had openly defined its stance as early as March 1997, calling, as seen, for –15% by 2010, and later also for –7.5% by 2005 (compared to 1990 levels, for a basket of three gases). Before AGBM 8, Japan (–5% and differentiation between Annex I countries), the G-77/China (support for the EU's approach for each gas individually, plus a 35% reduction objective for 2020) and the US (stabilization of GHG emissions at 1990 levels over the period 2008–2012, linked to flexibility) clarified their stance. With these proposals, the number of options regarding the issue of an overall target was essentially limited to three alternatives: the maximalist EU and G-77/China proposal, the US minimalistic approach, supported by JUSSCANNZ, and the Japanese middle-ground proposal. Only a compromise between the extreme poles would allow for reaching agreement. The fact that the EU's proposal remained fully on the table demonstrates that the Union had exerted influence with regard to the core norm of the regime up to and beyond this important watershed. Checking the constitutive dimensions of influence shows that the EU had intentionally attempted to set the agenda on this issue (*purposive behaviour*), had reached out to promote it *vis-à-vis* other parties (*interaction*) *prior* to all other major players (*temporal sequence*) and had attained its (intermediate) goal of having its proposal considered among the final ones (*goal attainment*). Going into the ultimate stages of the talks, the EU had thus managed to substantially influence the agenda on this crucial item. With its arguably central influence attempt in the whole negotiation process, it had altered the behaviour of the G-77/China and JUSSCANNZ. While it may have been fairly self-evident that the G-77/China bloc would adopt a position similar to the EU's due to internal differences rendering a support for the even more ambitious proposal by AOSIS impossible, JUSSCANNZ members were forced to accept the EU's proposal as valid proposition, even though they did not agree with it (test on *absence of auto-causation*). The decision on what would become the core provision of the Kyoto Protocol, however, had to await COP 3.

With few final decisions taken, but some pathways excluded, COP 3 marked a *fourth* turning point at which parties had to determine the main features of the ultimate agreement via the conclusion of a package deal. Three parties (and Chair Estrada) would prove to be central to elaborating the final compromise: the US, Japan and the EU (Depledge 2005; Schröder 2001; Oberthür/Ott 1999; Grubb *et al.* 1999; Interviews EU, US representatives 5, 17). The entire agreement hinged on the decision on emission reduction targets for Annex I countries. In the course of the negotiations, this issue had become broader than "just" the definition of numerical targets. Unsettled questions at the beginning of the second week in Kyoto concerned not only the numbers, but also the modalities of emissions reductions (gas coverage, budget periods or target year, differentiation). Further, the issue of targets had – ever since the US change in position at COP 2 – been inextricably linked to the notion of flexible mechanisms, supported by JUSSCANNZ, but moderately opposed by the EU and strongly refuted by the G-77/China. Each of these issues merits brief discussion. The question of *gas coverage* opposed essentially the US (proposing six gases) to the EU and Japan (in favour of three gases). The US proposal for comprehensive coverage of all gases on the basis of the Berlin Mandate finally prevailed in Annex A of the Kyoto Protocol because the EU and Japan accepted US arguments, and successfully obtained as a concession that 1995 instead of 1990 was stated as a base year on the three additional gases (HFCs, PFCs,

SF_6) (Yamin 1998: 118, for the full story: Grubb *et al.* 1999: 69; 75–76; Oberthür/Ott 1999: 120). Concerning the *base year* and whether to adopt a target year or a *budget period* (UNFCCC 1997a: Art. 3.1, alternatives A and B vs. C), consensus was quickly found on the idea of 1990 as a base year, which had been a reference point in the discussions for a long time. With regard to the end date, the Berlin Mandate and earlier proposals by AOSIS, the EU and others had clearly favoured single target years (2005, 2010, 2020), whereas the US had argued that spreading the risk over budget periods (e.g. 2006–2010) would be a fairer solution. In the course of the negotiations, other parties, including the EU, had apparently come to accept the rationale of this argument (Oberthür/Ott 1997: 119). In the last versions of the negotiation text, the idea was thus adopted and only the name changed: from budget to "commitment" period. The Chair's proposition of 2006–2010 as the initial period was later changed, with the EU's explicit approval, to the original US proposal for 2008–2012. Hence, on these technical issues, the US position almost completely prevailed, with the EU bowing voluntarily to what it perceived as the better arguments in the context of an overall compromise regarding the emission reduction obligations. On the matter of *differentiation*, the EU was influential *malgré elle*. It had opposed the idea of differentiation other than for REIOs throughout the entire talks. Its own burden-sharing approach had, however, weakened its argumentation strategy. Surprisingly, the US was not among those in favour of differentiation at first, as it had accepted EU differentiation as a form of "zero-cost emissions trading" (Grubb *et al.* 1999: 86). This changed radically at COP 3 when the Americans suddenly signalled openness to differentiation on the basis of a Russian proposal. This, together with the fact that the Japanese were also advocating differentiation, heightened the pressure on the EU to abandon its resistance. Differentiation became thus the third element of the discussion on emissions reductions targets on which the EU had to give in, delivering – through its own "bubble" – itself the best argument for its opponents. The final component of the decision on the core norm of the regime, the *numerical targets* themselves, was negotiated between the EU, the US and Japan during the last few days of COP 3. A first quantified proposal by Chair Estrada, assigning 10% reductions to the EU, 5% to the US, and 2.5% to Japan, was refuted as unacceptable by the EU (Schröder 2001: 79). It was then the UK, as member of the Troika, that – as seen, in informal talks and with high-level telephone diplomacy going on in the background – successively negotiated more acceptable targets for the EU, compensating the Union for its concessions on other points, especially the flexibility mechanisms (Cass 2007: 79–80).[14]

---

[14] The UK was criticized in the aftermath of COP 3 for having given up on the EU's ambitious targets too early, despite the other member states', notably Germany's opposition (Delreux 2008: 147). Before and during the negotiations, many observers had actually

The final outcome, a 8–7–6% cascade for the EU, US and Japan respectively, can be interpreted as a classical compromise and very much the median between the extreme positions taken by the US (0%) and the EU (–15%), even if it meant very different efforts to the three parties (Interview EU representative 5).[15] For the EU, this outcome was acceptable, since it implied almost no differentiation between the large three GHG emitters, and was within the range of 10% reduction which the internal burden-sharing covered. Although the overall target of 5.2% reductions came close to their original proposal, it was Japan as much as the US that had to deviate from their preferred outcomes on this item. The other JUSSCANNZ countries, in return, would make arbitrary "voluntary pledges" based on their "willingness to pay" (Oberthür/Ott 1999: 120).

In the final analysis, the EU had thus managed to exert considerable influence on the core norm of the reformed climate regime: it had been the first major actor to proactively state a numerical target, backed up by internal burden-sharing (*purposive behaviour, temporal sequence*), and to demand comparable efforts of other countries (*interaction*), thus not only setting the agenda, but also determining the scope of what was to be discussed in the further talks. Its *goals* were partially *attained*: it managed to pull others away from their minimal positions to reach a fairly ambitious, legally binding overall target as well as fairly high individual targets for the other two major industrialized emitters. In counterfactual analysis, checking the negative pole of the concept of influence, the outcome may have looked much different without the Union: "Without the 15% figure and the determination of the EU to push the US and Japan to a comparable reduction (…), the targets contained in the Kyoto Protocol may well have been much weaker" (Yamin 2000: 55). Especially striking is the fact that the EU had gained leverage over emissions in both the US and Japan by effectively managing to change these countries' behaviour – not through argumentation, but bargaining on the basis of a comprehensive position on its target. As concession, and this was the non-negligible downside of its successful influencing, it did have to give up on most of its ideas on binding policies and measures, and on its initial opposition to the flexible mechanisms which had been a *conditio sine qua non* of the US position.

---

felt that the final outcome would hinge on "how hard-nosed the EU is prepared to be" at COP 3 (ENB 1997k: 16). To the surprise of some, including Chairman Estrada and Dutch minister de Boer, it appeared as not having been "hard-nosed" enough in insisting on its –15% target (Oberthür/Ott 1999: 121). Others have pointed to the fact that the target was considered, by the majority of member states, as a negotiation position anyway (Cass 2007: 78).

[15] For Japan, with an already comparatively energy-efficient economy at the time, this outcome arguably meant a far greater effort than for the EU or the US.

When extrapolating from the influence on the two core pillars of the regime, the EU's overall influence on the talks can be determined. As the Union almost entirely achieved its apparent real aim of 10% legally binding reductions for key emitters, its influence on the norm of the regime (the overall magnitude of the emission reduction target) must be regarded as fairly high, while leverage over the crucial principle of responsibilities was not discerned. Over many other issues not explicitly analysed in depth, its influence seemed equally limited. Assuming an equal weight of the two key analytical units, the Union's overall *degree of goal attainment* was thus medium. So was the *degree of durability* of the final outcome: although the Kyoto Protocol represents an international treaty, it embodied legally binding emission reduction obligations only for a clearly delimited and comparatively short time period. Altogether, the EU's overall influence on the Kyoto Protocol negotiations must therefore be assessed as medium. This observation is confirmed by the existing secondary literature specifically on the EU in the Kyoto Protocol talks (Yamin 2000; Gupta/Grubb 2000). It is also confirmed by the reputation analysis carried out for this time period (Interviews EU representatives 5, 21; non-EU representative 17, Observers 27, 3).

To further nuance the assessment of the Union's impact, it helps to set its performance into a broader perspective. A common evaluation of the Kyoto Protocol negotiations sums up the outcome as "The EU got their numbers, the US got their institutions, Japan gained some prestige and the developing countries avoided reduction commitments" (Andresen 1998: 28). From a longitudinal perspective, the EU's success on the numbers was, however, linked to a series of significant concessions to the other negotiating blocs, making it impossible to consider Art. 3 and Annex B in isolation from other elements of the Protocol. The Union's acceptance of the non-bindingness of policies and measures (Art. 2 KP) and of the insertion of various flexible mechanisms (Arts. 6, 12, 17 KP) set the regulatory tone of the regime for the post-COP 3 period: even though the new "institutions" (above all: emissions trading) obtained by the US were, just like the PMs, not legally binding, and even though they were not even properly specified in the treaty, the attraction of these cost-effective mitigation means would be much greater for most (Annex I) countries than the interest in PMs over the course of time. What is more, the EU's ambiguous stance regarding developing country commitments led to a difficult relationship with both the G-77/China and JUSSCANNZ in this period and the time thereafter. The G-77 did not appreciate the fact that the EU sometimes (e.g. at COP 1) supported developing country positions, but demanded at different points in time (before COP 1, in late 1997, at COP 3) future efforts from the most advanced developing countries. By contrast, JUSSCANNZ, notably the US, lacked an understanding of what in their view was a too timid support by the Union for demanding developing country actions, given the clear scientific diagnosis that emerging economies' emissions limitations were essential for the future efficiency of the climate regime. The collective inability of industrialized countries to convince developing countries of the necessity to – eventually – contribute to mitigation efforts would prove to become quite devastating for the success of the Kyoto Protocol in the long run. The Clinton administration delayed the submission of the Protocol for ratification in the Senate, as it obviously did not fulfil the conditions of the Byrd-Hagel resolution (Missbach 2000: 142). This contributed to the country's complete withdrawal from the Kyoto ratification process under the following administration. In hindsight, the EU's relative success on the targets must therefore be regarded as, at least in part, a Pyrrhic victory: in the short run, the EU had attained its aim of adopting a treaty with legally binding, albeit ultimately not enforceable numbers, but lost, for the medium term, possible leverage over US and developing countries' emissions.

Explaining the EU's Influence during the Period 1995 to 1997

To facilitate the explanation of EU influence, two sets of assumptions were made when designing the analytical framework for this study: causal mechanisms underlying the exercise of influence (bargaining, arguing), and potential variables that may qualify as conditions enabling or restraining EU influence were identified. Two analytical strategies can be employed that capitalize on these assumptions: pattern-matching and explanation-building (see Chapter 1).

The story of the negotiations indicates that their decisive final stages must be considered as a relatively unequivocal example of classical bargaining (Rowlands 2001), especially regarding the crucial issues of targets and responsibilities. For one, the outcome, a "package deal" linking numerous issues, represented a compromise par excellence: in simplified terms, the acceptance of targets by the US depended on other parties' preparedness to integrate flexible mechanisms into the treaty regime, while the approval of the latter by the G-77/China was conditional on the exclusion of new commitments for developing countries. A preference for bargaining was also visible in the approach to the negotiations taken by the Chair, who repeatedly emphasized that "nothing is agreed until everything is" (ENB 1997k: 6), and in most parties' tendency to "backload" proposals (Depledge 2005: 171). Finally, no key party changed its beliefs or preferences on the key issues during the negotiations, with the exception of the US on the legal bindingness of targets at COP 2, arguably under the impression of new scientific findings. Parties only modified their behaviour to allow for an agreement at the very last stage. This confirms the above assumption that the EU's exercise of influence must have been bargaining-based. In other words, the causal mechanism that led others to alter their behaviour and move into the Union's direction on the targets was bargaining rather than persuasion. The questions that remain are how and why this causal mechanism was brought to bear on the issue of targets, and not, or to a much lesser extent, on other issues. The EU's influencing strategy must have met with favourable conditions in this respect, while it proved to be less suited to the negotiation context (or the discussed issues themselves) on other agenda items. In search for the scope conditions that triggered the causal mechanism, all variables identified in Chapter 1 are briefly considered: the global context and significant events, the issue of climate change, the internal dynamics of the regime, including actors' beliefs and preferences, positions, capacities and their interaction.

No significant *external* events were detected during this time phase that could have or did impact the climate negotiation arena. By contrast, advances in the scientific findings on *the issue of climate change* in the second IPCC report of late 1995 marked the negotiations in the spring and summer of 1996, enhancing the willingness of parties to consider legally binding targets. As for the *regime dynamics*, the fact that negotiations were mostly based on a bargaining rationale made interests and strategic behaviour central to the final outcome. This makes it necessary to reconsider the evolution of the interests (and their formation), strategies and capacities of the major actors throughout the negotiations. For the US, the intricate internal institutional set-up and the highly conditional support of the Senate for an engagement in global climate talks made it extremely difficult to define a position in the first place. Nonetheless, once it had defined its stance, it managed to behave strategically on the basis of interests that were predominantly economically motivated: any acceptance of climate mitigation policies was to avoid putting US competitiveness at risk. Its overall clout in the negotiations was guaranteed above all by the fact that it had remained the largest emitter on the planet. Japan had comparable internal problems. The turf wars between environmentally and economically minded ministries were ultimately settled through an intervention by the foreign ministry and the Prime Minister, based on general foreign policy considerations rather than economic or environmental concerns. Its clout in the talks was primarily a result of its function as a host country of the decisive COP. As such, it assumed the role of a facilitator, trying to balance the two extreme positions on targets and regulatory approaches. The G-77/China, despite all its heterogeneity, derived its force from its unity during key moments and on crucial issues in the negotiations. It remained a very defensive actor on many points, protecting its own achievements of the past (Art. 3 UNFCCC) and demanding further developed country action. Finally, if the EU possessed actor capacity during this period, it varied in function of its internal decision-making and coordination processes. As a result of its at times dysfunctional "climate policy-making machinery" and its insufficient preparedness for the final bargaining round, extensive internal coordination was necessary, hampering greater outreach activities and a more strategic approach (not only) at the Kyoto COP (Yamin 2000: 61). In contrast to the US, which had difficulties defining a position, but was more effective once it had one, the EU possessed elaborate positions quite early, but was rather inflexible in the actual talks. These positions and actions were, to a larger extent than in the case of the US and Japan, motivated by environmental and, on the issue of responsibilities, equity concerns (Interviews EU, US representatives 17, 5; Van Schaik/Schunz 2012). In its approach towards the target, it acted on the basis of the precautionary principle, and made the strategic choice to adopt a position that would force other industrialized countries to react in order to come to a fairly ambitious overall target with every Annex I party on board. In its strategy regarding developing country commitments, it regularly let moral concerns prevail and did not insist on a more active implication of advanced non-Annex I countries in mitigation policies. When these incommensurable approaches and interests clashed in a negotiation context necessitating decision by consensus, only strategic interaction and a package deal could be the outcome. The sole point of convergence of all (major) actors in the negotiations, at least after COP 2, concerned the necessity, in light of the political pressure that had built up, to agree on a new treaty with legally binding commitments. Capitalizing on this agreement, the necessary space for overcoming differences by extensive concession-making opened up during the final days of the talks.

This review of the potential explanatory factors of EU influence on the climate regime during the studied period thus points to a mixture of internal and external scope conditions enabling the EU's influence through bargaining, while also highlighting several restraining factors.

Turning to the *enabling conditions* first, the Union's own strategic approach around the issue of targets was certainly helpful: on this issue, it positioned itself proactively, managed to gain support from the G-77/China (and civil society actors), and remained firm until the final days of the talks. Moreover, it was able to organize its foreign policy implementation around this position, with a very engaged member state (the UK) taking the lead as key representative in the final stages of the negotiations. External factors that enabled its influence were arguably the goal-specific preference convergence between major actors (an outcome had to be achieved at COP 3 under the pressure of global public opinion), a level playing field among them, and the strong preferences of other key actors for approaches that the Union had refused from the start of the talks (like the notion of flexibility and differentiation), which opened pathways for mutual concessions. In these circumstances, its influence attempts were able to trigger "bargaining" as the causal mechanism leading to change in the behaviour of other players.

By contrast, the findings also allow for identifying several *hindrances* to the EU's capacity of exerting influence. Regarding internal factors, the negotiations on the issue of responsibilities indicated the need for internal coordination in the face of conflicting preferences among member states on strategic (as opposed to substantial) choices. As a result of this and a clear lack of "political will" of the member states, the Union's foreign policy implementation was rather limited and ambiguous on key issues (Yamin 2000; Sjöstedt 1998: 238–239). Concerning external determinants of the Union's influence, the latter was considerably restrained by other actors' power and their capacities to convert this power into influence. These, in turn, depended to a large extent on the internal institutional set-up, interest and belief constellations within third countries and negotiating coalitions. In some cases, as with the G-77/China on the issue of responsibilities or with the US on the issue of flexible mechanisms, the relative power of these actors – rooted in material and immaterial capacities including emission profiles (for the US) and the capacity to build coalitions (for the G-77/China) – was so significant that they created an uneven playing field, to the detriment of the EU. As a result, these actors' negotiation positions absolutely had to be accommodated to ensure the overall aim of reaching an agreement.

CHAPTER 4

From the Buenos Aires Action Plan to the Year 2007 (1998–2007)

EU Influence on the Consolidation of the Global Climate Regime

Chapter 4 analyses the EU's activities and impact in the context of the global climate negotiations during a time period that marked the transition between two phases of active regime reform attempts (1995–1997 and 2007–2009). Although talks during this period were not primarily concerned with the major topics studied in this work, i.e. a new interpretation of the core norm and principle of the regime, the ten-year span needs to be given consideration for two main reasons. First, the debates on the concrete modalities of the operationalization (1998–2001) and ratification (2002–2004) of the Kyoto Protocol were quite essential for the overall development of the climate regime. Their purpose lay in the transformation of the – only politically adopted – provisions of the Protocol, including its central norm (i.e. the legally binding target of 5.2% for Annex I parties), into international law. Second, in parallel to these ratification discussions and especially from 2005 until late 2007, debates on the future of the regime in the large sense of the term were gradually started (see also Annex II).

COP 4 to COP 7: From the Buenos Aires Action Plan to the Marrakech Accords (1998–2001)

In the final hours of COP 3, the negotiation skills of Chair Estrada had proven crucial for ensuring a deal all parties could undersign (see Chapter 3). Essential for his success in ensuring the adoption of the Kyoto Protocol had been the artful postponement of crucial decisions to a post-Kyoto follow-up process. As a result, the Protocol contained demands on the "Conference of the Parties serving as the meeting of the Parties to this Protocol" to determine "at its first session or as soon as practicable thereafter" inter alia the modalities of joint implementation (Art. 6 KP), the clean development mechanism (Art. 12 KP), emissions trading (Art. 18 KP) and compliance (Art. 18 KP), as well as the counting of sinks (i.e. processes or activities removing greenhouse gases from the atmosphere such as forest or land management) (Art. 3.3 and 3.4 KP). The Protocol provisions themselves imposed thus, to a large extent, the agenda for the immediate post-Kyoto period: if the parties to the UNFCCC were truly determined to make the new treaty operational, they needed to take crucial decisions either directly at COP 4 or in the course of a longer process leading up to the first meeting of the parties to the Protocol. This section focuses on the negotiations between COP 4 and 7, which accomplished the operationalization of the Protocol.

The Context: Major Developments in Global Politics and Climate Science

Significant events outside the UN climate arena concerned above all the US during this period. In January 2001, George W. Bush became the 43rd President and his administration immediately began to alter the context for global politics through a decidedly hostile stance towards multilateralism. Later that year, the terrorist attacks of 11 September – striking the US, but inciting an almost global response – sparked speculations about a potential US re-engagement in multilateralism, but in fact led to further unilateral activity (Dessai *et al.* 2003: 192–193).

Major scientific advances were incorporated in the IPCC's Third Assessment Report (TAR) published in 2001. Although it would not attain the same degree of importance in the regime negotiations as its predecessors, the TAR presented again increased evidence for global warming and highlighted the discernible human influence on climatic variations, bound to become more significant in the future (IPCC 2001a, b).

Key Actors in the Global Climate Regime and their Negotiation Positions

Many parties to the negotiations at COP 3 returned home with the perception of having made gains from the Kyoto Protocol (Flavin 1998). This was particularly true for several members of the JUSSCANNZ coalition, who realized that an extensive use of flexible mechanisms and GHG sinks would considerably facilitate the task of meeting their commitments. A shared interest in exploiting the Kyoto mechanisms also led to the emergence of a new coalition: the *Umbrella Group*, bringing together the members of JUSSCANNZ and major economies in transition (Japan, the US, Canada, Australia, New Zealand, Russia, Ukraine, Kazakhstan, Norway and Iceland) (Ott 2001a: 280).[1] Common to all

---

[1] In 2000, another small and untypical coalition emerged under the term *Environmental Integrity Group*. It regrouped Annex I and non-Annex I parties (Switzerland, Mexico, South Korea, Liechtenstein, Monaco) (Romero 2004: 10). The group's overarching aim and founding rationale was to protect the environmental integrity of the

these countries was a preference for thoroughly operationalizing the new rules of the regime prior to ratifying the Protocol. Their central objective was to obtain maximum flexibility in the application of the mechanisms, the compliance system (with the US advocating a more rigid approach on this item) and the use of sinks (ENB 1998; Pallemaerts 2004: 38, 45). Further, several of the long-standing JUSSCANNZ members, especially the US and Australia, continued to display a special interest in genuine regime development by encouraging major developing countries to take on emissions reduction commitments. In the US, given the political constellation in the Senate (see Chapter 3), this had even become a *condition sine qua non* of the country's ratification of the Kyoto Protocol (Tangen 1999). In the course of the talks, these positions would gradually harden (Grubb/Yamin 2001).

The majority of countries from the other big negotiating bloc, *G-77/ China*, was equally in favour of an operationalization of the Protocol before its (swift) ratification, but refused to consider emissions reductions obligations of its own, arguing that the implementation of existing provisions, in line with the CBDR principle, had absolute priority (Tangen 1999). The G-77/China preferred less flexibility, a limited use of sinks, a strong compliance regime and the quick operationalization of the financial mechanisms (ENB 1998; Torvanger 2001: 2–4).

For the *European Union*, the immediate post-Kyoto phase produced a hang-over of sorts when it realized that the outcome of the regime reform process was not as favourable as it had initially sounded, and that much work still lay ahead before it would attain its primary aim, the entry into force of the Protocol (Interview EU representative 21). After a long period of concentration on the global negotiations, it also had to recognize that it had largely neglected its internal climate policies, complicated by diverging preferences among member states (Pallemaerts 2004: 44–45; Jordan/Rayner 2010: 65–66). The first important decision aimed at rectifying this was taken in June 1998 when the Council forged a political agreement on the sharing of the EU's 8% reduction obligation imposed by the Kyoto Protocol, accompanied by a call to develop a range of climate policy measures within the EU (Pallemaerts 2004: 44–45).[2] In the period that followed, the Union gradually developed a series of incentivizing and coordinating measures comprising inter alia a volun-

---

negotiation process by proposing consensus positions acceptable to the different other groups (Romero 2004: 20). As the group would not play a major role in the negotiations analysed in this study, it is only explicitly mentioned whenever it or one of its members contributed significantly to the regime talks.

[2] The political agreement on burden-sharing was, however, not translated into a legal act until a Council decision of June 2002, which, at the same time, constituted the EC's ratification of the Kyoto Protocol (Council 2002).

tary agreement with the European Association of Car-makers on CO_2 emission standards (1998) and a directive aimed at promoting electricity from renewable energy sources (EP/Council 2001; Pallemaerts 2004: 46–48; Jordan/Rayner 2010: 64–68). Further, the Commission launched a European Climate Change Programme in 2000 in order to identify and elaborate policies necessary for a sound implementation of the Kyoto Protocol (European Commission 2008a). While growing clarity about its internal climate policies through secondary legal acts was bound to strengthen its actor capacity, the Union's (primary) legal and institutional set-up remained, despite the entry into force of the Treaty of Amsterdam in May 1999, largely unmodified. One significant exception has to be signalled, however: to guarantee greater continuity in the EU's foreign policy approach, a specific "Climate Troika" was formed after COP 6 in The Hague, composed of the current Presidency of the Council, the future Presidency, and – as institutional memory of sorts – the Commission (Grubb 2001: 10).[3] The Union's negotiation position had, finally, not considerably evolved after the Kyoto COP. On the issue of responsibilities, the EU still wanted to loosely incite major developing countries to begin considering emissions reduction obligations in the medium term (Tangen 1999: 176). Regarding the issues that needed to be resolved before the Protocol could be ratified, the EU favoured a sound operationalization of the flexible mechanisms (including a demand for "supplementarity", i.e. the definition of a cap on non-domestic measures counting towards the fulfilment of a party's Kyoto target), a strong compliance system, and a limited use of sinks (Torvanger 2001: 2–3).

The Negotiation Process and the EU's Influence Attempts

Held between 2 and 13 November 1998 in Buenos Aires, **COP 4** was to provide a first opportunity to prepare decisions that were needed to accelerate the ratification of the Kyoto Protocol. Besides the aim of elaborating a work plan to that end, one other issue – directly related to a core pillar of the climate regime – was heavily debated during the first days of talks (EPL 1999: 3). During the opening plenary, the Argentinean hosts proposed to include voluntary commitments by developing countries as an additional item on the agenda,[4] coupled with the promise by President Menem that Argentina would itself soon adopt a voluntary quantified

---

[3] It replaced a Troika consisting of the former, current and future Council Presidencies. This change was also due to difficulties in the EU's outreach strategy during the negotiations at COP 6, as this chapter explains.

[4] To recall, during the Kyoto talks, it had been AOSIS and the AGBM chair Estrada, himself from Argentina, who had already championed a provision to allow for voluntary commitments of developing countries (see Chapter 3).

emission reductions target (Ott 1998: 186; ENB 1998: 10). The suggestion was met with fierce resistance from the majority of the G-77/China bloc, notably India and China (ENB 1998: 3). For that reason, the issue did not formally make it on the agenda, but was nonetheless informally debated. A heated dispute developed between the US, whose Senate still regarded voluntary commitments of major developing countries as an essential precondition for its ratification of the Protocol, and China, the major emitter among the developing countries (Flavin 1998; Tangen 1999). In these debates, as in the COP on the whole and during previous negotiation sessions, the EU found itself somewhere in between the conflicting parties. It appeared to observers as impaired by internal problems and "rather passive", despite taking the stance that "broadening commitments in the long term is necessary and unavoidable" (Tangen 1999: 176; ENB 1998: 3, 14). In spite of Argentina's proposal, followed a day later by the – highly symbolic – signing of the Protocol by US President Clinton (ENB 1998: 13), the G-77/China managed to ultimately fight off any decisions on voluntary developing country actions. Slowed down by conflicts about this issue, the COP either did not engage in substantive discussions on agenda items directly concerned with the operationalization of Protocol rules or, if it did, was unable to reach decisions (ENB 1998: 14). During the final days, the main aim of the negotiators was thus to ensure at minimum a decision on the further proceedings of talks. The resultant Buenos Aires Plan of Action (BAPA) did not settle any pending issues, but rather set COP 6 in 2000 as a deadline to forge agreements on, above all, (i) financial mechanisms to assist developing countries in their efforts against climate change, (ii) development and transfer of technologies; (iii) rules governing the Kyoto mechanisms, including on the issue of supplementarity, with priority given to CDM; (iv) rules and procedures on compliance; and (v) preparations for MOP 1 (UNFCCC 1999a; Bollen/van Humbeeck 2002: 100; ENB 1998).

COP 5 (25 October–5 November 1999, Bonn) began with a high-level plenary meeting during which the host country's Chancellor Schröder expressed his hope for celebrating the entry into force of the Kyoto Protocol at the 10th anniversary of the Rio Earth Summit in Johannesburg in 2002 (ENB 1999: 2). This deadline was supported by the EU and numerous others, including Japan and AOSIS (ENB 1999: 12; von Seth 1999: 227). Central agenda items of the talks were the Kyoto mechanisms and the elaboration of a detailed work plan towards COP 6, even though other items stipulated in the BAPA were also taken up (compliance, technology transfer, sinks) (Bollen/van Humbeeck 2002: 100; von Seth 1999). Regarding mechanisms, debates concentrated on a synthesis of proposals by parties on principles, modalities, rules and guidelines (UNFCCC 1999b, c). During these deliberations, the EU emphasized the need to link the mechanisms to strong monitoring and reporting requirements, while other industrialized countries spoke in favour of maximum flexibility (ENB 1999: 8–9). Preliminary decisions on the mechanisms were taken and forwarded to preparatory meetings before COP 6 (ENB 1999: 9). A work plan foresaw several intersessional rounds of talks and an acceleration of their pace (von Seth 1999: 228; ENB 1999: 8; UNFCCC 1999d). All in all, the conference made thus modest advances on technical issues, keeping the negotiations on track towards the crucial sixth conference (ENB 1999; von Seth 1999: 231). Major events of relevance for the development of the regime were Argentina's specification of its voluntary emission reductions pledges made at COP 4 (2–10% below business as usual during 2008–2012) as well as Kazakhstan's demand to be included in Annex I (von Seth 1999: 228; ENB 1999: 13). This was welcomed by the US, Japan and Australia, while the EU stated its by then well-known position that a "possible way of making all countries limit their GHG emissions is to agree on increasing global participation *after* the first commitment period" (ENB 1999: 13, emphasis added). This was refuted by China and India, pointing to the "main responsibility" of developed countries (ENB 1999: 13). In the face of this opposition, the announcements by Argentina and Kazakhstan could not prevent the emergence of debates on the possibility of non-ratification of the Protocol by the US (von Seth 1999: 233).

COP 6 (13–25 November 2000, The Hague) was to bring the showdown in the two-year operationalization process of the Kyoto Protocol kicked off in Buenos Aires. It had been prepared in several meetings during the year 2000, of which observers remarked "the distinct lack of urgency, (…) not only in the conference halls (…), but also in the upper echelons of politics" (Ott 2001a: 280). Preparatory talks had covered all issues of the BAPA, but yielded little results. The COP split into several contact groups to consider the issues enumerated in the BAPA (ENB 2000). At the end of its first week, little had been achieved. In the face of the large number and complexity of undecided issues, parties seemed to wait for a compromise text from the Dutch President, Environment Minister Pronk (Ott 2001a: 281; Grubb/Yamin 2001: 268; Dessai 2001: 141). The finally introduced "Pronk paper" identified a total of 39 (!) "crunch issues" requiring decisions at the highest political level. The themes were regrouped into four boxes (Box A: developing country issues such as financing and technology transfer; Box B: Kyoto mechanisms; Box C: sinks – discussed under the term "land-use, land-use change and forestry" (LULUCF); Box D: policies and measures as well as compliance issues) (UNFCCC 2000: 2–14). In spite of all preparatory work, the paper had remained political rather than technical in nature, introducing new ideas and clearly attempting to accommodate the preferences of the Umbrella

Group (Ott 2001a: 281; Dessai 2001: 142; Grubb/Yamin 2001: 268–270). This latter tendency was particularly noticeable with regard to the Kyoto mechanisms and the use of sinks, two central concerns on which the US-led coalition clashed with the EU's preferences (UNFCCC 2000: 6–11; Ott 2001a: 281). Together with the COP President's unusual approach for tackling the final talks with such a political (rather than a legally worded) text, it transformed the talks into an open conflict between the major industrialized coalitions, during which the G-77/China was effectively sidelined (ENB 2000: 18–19; Ott 2001a: 283; Grubb/Yamin 2001: 269; Dessai 2001: 142). In the US-EU exchanges that followed, the two most contentious issues were the supplementarity of the flexible mechanisms and the modalities of the use of sinks (Grubb/Yamin 2001: 271–272). On the first issue, which turned around the question to what extent activities to reduce emissions carried out through flexible mechanisms should be restricted, the Pronk paper suggested, rather vaguely: "Annex I Parties shall meet their emission commitments *primarily* through domestic action since 1990" (UNFCCC 2000: 7, emphasis added). This left much leeway to the Umbrella Group, seeking flexibility, to the detriment of the EU's desire to set a clear quantitative ceiling on the use of the mechanisms (Ott 2001a: 283; Grubb/Yamin 2001: 272). The second issue would, however, reveal to be even more problematic: the US and other members of the Umbrella Group (Canada, Australia), had made it clear that they not only desired a maximum use of forest sinks, foreseen in Article 3.3 KP, but also sought to receive credits for carbon absorption through other managed lands under Art. 3.4 KP (Grubb/Yamin 2001: 271). This was heavily opposed by the EU, the developing countries and the environmental NGOs, who considered far-reaching use of existing management measures towards meeting emission reduction targets as unacceptable (Bollen/ van Humbeeck 2002: 103). The Pronk paper did not, however, take the arguments of the opponents to the US approach into account when stating that "a Party may include the following activities: grazing land management, cropland management and forest management" (UNFCCC 2000: 10). While some members of the G-77/China were seemingly prepared to grant the US derogations on this point, the EU was more reluctant to compromise, fearing that "ordinary business-as-usual activities in the agricultural sector" would count as "climate protection measures" (Grubb/ Yamin 2001: 271; Ott 2001a: 282). Bilateral talks during the very final hours of the COP, involving the UK's Deputy Prime Minister Prescott[5] and the US lead negotiator did, against all expectations, produce a compromise formula. As part of a broader package, it foresaw the opportunity

---

[5] Prescott had already played an important role during the final days of negotiations at the Kyoto COP when he represented the EU as part of the Troika (see Chapter 3).

for the US to claim but a limited amount of reductions from its own sinks (75 million tons of CO_2); in return, the EU gave up on its insistence on a quantitative cap on mechanisms (Dessai 2001: 142). The Troika, led by France, brought this compromise to the EU-15 coordination meeting, where it encountered strong opposition by the Scandinavian countries as well as Germany, arguing that the UK never had a mandate to make such concessions (Jacoby/Reiner 2001: 302; Yamin/Grubb 2001: 263; ENB 2000: 18). This forceful opposition within the Union had repercussions for its overall position and effectively led to a breakdown of negotiations at COP 6 (Dessai 2001: 142). The conference ended therefore in a face-saving formula to suspend talks and reconvene later on the basis of what had been achieved so far (ENB 2000: 18–19; Jacoby/Reiner 2001: 302). All stories told about these negotiations emphasize, besides the unfortunate Pronk paper and the tough stance of the US, the less than optimal role of the EU (Ott 2001a; Grubb/Yamin 2001; Dessai 2001; Jacoby/Reiner 2001; Dessai *et al.* 2003; Vogler 2005). The Union's performance will be subject to more detailed analysis, set into a broader context, when it comes to assessing its overall influence below.

The **period after the failure of COP 6** saw several attempts by the Umbrella Group and the EU to prepare a deal on the basis of the partial agreements reached in The Hague. Following an initiative by US President Clinton, whose final term in office was drawing to a close, a meeting was convened in Ottawa in December 2000. Yet, negotiators could not reach common understandings on key issues, even re-opening some of the agreements of COP 6 (Bollen/van Humbeeck 2002: 104). The originally planned follow-up meeting at ministerial level was therefore cancelled (Ott 2001a: 284; Jacoby/Reiner 2001: 303). Cooperation was further complicated after George W. Bush had been sworn in as US President. In March 2001, he indicated in a letter to a group of Senators that his administration opposed the Kyoto Protocol because "it exempts 80% of the world, including major population centers such as China and India, from compliance" (White House 2001a). On the basis of this and related arguments, including concerns about competitiveness and loss of jobs, the US thus effectively withdrew from the Protocol ratification process (Bollen/van Humbeeck 2002: 104). The EU and other governmental actors (such as Japan and the G-77/China), but also a whole range of civil society organisations, attempted to influence the US administration to reconsider its stance through sending envoys and letters to Washington, organising protests and issuing open appeals – all in vain (Bollen/van Humbeeck 2002: 104; Dessai *et al.* 2003: 188).

When it became obvious that the US position would remain unchanged, the EU, led by the Swedish Council Presidency, made clear that Kyoto was "the only game in town" (ENB 2001a: 13) and engaged in a "diplomatic tour" to, inter alia, Iran (the G-77 Presidency at the time), Russia, Japan and China, in order to gather support from crucial parties for the further ratification process (Grubb 2001). This yielded positive effects: almost all parties assured their commitment to the Protocol (Dessai *et al.* 2003: 188; Gupta/Ringius 2001; ENB 2001a: 3). With the US appearing isolated in its opposition to "Kyoto", President Bush announced on 11 June 2001 that his country would not obstruct the further negotiation and ratification process (White House 2001b; Bollen/van Humbeeck 2002: 105). At an EU-US summit on 14 June 2001 in Gothenburg, the two parties agreed to disagree on the further negotiation process and the Union cautioned its partners not to interfere with future UN climate talks (Dessai *et al.* 2003: 190).

COP 6bis, held in Bonn between 16 and 27 July 2001, broke with the logic of its predecessors: a high-level segment came together already during its first week. After three days of ministerial consultations, COP President Pronk introduced, on 21 July, a consolidated and unbracketed negotiating text he had prepared together with the UNFCCC secretariat (ENB 2001a). This document, product of the agreements reached at COP 6 and further informal talks, proved to be a sufficient basis for reaching a deal this time around. Parties could agree on all items except the sections on compliance, where previous agreements were re-discussed and the EU's preferences for a strong mechanism clashed with the flexibility concerns of Japan, Russia and Canada (ENB 2001a: 4, 14; Dessai *et al.* 2003: 190–191; Ott 2001b: 470). After the decision on this issue had been effectively shifted to a later stage, the path was cleared for an adoption of the "Bonn Agreement" on 23 July (Ott 2001b: 469–470). The Agreement contained several key decisions on issues raised in the BAPA: regarding financial assistance and technology transfer to developing countries, three new funds were introduced (Special Climate Change Fund, Least Developed Countries Fund, Adaptation Fund), all of which were to be managed by the Global Environmental Facility (UNFCCC 2001a: 2–5). Concerning the Kyoto mechanisms and supplementarity, "the European Union and others lost their battle to have a quantitative cap" (Dessai *et al.* 2003: 191). On sinks, the EU and the G-77/China also made considerable concessions, as a longer list of activities, including those proposed by the US and in the Pronk paper at COP 6, were added to Article 3.4 KP. This augmented the range of possibilities for parties who wanted to reduce emissions by counting also various land management measures (UNFCCC 2001a: 10–13; ENB 2001a: 13). It was thus essentially through substantive EU concessions that the deal was ensured (ENB 2001a: 13; Jordan/Rayner 2010: 68).[6] The remainder of the con-

---

[6] This obviously raises questions about the EU's influencing strategy and impact in function of US presence or absence during climate talks. They are addressed in the concluding discussion of its influence during this period.

ference proved too short to translate these political agreements into (a) coherent legally worded document(s). Notable sticking points were the issues of mechanisms, sinks and compliance (Torvanger 2001: 14; Dessai *et al.* 2003: 192; Bollen/van Humbeeck 2002: 107). The finalisation of the BAPA thus had to await COP 7.

COP 7 in Marrakech (29 October–10 November 2001) had the delicate task of translating the Bonn Agreement, a package of draft decisions, into legal language. Despite existing political compromises on almost all issues, the conference turned out to be more politicized than expected: where the EU considered it as a final formal step towards concluding the BAPA and operationalizing the Kyoto Protocol, some Umbrella Group members, especially Japan, Russia and Australia, sought further gains for their positions, while the US took "little overt part" (Schneider/Wagner 2002: 3; ENB 2001b: 14–15). Talks focused on three outstanding issues from the BAPA (ENB 2001b; Dessai *et al.* 2003: 194–195; Bollen/van Humbeeck 2002: 109–113): compliance, the main left over from COP 6bis, and the two topics that had broken the deal at COP 6: sinks and flexible mechanisms. Regarding the issue of compliance, outstanding problems could be resolved at expert level in a package deal between the EU and the G-77/China on the one hand and the Umbrella Group on the other hand (ENB 2001b: 15, 6–8). The decision finally adopted foresaw a comparatively strong mechanism, which reflected EU preferences, notably with regard to its major institutional novelty, a two-branch compliance committee (ENB 2001b: 15; for an in-depth discussion: Schneider/Wagner 2002: 10; Dessai/Schipper 2003: 151–152). The other two issues were not resolved until the last two days of the COP when parties decided on very technical issues related to LULUCF reporting and the eligibility criteria for participation in the flexible mechanisms (ENB 2001b: 15; Dessai *et al.* 2003: 194). On these items, the EU gave in to various demands of the Umbrella Group (Dessai/Schipper 2003: 151; Schneider/Wagner 2002: 6–10). Ultimately, the COP reached its overarching aim: the Marrakech Accords, a package of 245 pages of decisions adopted on 10 November 2001, ended the BAPA process (UNFCCC 2001b).

The Outcome: the Marrakech Accords – Clearing the Way for Ratifying the Kyoto Protocol

Exactly three years of negotiations had been necessary to detail the rules of the reformed climate regime. Although this time period was not per se concerned with regime development regarding its core pillars,[7] talks were essential for the continuity of the climate regime under the UN

---

[7] The major exception was the described debate about developing country commitments at COP 4.

umbrella, and thus also for the EU's objectives on the further evolution of the regime. In this regard, the Marrakech Accords proved crucial for clearing the path for ratification of the Kyoto Protocol. A critical reading could characterize these Accords as the "Marrakech dilution of the watered down Bonn agreement to the fatally flawed Kyoto Protocol to the UNFCCC" (Dessai et al. 2003: 194). More nuanced interpretations, by contrast, regarded the compromises on some of the crunch issues as necessary steps in the development of a regime that had to provide enough incentives for all actors to participate, even keeping the door open for the US to eventually re-join the club (ENB 2001b: 16). To enable such a catch-all agreement, the "EU and the G-77/China had been compelled to concede to many of the demands of key Umbrella Group countries" in the "Marrakech bazaar", but also beforehand at COP 6bis (ENB 2001b: 15; Dessai et al. 2003: 193). With the exception of the compliance system, "unique in the world of environmental law" because of its level of detail with regard to both institutional set-up and penalties (Dessai/Schipper 2003: 151; UNFCCC 2001b: point L.), many key decisions in the Marrakech Accords reflected thus the preferences of the non-EU industrialized countries for greater cost-effectiveness – to the detriment of environmental integrity. This is particularly the case regarding the use of sinks, with the inclusion of all land management activities (Dessai/Schipper 2003: 151; UNFCCC 2001b: point K.), and the modalities of the flexible mechanisms. On this issue, supplementarity, so important to the EU, had become an "almost meaningless item within the Accords" (Schneider/Wagner 2002: 5–9; UNFCCC 2001b: point J.; Dessai et al. 2003: 195).

The EU's Influence on the Negotiations Leading to the Marrakech Accords

A brief assessment of the EU's influence during the talks that led to the Marrakech Accords, based primarily on testimonies of observers, allows for maintaining the longitudinal perspective of the study. To that end, the Union's main influence attempts and their effects at major turning points are assessed, before tentatively explaining its influence.

There is no evident coherent pattern of EU influence attempts that can be extracted from the discussion of this time period. The Union was active as a defender of the idea of prompt operationalization of the Kyoto Protocol, as part of its broader objectives related to regime development, but rather passive regarding the concrete substance of talks at COPs 4 and 5, balancing between the developing countries and the Umbrella Group (Tangen 1999; Ott 1998). Gradually, it would define, in a similar vein as during the Kyoto Protocol negotiations, a very detailed position and submit proposals on all key issues, with emphasis on sinks, compliance, and, importantly, the supplementarity of the flexible mechanisms (Torvanger 2001). This position tended, however, in another parallel to the period 1995–1997, to be quite inflexible: at COP 6, the EU was unprepared to make concessions to the US, thus contributing to the failure of the talks (Ott 2001a: 285; Grubb/Yamin 2001). It was only in the aftermath of the US withdrawal from the process, that it altered both its position and behaviour. Not only did the EU engage in more wide-reaching diplomatic activities, but it also displayed, under pressure, greater willingness to compromise with the Umbrella Group.[8]

Three relevant turning points can be identified during this period. The first one touched upon the maintenance of the norm, the other two on the modalities of its implementation.

Turning point 1, the US withdrawal from the process, came after a first major failure in the history of climate negotiations (COP 6). It led to a partial convergence of preferences among the remaining key actors, who agreed that the operationalization and ratification process of the Protocol should be pursued. The EU may claim credit for having actively – and successfully – led the way to forge this consensus on further regime development both rhetorically and by proactively searching the dialogue through sustained diplomatic efforts (and thus fulfilling four necessary conditions for influence: *purposive behaviour, interaction, temporal sequence, goal attainment*). It is however difficult to assess the *degree of auto-causation* of the other players. The remaining Umbrella Group members probably realized that the US withdrawal would enable them to impose the terms of the agreement. This made a participation in the regime even more appealing to them, as they could seek symbolic (gain international profile, e.g. Japan) and material benefits (via flexible mechanisms, e.g. Russia). Counterfactual reasoning may help to decide whether the condition "absence of auto-causation" was, at least partially, fulfilled: had the EU not "led the way", would the other Umbrella Group members have behaved the way they did? Probably not, as they might have been too concerned about their relations with the US (e.g. for Japan, Grubb 2001). It did take the clear signals from the economic heavyweight EU to convince them to pursue with the negotiations at that stage. In the final analysis, it can therefore be plausibly argued that the Union did exert a – medium – degree

---

[8] Quite a few studies that have examined the Union's external climate activities from a leadership perspective have actually focused precisely on its performance after the 2001 withdrawal of the US from the Kyoto Protocol ratification process to attribute the EU an avant-garde role in the climate regime (see, for instance, Grubb 2001; Gupta/Ringius 2001; Oberthür/Roche-Kelly 2008). An influence analysis perspective demonstrates that its alleged leadership came at a high cost: by making numerous concessions to the other Umbrella Group members, the EU effectively gave up its environmental integrity concerns in order to ensure the continuity of the regime.

of influence at this major turning point: without its commitment to the multilateral climate regime, the latter may have slipped into a longer crisis, and the commitments embodied in Article 3.1 and Annex B of the Kyoto Protocol might never have gained any significance whatsoever. EU impact thus ensured the maintenance of the norm.

The second and third turning points in Bonn (COP 6bis) and Marrakech (COP 7) concerned final decisions on various concrete technical issues regarding the modalities surrounding the issue of emissions reduction targets. Had these discussions failed completely, the norm on which the EU had been so influential in Kyoto would never have come into force. This was avoided, but the rules around the emission reduction targets were nevertheless severely watered down in a process in which the EU unsuccessfully tried to protect the environmental integrity of the Protocol (*purposive behaviour, interaction*). As the Umbrella Group exploited its bargaining power to the fullest to gain maximum flexibility regarding the use of sinks and mechanisms, the Union felt compelled to adapt its previously very rigid position and to give in to almost all demands, with the partial exception of compliance, in order to ensure regime development.[9] In failing to reach its objectives, it thus clearly did not fulfil a crucial necessary condition for exerting influence (*no goal attainment*). No additional EU influence over the specific contours of the regime is thus discerned at these two turning points. All in all, the EU's influence attempts were oriented towards the long term, and had enduring effects: the Union ensured that the decisions necessary for the Protocol to come into force – and the regime to further develop – were taken. It had, however, little influence on their substance.

To account for the EU's influence during this period, endogenous/ actor-related and exogenous variables must again be considered. In terms of *actor-related variables*, the EU's actor capacity can be largely taken for granted, but its influence attempts and overall foreign policy implementation may partially explain why it did not fare well during most of this period. Observers of COP 6 in The Hague noted its "rather weak performance" as one of the reasons for the conference's failure, attributing it to "the uncoordinated, reactive and fragmented style of European diplomacy" (Ott 2001a: 285; Grubb/Yamin 2001; Vogler 2005: 840). The key problem seemed to be a profound divergence of preferences regarding substantial and strategic policy choices among the member states.[10] In

---

[9] This marks a striking parallel to the Kyoto Protocol negotiations, where the EU also gave in to demands by the US (and Japan) on almost all items to attain its aims regarding the emission reductions target.

[10] The "blame game" that the UK negotiators and the French EU Presidency engaged in immediately after COP 6 may be one indicator that apparent problems were rooted in divergent preferences (Grubb/Yamin 2001: 274; Dessai *et al.* 2003: 186–187).

an institutional context requiring unanimity, such differences resulted in slow and cumbersome processes of foreign policy decision-making and coordination as well as suboptimal outreach (Sprinz 2001: 7). As in previous rounds of talks, the EU spent "so much time negotiating with itself, and secondarily focusing on its position *vis-à-vis* the United States, that very little investment [wa]s made with respect to other countries" and "little or no account of the realities in the rest of the world" was taken (Grubb/Yamin 2001: 274). The resulting inflexibility was best exemplified in its stance on the issues of sinks and supplementarity at COP 6, of which – according to Grubb/Yamin (2001: 272) – "many EU members had known for a year" that they were "unacceptable and probably unworkable" for key third countries. These positions reflected a lack of strategic thinking on realistic fallback positions that would have enabled the EU to seal a deal as early as COP 6, avoiding the delay and painful experiences of the year that followed. Strikingly, the Union immediately adapted its position and behaviour after the failed COP 6 and the US withdrawal from the Protocol ratification process (Grubb 2001: 10), subscribing to compromises that would, arguably, "have satisfied the US in The Hague" (ENB 2001a: 13).[11] This adaptation of its position may be regarded as the key positive explanatory factor for its influence on the overarching objective of ensuring regime continuity, but it came at high cost regarding its environmental integrity. Turning to the *external determinants* of EU influence, the US withdrawal from the Kyoto Protocol ratification process constituted the most significant factor. It considerably impaired the Union's chances of exerting influence on the regime. As the entry into force of the Protocol required the ratification of 55 parties to the UNFCCC, including Annex I parties representing 55% of the total emissions of all Annex I parties (Art. 25.1 KP), US disengagement meant that virtually all other industrialized actors needed to ratify if the EU was to reach its aim of getting the treaty into force. This provided the bigger members of the Umbrella Group (Japan, Russia, Canada, Australia) with de facto veto power (Dessai *et al.* 2003: 190, 197; Bollen/van Humbeeck 2002: 106). Since their diverging preferences were hard to accommodate, the Union had to bow to virtually all their wishes to attain its objectives. In a context of preference heterogeneity, the structural features of the negotiations thus did not play out to the advantage of the industrialized party who most wanted the due enactment of the Protocol (Yamin/Grubb 2001).

---

[11] The strong focus on the US, displayed at COP 6 (Grubb/Yamin 2001: 274), was, however, continued in a different manner thereafter. Indicating how some players within the EU perceived the climate negotiations, the successful ending of the resumed sixth conference, in which the US actually took little active part, prompted EU Environment Commissioner Wallström to comment: "I think something has changed today in the balance of power between the US and the EU" (ENB 2001a: 14)!

After the Marrakech Accords: Ensuring Ratification of the Kyoto Protocol (2002–2004)

With the technical hurdles for ratification of the Kyoto Protocol overcome, climate regime talks were to focus on the preparation of the first COP/MOP and the first review of the Protocol, scheduled for seven years before the end of its first commitment period (i.e. 2005, Art. 3.9 KP). It all came a bit differently, though: US disengagement from the climate regime made the ratification of the treaty essentially dependent on Russia, with its 17% share of Annex I country emissions (UNFCCC 1997e: 60), and Moscow sought to exploit this situation to its advantage by delaying ratification. In the struggle for Russian ratification, but also in the few discussions pertaining to regime development that emerged during this period, the EU became a key player.

The Context: Major Developments in Global Politics and Climate Science

During this period, no significant events outside the UN climate regime had an impact on the climate negotiations or the EU's influence on these, although the effects of US unilateralism and the 2001 Third IPCC Assessment Report persisted.

Key Actors in the Global Climate Regime and their Positions

After their bargaining successes in Bonn and Marrakech, most remaining *Umbrella Group* members were above all concerned with the ratification and implementation of the Protocol. Japan, Canada and New Zealand actually ratified the Protocol before or shortly after COP 8 (Ott 2003: 2). In this process, Russia, although it had openly declared its commitment to ratification, counted on taking advantage of its special role as a veto power of sorts (European Parliament 2003; Douma 2006: 54). Last but not least, the US stuck with its disapproval of the ratification process. Its pledge not to obstruct the international negotiation process would also soon come under strain (White House 2001b).

The *G-77/China* considered the decisions on finance and technology transfer in the Marrakech Accords as only a partial success and urged not only the swift ratification of the Protocol, but also a further strengthening of Annex I party commitments to aiding developing countries in adaptation to climate change via a stronger focus on sustainable development (Najam *et al*. 2003; Ott 2003: 2). Linked to this shift from mitigation to adaptation, the majority of the group (except for AOSIS) was also hostile to starting talks on a reform process for the period after 2012, since it wanted to avoid being drawn into debates about own GHG reduction duties (Watanabe 2003: 19; ENB 2002: 11–12).

After its efforts spent on ensuring that the formal prerequisites for ratification of the Kyoto Protocol would be fulfilled, the *European Union* was above all concerned with the prompt enactment of the treaty. The member states contributed their share to this by adopting, on 25 April 2002, a Council decision containing the 1998 political agreement on internal burden-sharing (Council 2002). Together with their own ratification documents, the EC ratification of the Protocol was deposited at the UN headquarters on 31 May 2002 (Pallemaerts 2004: 49). Internally, the EU's joint ratification sparked the adoption of another series of measures to prepare its fulfilment of the target for the first commitment period. Directives were adopted, inter alia, on energy efficiency of buildings and the promotion of biofuels (Pallemaerts 2004: 49; Jordan/Rayner 2010: 68–69). The arguably most important measure introduced during this period was, however, the October 2003 directive on the creation of a European Emissions Trading System (ETS) (EP/Council 2003). Originally opposed to flexible mechanisms, the EU gradually, and for various reasons discussed elsewhere (Damro/Mendez 2003), warmed to the idea of applying such cost-effective tools for reducing emissions. In place since 2005, the ETS would soon become the cornerstone of the Union's internal climate regime. While these measures were to strengthen the EU's credibility, its actor capacity was bound to come under threat by enlargement, foreseen for January 2004. It was assumed that the input of ten new countries with diverse energy systems and environmental policy approaches would render the already problematic decision-making and coordination processes on external climate policy even more protracted. To cushion the impact of enlargement, but also to generally improve the performance of the EU in global climate negotiations, the Union engaged in a reform process of its institutional set-up under the Irish Council Presidency in the first half of 2004 (Interview EU representative 20; Lacasta 2008). In this process, the structures of the Ad Hoc Working Group on Climate Change that had been created in 1994 under the Environment Council – and had since 2000 been re-baptized as the Working Party on International Environmental Issues-Climate Change (WPIEI-CC) – were modified (Oberthür/Roche-Kelly 2008: 38; Costa 2007: 20; Lacasta 2008). Gradual reforms over the years had already led to the development of several expert groups under the WPIEI-CC. The novelties of the 2004 reform concerned then, primarily, the introduction of the position of "lead negotiators", i.e. individuals from any member state or the Commission, who would be in charge of negotiating on behalf of the EU in particular groups in the international arena for long periods of time (Interviews EU representatives 20, 13; Oberthür/Roche-Kelly 2008: 38; van Schaik 2008). Further, "issue leaders" were designated, who together formed small groups that would join the lead negotiators in designing and promoting EU positions in cooperation with the expert groups (Interview EU representative 20; Oberthür/Roche-Kelly 2008: 38; Lacasta 2008). This system, it was believed, would not only enhance the continuity of EU external climate policy with regard to know-how, but also improve its outreach capacities by ensuring continuous contacts with third country negotiators. Regarding its negotiation position, the EU began turning to the period after 2012: it intended to initiate a thought process on the development of the climate regime beyond the first Kyoto Protocol commitment period by addressing the questions of how to fulfil the objective of Article 2 UNFCCC and broaden the scope of parties with binding commitments (Ott 2003: 2; ENB 2002: 11–12).

The Negotiation Process and the EU's Influence Attempts

COP 8 in New Delhi (23 October to 1 November 2002) was meant to be a "working conference", focusing on technical issues related to, e.g., the CDM and the new funds created with the Marrakech Accords (Ott 2003: 2). During the opening plenary, observers noted that the grounds for a strong politicization of talks had already been laid out. The Indian hosts wanted to embed climate negotiations into broader sustainable development debates so as to underscore the development concerns of the G-77/China members by addressing, above all, issues related to adaptation (ENB 2002: 3). In the course of the talks, the EU made it very clear that this went against its own preferences, which were strongly concentrated on mitigation in a long-term perspective. Consequently, it invited parties to start a dialogue on the further commitments under the Kyoto Protocol (ENB 2002: 11–12; Ott 2003: 2–3; Watanabe 2003). This, in turn, was met with fierce resistance from the entire G-77/China bloc, arguing that the main focus of the COP should be the implementation, not the development of the Protocol (Ott 2003: 3). As the conference evolved, this conflict led to the emergence of an unusual coalition between the developing countries and the US: while the former, notably China and India, were fearful of having to take on GHG reduction obligations that would hamper their development, the US saw its support for those countries as an opportunity to control and slow down the regime development process by focusing talks on the UNFCCC rather than the Kyoto Protocol (Watanabe 2003: 19; Ott 2003: 3).[12]

After a first week of mostly technical discussions, the COP Presidency presented an informal draft for a final "Delhi Declaration", which clearly reflected the hosts' desire of more strongly linking climate change and

---

[12] The US change in tactics, breaking its June 2001 promise not to obstruct talks, came in parallel to the administration's more wide-reaching attempts to forge bilateral energy-related partnerships, e.g. with India, outside the UN context (Ott 2003: 7–8).

sustainable development (UNFCCC 2002a). By contrast, the draft did not contain any reference to the Kyoto Protocol, the latest scientific findings summarized in the Third Assessment Report or the future development of the UN climate regime (UNFCCC 2002a; ENB 2002: 14; Ott 2003: 3–4). As this was not much to the liking of the EU and some Umbrella Group members (Japan, Canada) (ENB 2002: 14–15; Ott 2002: 3–4; Watanabe 2003), the final days of the COP were spent on intense talks over this text (ENB 2002: 14; Watanabe 2003). Backed by Japan and Canada, the EU vocally pursued its objective regarding the division of responsibilities within the regime, but "fought in vain for over 15 hours to secure a reference to 'wider participation' after 2012, which would have invited emerging countries to phase in commitments" (Watanabe 2003: 19). The ultimately adopted Delhi Declaration did not contain any mention of further development of the regime. It did however comprise references to the TAR as well as to the need for the swift ratification of the Protocol (UNFCCC 2002b).

COP 9 in Milan (1–12 December 2003) had originally been planned as the first conference after the ratification of the Kyoto Protocol, to be held in parallel to MOP 1. Pending ratification by Russia transformed it into technical talks under a veil of uncertainty, which, according to observers, provoked a non-negligible lack of motivation (Fodella 2004: 24). The talks were concerned with further consideration of the CDM, modalities of sinks and the funding mechanisms (Dessai et al. 2005; Fodella 2004: 25). Showing signs of its oft-remarked "bunker mentality", the EU's tough stance on the latter issue hampered the search for compromises with the G-77/China bloc (Dessai et al. 2005: 111; Interview EU representative 20). Further regime development was clearly not among the key concerns, even though opportunities for debating long-term issues did open up, notably in the framework of discussions of the TAR. The EU's attempts to raise this issue, presented less forcefully than in New Delhi, encountered "G-77/China's strong resistance to adopting a COP decision on the TAR, and in fact to discussing anything beyond procedures for further consideration of this issue" (ENB 2003: 17). This was interpreted as a "clear reflection of the group's determination not to allow negotiations to head anywhere close to the issue of developing countries' future commitments" (ENB 2003: 17; Fodella 2004: 25). The decision on this issue prompted the elaboration of an uncontroversial – and equally vague – technical work programme (ENB 2003: 4). Other major decisions touched upon the guiding principles for the Special Climate Change Fund as well as the Least Developing Countries Fund and the modalities of LULUCF (UNFCCC 2003). Altogether, COP 9 thus confirmed the tendencies observed for its predecessor: technical progress, but paralysis regarding the further development of the regime – and a smouldering conflict around the latter between the EU, supported at times by some Umbrella Group members, and the G-77/China, sometimes in liaison with the US (Dessai *et al.* 2005: 119–121; Interview EU representative 20).

Between COP 9 and COP 10, reinforced diplomatic efforts were undertaken, notably by the EU, to ensure Russian ratification of the Kyoto Protocol. Without going into the details of the debate in Russia, the role of the EU in encouraging this ratification should be highlighted (for insightful discussions, see European Parliament 2003; Douma 2006). In the face of a range of Russian concerns about the Protocol, among them the fear of being alienated from the US,[13] the EU repeatedly made moves to ensure the country's support by strengthening bilateral ties. An extension of the 1997 Partnership and Cooperation Agreement (PCA) was to serve this purpose, but also to clear the path for a Russian membership bid to the World Trade Organization, which the EU promised to support (Douma 2006: 61; Buchner/Dall'Olio 2005). The final hurdles for the extension of the PCA were taken at the EU-Russia summit of May 2004, followed a few months later by the ratification of the Protocol in the Duma (Douma 2006: 61–62). Without giving in to all Russian demands (Douma 2006), the EU had thus, for the first time in a climate negotiation context, successfully employed economic tools, using the PCA reform and the promise of support for WTO membership as incentives for Russia to become a party to the Protocol.

Reportedly, the EU did not overtly celebrate this success at **COP 10** (6–18 December 2004, Buenos Aires), but immediately turned its attention towards the future (Ott *et al.* 2005: 84). In terms of regime development, it would become a key player in an intriguing debate developed around an issue introduced by the Argentinean Presidency. At the beginning of the session, the latter had suggested holding two seminars in the course of 2005 to exchange views on negotiations about the post-2012 period (Ott *et al.* 2005: 85–86; ENB 2004b: 2). This proposal was opposed by the US and by the majority of the G-77/China, either fearful of debates about their own emission reduction targets (China, India) or concerned with slowing down regime development altogether (OPEC). The EU, supported by AOSIS, thus found itself in the role of the only major

---

[13] Although President Putin, among others, had promised Russian ratification of the Protocol several times, notably at the World Summit on Sustainable Development in 2002, the Duma pondered its approval of the treaty for a long time (Douma 2006: 54). Besides a desire to remain on good terms with the US, many Russians did not believe in the benefits of reducing emissions or contested the science (European Parliament 2003). By contrast, they saw an opportunity to exploit the ratification of the Protocol to shape the relationship with the EU to its advantage, accede to the WTO, and improve its overall international standing (European Parliament 2003).

defender of Argentina's proposal (Ott *et al.* 2005: 85). The compromise that was agreed during the final night of talks was to hold one seminar for the purpose of information exchange that would be "without prejudice to any future negotiations, commitments, processes and frameworks" (ENB 2004a: 1). Other decisions require no further discussion here, as they were mainly technical, relating, e.g., to adaptation, LULUCF and CDM (ENB 2004a). The fact that even a comparatively innocent issue like the organisation of a seminar would spark extensive debates set the tone for the talks that would follow at the first joint session of the COP and the MOP (ENB 2004a: 15).

The Outcome: the Ratification of the Kyoto Protocol

The outcomes of these three years of further regime negotiations were rather meagre. Besides technical progress, mostly concerning the crunch issues of both the Kyoto Protocol and the Marrakech Accords (mechanisms, sinks, funding), the major advancement in terms of immediate regime development was the Russian ratification of the Protocol, which allowed for its entry into force on 16 February 2005. Finally, COP 10 saw the timid beginnings of a post-2012 regime dialogue, a topic which would occupy negotiators in the period to come.

The EU's Influence on the Ratification of the Kyoto Protocol

In search of EU influence attempts, the most striking feature of EU foreign policy activity within the climate regime throughout this period was its desire to quickly move ahead with regime development. In Delhi, it alienated many parties, notably the majority of developing countries, when it made proposals for considerations of wider post-2012 mitigation efforts at a time when the G-77/China was more interested in addressing development concerns. As a result of the EU's impatience, a climate of mistrust formed between many G-77 countries and the Union (Ott 2003: 8). Rather than, at least strategically, responding to some of these countries' concerns, the EU let its own agenda prevail, testifying once again to an often fairly self-centred approach and deficient outreach (Najam *et al.* 2003; Ott 2003: 8). This tendency continued into COP 9, although the EU did temper its impatience somewhat, realizing that the many technical issues related to the operationalization of the funding mechanisms were of crucial importance to the G-77/China (Dessai *et al.* 2005: 111, 119–121). In the period that followed the Milan conference, its main influence attempts lay outside the direct regime context. In an unprecedented move, the Union managed to use economic tools, linking climate change to issues originally unrelated to environmental politics in order to influence Russia. Finally, back in the UN arena, the EU turned again towards the future of the regime at COP 10, pursuing its own agenda in spite of a lacking international mandate to start post-2012 talks (Ott *et al.* 2005: 90). Continuing patterns from the two previous COPs,[14] it alienated both the G-77/China and the US with this approach, creating a particular challenge for the period that followed: if it wanted to advance the regime talks, the Union would need to convince the major developing countries, whose emissions were rising steeply, just as much as the major industrialized emitter (Dessai *et al.* 2003: 201; Ott *et al.* 2005: 90–91). Balancing between these two players would therefore represent the major task for the future (Biermann 2005).

The only key event and a turning point of sorts with regard to the development of the climate regime during the analysed period was the Russian Kyoto Protocol ratification, crucial to the successful completion of a reform process begun ten years earlier at COP 1 in Berlin. Although it was ultimately the decision of a sovereign country, the EU did exert medium influence over Russia's choice: it approached the Russians with concrete proposals that were openly linked to the Kyoto Protocol ratification process (*purposive behaviour, interaction, temporal sequence*). These proposals and subsequent agreements allowed the Union to *attain its aim* (ratification of the Protocol). Finally, even though the Russians may have decided to ratify the Protocol without the Union's interference, the likelihood that the incentives provided by the EU were essential for it to decide the way it did is high (*partial absence of auto-causation*) (Douma 2006, who cites observations by Russian politicians supporting this interpretation). The EU was thus able to set forth its role as a decisive player when it came to ensuring the continuity of the multilateral regime. Concerning the attempts to initiate post-2012 talks, it booked, however, together with AOSIS, only a very limited success against an opposition of developing countries and non-Kyoto ratifiers (US, Australia) at COP 10.

Without engaging in wide-reaching explanatory efforts of its influence, the case of Russian ratification demonstrated that (i) the EU generally has more foreign policy instruments at its disposal in a climate policy context than it had been willing to employ before that, and that (ii) these tools, used for bargaining, hold the capacity to enhance the Union's chances for exerting influence. By contrast, its impatience and insensitivity towards other partners' interests, a recurring pattern, at times strongly alienated third parties. What may appear as an attempt at leading carries thus, in fact, the risk of isolating the EU in the climate arena, limiting its prospects for exercising influence.

---

[14] Ott *et al.* (2005: 90) note an interesting parallel between the EU's constant desire of developing the climate regime and its own evolution as an entity (and foreign policy actor) with a range of treaty reforms in a relatively short time span: 1987 Single European Act, 1993 Maastricht, 1999 Amsterdam, 2001 Nice, 2009 Lisbon.

Towards a Post-2012 Regime: Loose Talks on the Way to Bali (2005–2007)

With the Protocol finally ratified and first steps undertaken towards a post-2012 dialogue, the beginning of a new era of climate regime development seemed around the corner.

The Context: Major Developments in Global Politics and Climate Science

Important changes to the broader context into which the regime discussions were embedded occurred towards the end of the time period analysed in this section. Considering the broader global politics of climate change, while the focus had clearly been on the UN regime as long as the Kyoto Protocol was not yet ratified, numerous processes were initiated outside the UN framework in 2005. This was primarily the result of activities by the United Kingdom, which held both the EU and the G-8 Presidency in the second half of that year (Vogler 2008: 21). Part of the reason for promoting the use of fora outside the UN context was the desire to re-engage the US, which was considered as indispensable for ensuring the environmental effectiveness of the climate regime, but remained sceptical of both the multilateral process and the science of climate change (Afionis 2008). At the G-8 summit in Gleneagles in July 2005, a communiqué was nonetheless adopted in which all leaders, including the US President, underscored that "climate change is a serious and long-term challenge" and that it was "in our global interests to work together, and in partnership with major emerging economies" to reduce GHG emissions (G-8 2005: points 1 and 3). Further, the G-8 reaffirmed that "the UNFCCC is the appropriate forum for negotiating future action on climate change", but still agreed to begin an extra-UN, G-8 "Dialogue on Climate Change, Clean Energy and Sustainable Development" (G-8 2005: points 14, 9). The Dialogue's aim was to share best practices and promote the transformation of energy systems; it was open to "other interested countries with significant energy needs", notably the five major emerging countries (China, India, Brazil, Mexico, South Africa) already present in Gleneagles (G-8 2005: point 9; Afionis 2008: 3–5). The G-8 issued a work plan to that end, creating several working groups (Afionis 2008: 4). The EU, represented by its big member states and Commission President Barroso at the summit, endorsed this plan (Afionis 2008: 4). Only months later, before a first Gleneagles dialogue meeting, the US itself co-initiated another small multilateral forum, the Asia-Pacific Partnership on Clean Development and Climate Change (APP, US, China, India, Australia, Japan, South Korea). The aim of this partnership was equally to advance cooperation on clean energy technologies outside the

UN process (APP 2009). Both the G-8+5 and the US-led initiative would diversify the landscape of global arenas in which climate change would be discussed from the mid-2000s on. These trends would increase a feeling of uncertainty about where and how best to pursue global debates.

Regarding the issue itself, 2006 has been referred to as the "year of climate change", as it saw a rise in public and political interest in the topic, with the effect that important findings by climate scientists slowly began to transfer into "popular knowledge" (Sterk *et al.* 2007: 139; Davies 2006). The attention of the public was drawn to the issue, for instance, by a documentary film written by and starring former US Vice-President Gore. "An Inconvenient Truth" became a public success in many parts of the world. Further, a study on the economics of climate change, commissioned by the UK government and conducted by a team led by Nicholas Stern, appeared in late October 2006 and became a key reference for politicians and the interested public. The central message of the Stern report (2007) was that early action to mitigate climate change would come at lower cost than late adaptation and mitigation measures. Finally, the year 2007 would see the gradual release of the Fourth IPCC Assessment Report, discussed in detail in the Introduction to this work.[15] All in all, the degree of attention climate change attracted from 2006/2007 on was bound to have an impact on the emerging discussions about the further reform of the climate regime.

Key Actors in the Global Climate Regime and their Positions

Prior to the first COP/MOP, the *Umbrella Group* was split on how to proceed with regime development. While the US and Australia were not particularly inclined towards discussing further reductions at all, the other members of the group made it clear that they considered it necessary that major developing economies engage in GHG reduction measures, but divergences existed regarding the modalities of such efforts (ENB 2005b: 13–14; Germanwatch 2006: 5). By contrast, the *G-77/China* camped on its old position: it wanted by all means to prevent debates about own targets, preferring to focus discussions about regime development on the review of the commitments for Annex I countries stipulated in Article 3.9 KP (ENB 2005b: 13; Kasa *et al.* 2008: 116).

Since 2004, the *European Union* was operating with the described new internal division of labour in the climate change domain. The lead negotiator and issue leaders arrangement was bound to show its real impact on the Union's actor capacity during the more complex, two-track

---

[15] As the concrete findings of the report would really only have specific consequences for the talks from the 13th COP in Bali on, they are recalled in Chapter 5.

talks that were to come with the COP/MOPs. One sign of a greater interest for strategic thinking in the Union's foreign climate policy approach was a major transformation of its outreach activities. In reaction to the proliferation of fora addressing climate change and the increased importance of mini- and bilateral partnerships (ENB 2006: 19), but also to generally boost its foreign policy strategy, the EU sought to improve its bilateral relations with individual countries or groups of countries during the years 2005 and 2006 (European Commission 2005: 10). Further, the Union stepped up its diplomatic efforts for generalized outreach by exploiting the Green Diplomacy Network (GDN) to promote its positions on climate change. In terms of bilateral relations, the EU institutionalized talks with both Annex I and non-Annex I countries. Older relations with industrialized countries were reinvigorated. On top of long-standing relations with Canada, the Union had created, as early as 2002, a US-EU High Level Representatives Dialogue on Climate Change to keep the channels of transatlantic communication open despite substantive disagreement on how to deal with climate change (European Commission 2009b, 2009c). Moreover, EU-Japan relations were further strengthened: since 2001, they were based on a rolling work plan in the environmental domain, foreseeing regular high-level exchanges between the Troika and interlocutors from the Japanese ministries of environment, economy (MITI) and foreign affairs (Interview EU representative 24, European Commission 2009d). New and more wide-reaching efforts to strengthen cooperation were undertaken *vis-à-vis* the two major emerging countries, China and India. In 2005, the EU-China Summit adopted a Joint Declaration on Climate Change, establishing a partnership covering concrete cooperation in the fields of carbon capture and storage and clean energy technology, but also ensuring regular political dialogues (Interviews EU representatives 1, 31; European Commission 2009e; Joint Declaration 2005). In the Indian case, a joint action plan and a 2005 EU-India Initiative on Clean Development and Climate Change were to ensure a collaborative promotion of public-private partnerships for research and development of clean technologies as well as continuous dialogue on climate change (Interview EU representative 31; European Commission 2009f). Furthermore, the EU also attempted to forge stronger relations with least developing countries. In 2004, the Council adopted an "Action Plan on Climate Change in the Context of Development Cooperation" (European Commission 2007a: 10). It intended to help developing countries in building capacities to cope with mitigation and adaptation, and was meant as a basis for more concrete partnerships with LDCs. Finally, the Union also started to integrate debates about climate change into existing bilateral dialogues, e.g. the Asia-Europe Meeting (ASEM) (European Commission 2009g). In terms of wider diplomatic efforts, the EU began to undertake attempts to further coordinate its outreach via the Green Diplomacy Network, an informal network of environment experts within the foreign ministries and embassies of the 27, supported by the Commission (Interviews EU representatives 12, 1, 26, 24, 30). This network had carried out a first series of démarches in key countries before COP 10 in Buenos Aires and would repeat this activity before each of the subsequent COPs (European Commission 2009a: 2). The démarches, conducted about three weeks before the conferences in countries considered central for the outcome of the UN climate negotiations, consisted in an exercise of informing negotiation partners of the EU's position and an exchange of views (Interviews EU representatives 12, 1, 26). Based on instructions agreed in the WPIEI-CC, the local GDN would normally target the message to the specificities of its host country (Interviews EU representatives 12, 1, 26). This permitted the EU to explain its position more thoroughly than ever and, at least in theory, to gain an understanding of the concerns in partner countries so as to strive for an adaptation of its position and behaviour. Regarding the Union's internal climate regime, a policy decided earlier would come to full bloom in 2005: the ETS began its first trial period. In the same year, the Commission equally launched its second European Climate Change Programme, focusing on topics like emissions from aviation, cars and carbon capture and storage (European Commission 2008c). Concrete new legislation was added to the EU's acquis in 2006, notably a directive on energy efficiency and energy services (EP/Council 2006) (see Jordan/Rayner 2010: 68–71). Finally, the Union's negotiation position on issues related to regime development had become more concrete: in early 2005, the Commission communication "Winning the battle against climate change" (2005: 10) had demanded

> to establish a multilateral climate change regime post-2012 with meaningful participation of all developed countries and the participation of developing countries which will limit the global temperature increase to 2°C (...). The reduction commitments that the EU would (...) take (...) should depend on the level and type of participation of other major emitters.

The March 2005 European Council endorsed this EU "blueprint for a post-2012 world" (Jordan/Rayner 2010: 69), reiterating its "determination to reinvigorate the international negotiations" (European Council 2005: 15). Without specifying a target, the heads of state and government stated that they aimed at "reduction pathways for the group of developed countries in the order of 15–30% by 2020" compared to 1990 levels (European Council 2005: 16). The Union was thus determined to start talks on the future of the climate regime as soon as possible, and sought ways to promote broader participation in emissions reductions efforts by engaging major emerging economies and the US.

The Negotiation Process and the EU's Influence Attempts

The seminar on the future of the UN climate regime that had been the subject of much controversy at COP 10 was organised in Bonn in May 2005 and proved to be an "open, frank and broad-ranging" dialogue (ENB 2005a: 7). It had, however, not succeeded in overcoming key conflicts, as the general impression was that persisting differences "could soon translate once more into heated discussions and intransigence once formal negotiations resumed" (ENB 2005a: 7). Proceedings at **COP 11/ MOP 1** (28 November to 10 December 2005, Montreal) would therefore become crucial with regard to further regime development. Against the will of the US and Australia, the majority of countries had been in favour of holding the two meetings in parallel. While all parties to the UNFCCC, including the US, were gathering as COP, parties who had not ratified the Kyoto Protocol were granted observer status under the MOP. This entailed limited speaking and no voting rights. To structure talks, the Canadian COP Presidency had identified three broad agenda items: implementation, improvement and innovation (known as "the three I's") (Bausch/Mehling 2006: 195).

Regarding *implementation*, the first meeting of the parties to the Kyoto Protocol proved to be productive: without much conflict, parties formally adopted the Marrakech Accords. Concerning the agenda item *improvement*, several decisions were taken on the modalities of the CDM and joint implementation as well as on the compliance committee (Wittneben *et al*. 2006; Depledge 2006; Bausch/Mehling 2006; ENB 2005b). Main agenda items of the COP concerned ongoing issues (deforestation, adaptation), but discussions did not result in major advances (ENB 2005b). As far as the last, and for this analysis most crucial agenda item was concerned, *innovation* through a further development of the regime, problems emerged.

The MOP had to deal with two related issues in this regard. The Protocol itself foresaw a review of the adequacy of Annex I parties' commitments seven years prior to the end of the first commitment period, i.e. in 2005 (Art. 3.9 KP). At the same time, it also called for a general review in light of the "best available" science, and "at regular intervals and in a timely manner" (Art. 9 KP) (Bausch/Mehling 2006: 196–197). In a contact group on "future action", the usual cleavages quickly resurfaced: while the EU and Japan jointly introduced the idea of linking these two reviews to start discussing broader commitments beyond a mere review of Annex I targets, the G-77/China submitted a proposal that did not mention Article 9 and refused categorically any debates about non-Annex I reduction obligations (Depledge 2006: 18; ENB 2005b: 13–14). An informal group was nonetheless formed to discuss the review process under Article 9.

In parallel, discussions were conducted in yet another group on a proposal by the Canadian Presidency foreseeing the initiation of a dialogue on long-term cooperative action under the Convention, introduced to provide a forum in which the US could be a full member (ENB 2005b: 14; Depledge 2006: 18). Reunited in a high-level informal meeting, members of the three groups deliberated during the last two days of the conference on the design of future regime reform talks (ENB 2005b: 14). In a reportedly dramatic move, the US walked out of this group when the discussions came – in their view – too close to considering new commitments for Annex I parties (Müller 2006: 12; Depledge 2006: 18; Germanwatch 2006: 5). The other parties, led by the EU, Japan, and major developing countries (China, India), pursued their debate and decided to draft a plan for a dialogue on long-term action under the UNFCCC (Germanwatch 2006: 5; ENB 2005b: 14). After the UK had used its special relationship to bring the US back to the table the next (and last) day (Müller 2006: 12), US negotiators had only minor changes to make to this plan (Germanwatch 2006: 5).[16]

The resultant structure for further regime development discussions was three-fold (Bausch/Mehling 2006; Depledge 2006; Germanwatch 2006: 4–6; ENB 2005b: 14): the COP decided to begin a "Dialogue" under the Convention, which would serve the purpose of exchanging views about an enhanced implementation of the UNFCCC in a "non-binding" manner "not open[ing] any negotiations" (UNFCCC 2005a: 2–3; Wittneben et al. 2006: 17–18; ENB 2005b: 14). Up to four workshops were scheduled until COP 13. Under Article 3.9 KP, the MOP decided to initiate talks on commitments for Annex I parties and created an open-ended Ad Hoc Working Group (AWG-KP) (UNFCCC 2005b: 1; Wittneben et al. 2006: 16–17; ENB 2005b: 14). The group was to complete its work in time to avoid a gap between the first and a second commitment period of the Protocol. Finally, to the disappointment of the EU and Japan, a decision on the Article 9 review was postponed: parties were invited to submit views on this item (Depledge 2006: 18).

Against the background of the increased attention given to the topic of climate change in 2006, **COP 12/MOP 2** in Nairobi (6–17 November 2006) apparently had a sobering effect. Observers remarked a lacking sense of urgency – with talks proceeding at "an almost surrealistic slow pace" – and the usual tendency to backload, resulting in little progress (Sterk et al. 2007: 139–140; Okereke et al. 2007: 32–33, 40–41). Apart from adaptation, which occupied a central place on the agenda, with the adoption of an ambitious "Nairobi Work Programme" (Okereke et al.

---

[16] The US had apparently hoped to present the Asia-Pacific Partnership as an alternative track to the UN regime. China, India and Japan chose, however, to stick with the UN as the prime arena to discuss climate change (Germanwatch 2006: 5).

2007: 3–10; UNFCCC 2006a; ENB 2006: 19), talks on regime development received the most attention. They were held under the two tracks created to that end at COP 11/MOP 1, with some parties, notably the EU, trying to extend them to a third forum.

Under the *AWG-KP* of Article 3.9 KP, which had met once before – in May 2006 – to elaborate a future plan of work (ENB 2006: 2), the EU and several other parties (Switzerland, Canada) again tried to construct a link with the discussions on Article 9 KP (ENB 2006: 11). Further, the Union and Australia declared that action only by Annex I countries would not suffice to mitigate climate change in a future regime arrangement (ENB 2006: 11). The EU in particular also attempted to insert its own temperature guideline, a reference to keeping global temperature increase below 2°C, into the negotiation texts, while arguing for a debate on a long-term vision of halving emissions by 2050 (Sterk *et al.* 2007: 141). In the face of G-77/China opposition, the AWG-KP did not reach agreement on these issues, nor on a specific timeline for its proceedings. Only very general conclusions were adopted, emphasizing the need for industrialised countries to lead and the future targets to be based on scientific analyses (ENB 2006: 11; Okereke *et al.* 2007: 14–15). An extended work programme was also endorsed: in 2007, parties were to proceed with analyses of mitigation potentials, followed by discussions on issues of means and actual reduction objectives for Annex I parties (UNFCCC 2006b; Sterk *et al.* 2007: 141). The second forum for regime development, the *Convention Dialogue*, was arranged as a series of workshops, focusing on broader topics. Workshop number two was conducted at COP 12, but did not yield any substantial advances beyond the exchange of views (ENB 2006: 4; Okereke *et al.* 2007: 15). Observers remarked a certain loss of momentum of the talks under this track (ENB 2006; Sterk *et al.* 2007: 142). A third opportunity to discuss the future of the climate regime was still available under Article 9 KP, which stipulated a periodic review of the Protocol: the EU tried, once again, to gain a profile as the champion of a thorough and comprehensive review, while the G-77/China attempted to limit the assessment of advances to this MOP, arguing that Article 9 KP foresaw a "review", not a "revision" of the Protocol (ENB 2006: 11; Okereke *et al.* 2007: 16; Sterk *et al.* 2007: 142). The debates did not end with any substantial reconsideration of the Protocol, but rather a compromise agreement on procedure: a new, comprehensive review process was scheduled for MOP 4 in 2008 (Sterk *et al.* 2007: 142). The G-77/China obtained an assurance that this review would not lead to new commitments for any party (UNFCCC 2006c). In sum, while "the focus of COP 12 and COP/MOP 2 was undoubtedly on the future – of the Protocol, the Convention, and longer-term action to combat climate change" – (ENB 2006: 19), not much was actually achieved in terms of regime development.

From the Buenos Aires Action Plan to the Year 2007 (1998–2007)

After the sobering effect of the Nairobi COP/MOP, the **year 2007** would witness a remarkable build-up of political momentum towards COP 13/MOP 3 in Bali, Indonesia, within and beyond the UN arena. To illustrate both the extension of the negotiations into non-UN bodies and the EU's expanding foreign climate policy activities, a comprehensive story of this transition year toward a period of intense post-2012 talks, discussed in the next chapter, is presented here.

In January 2007, **the EU** was the first major actor in the global climate arena to advance its conception of a post-2012 agreement when the Commission proposed the start of a negotiation process toward a comprehensive regime reform in its communication "Limiting Global Climate Change to 2° Celsius: The way ahead for 2020 and beyond" (European Commission 2007b). In March, the European Council conclusions endorsed the Commission communication, linked to the proposal that the EU would commit itself to 20% unilateral emissions reductions and 30% emissions reductions by 2020 if other industrialized countries were to adopt reductions in that same range (European Council 2007). Gradually completed, this proposal would form the core of the Union's position – constituting its major influence attempt – in the post-2012 talks.

After a spring during which debates on climate change were held for the first time in the UN Security Council (UNSC 2007), negotiations within the **UN climate regime** were set forth under both tracks in May 2007 in Bonn. Focusing on technology and adaptation, the *Dialogue Workshop* (16–17 May) did not produce any concrete advances in terms of regime development (ENB 2007a: 15). The summary of the meeting did, however, identify a number of elements that had emerged as the parties' favoured building blocks of a future climate regime, including a long-term goal "consistent with science", "national climate change strategies" on mitigation and adaptation, the continued use of market mechanisms and the possible introduction of sectoral mechanisms as well as enhanced consideration for adaptation and technology transfer (UNFCCC 2007c: 2). The *AWG-KP* meeting went into quite some detail to discuss mitigation potentials and emission reduction objectives for Annex I parties as well as the review of its work programme (ENB 2007a: 14). In the closing plenary, the EU, while acknowledging the importance of the group, highlighted the significance of linking its discussions to debates under the Convention. The conclusions of the session stated that work was to continue at and beyond Bali and that it should focus on further Annex I emissions reductions, while taking into account the "shared vision" of how to reach the objective of Article 2 UNFCCC (ENB 2007a: 14–15).

During the early summer of 2007, the **EU** became increasingly active in **bilateral relations** with key parties outside the UN framework. The

EU-China dialogue of 29 May provided an opportunity for an exchange of positions, but did not deliver any common vision of the future shape of global climate policies, with China refusing to discuss any binding emission reductions/actions for itself (Agence Europe 2007a). By contrast, the Union's meeting with Japan on 5 June delivered a "broad consensus" about the urgency of tackling climate change and the necessity to kick off a new negotiation round in Bali (Agence Europe 2007b). Further impetus to the UN process came from the G-8 in June: under the impulsion of Germany, holding both the EU and the G-8 Presidency in the first half of 2007, leaders concluded in Heiligendamm: "we will consider seriously the decisions made by the European Union, Canada and Japan which include at least a halving of global emissions by 2050" and "have agreed that the UN climate process is the appropriate forum for negotiating future global action on climate change. We are committed (...) to actively and constructively participate in [COP 13] with a view to achieving a comprehensive post 2012-agreement (post Kyoto-agreement) that should include all major emitters" (G-8 2007: point I; Agence Europe 2007c). For the first time, the Bush administration would thus loosely acknowledge in an international forum that substantial emission reductions (in the long term) were necessary. The declaration also sent a clear signal to the major emerging countries that all key industrialized countries, unlike during the Kyoto Protocol negotiations, were committed to demanding actions from them.

Back in the **UN arena**, further support for the EU's vision of a comprehensive post-2012 agreement came from a three-day UN General Assembly debate dedicated to climate change (31 July–1 August 2007), in which numerous countries expressed their growing concern about the problem and called for a new negotiation process to be initiated in Bali (UNGA 2007). In the **UN climate regime**, the final Convention Dialogue and the first part of AWG-KP 4 were held between 27 and 31 August 2007 in Vienna. The *Dialogue* dealt essentially with two topics: financial issues and the question of how to proceed with discussions beyond 2007, with the scenario note by the co-facilitators stating the hope that the abovementioned "building blocks" could be "welded together into an effective and appropriate international response to climate change" and that talks would prepare for the decision on a "road map" in Bali, an explicit request voiced, inter alia, by the EU (UNFCCC 2007c: 2; ENB 2007c: 6–9, 9). In its submission informed by the European Council conclusions of March 2007, the EU had suggested that "COP 13 and COP/MOP 3 in Bali need to lead to the engagement of all Parties in a comprehensive negotiation process including both the Convention and Kyoto Protocol tracks agreed at COP 11 and COP/MOP 1, with a view to reach a global and comprehensive post-2012 agreement by 2009 at COP 15 and

COP/MOP 5" (UNFCCC 2007d: point 3). The paper further identified one key objective and eight building blocks that the EU wanted to feature in the future negotiation process after Bali: "the need to limit the global average temperature increase to not more than 2°C above pre-industrial levels and a 2050 goal/yardstick of low carbon development to reduce global emissions to at least 50% below 1990 levels" necessitated discussions on (i) a shared vision, (ii) deeper absolute emission reductions by industrialized countries, (iii) "further fair and effective contributions by other countries", (iv) an extension of the carbon market, (v) increased attention to technology cooperation and transfer, (vi) adaptation measures, (vii) the reduction of emissions from international aviation and maritime transportation, and (viii) the regulation of deforestation and land-use (UNFCCC 2007d: points 3, 2). Many of these elements, but also numerous suggestions of other parties were contained in the final report that the co-facilitators transmitted to the COP, without, however, privileging any option, a request explicitly voiced by developing countries (ENB 2007c: 9). Highlighting that broad agreement existed on continuing the Dialogue to come to "an effective global response to climate change", the report proposed various possibilities as "next steps": extend debates either (i) under the existing Dialogue or (ii) under the COP; (iii) open a "negotiating process" in a working group with a clear mandate and timetable either separately from the AWG-KP or "fully integrated" with the existing process under the KP (UNFCCC 2007h: points 14, 61, 64). A decision on this was shifted to the Bali COP (UNFCCC 2007h: point 69).

In a second submission transmitted to the UNFCCC secretariat *after* this fourth session of the Dialogue, the EU further elaborated on each of its eight building blocks (UNFCCC 2007f). Compared to the short and often vague declarations by most other parties that had made submissions at that stage (China, Australia, Canada, Russia, South Africa, Saudi Arabia), the EU – together with AOSIS, which also provided a very detailed blueprint of its ideal outcome[17] – was the only player to present a concise vision of its positions.

---

[17] The AOSIS proposal formed the basis of its positions in the post-2012 negotiations and therefore merits brief consideration regarding the two issues under analysis here. On the targets to be adopted under a post-2012 regime, AOSIS demanded that "long-term temperature increases are stabilized well below 2 degrees Celsius" (UNFCCC 2007e: 3). To achieve this, "global GHG emissions must peak within the next 10–15 years, and be followed by reductions of at least 50–80% of 2000 levels by 2050" (UNFCCC 2007e: 3). On the issue of responsibilities, AOSIS stated: "the largest historical emitters must now take aggressive action under the Convention (…) All Annex I Parties should take on hard quantified emission limitation or reduction targets in the Post-2012 period and through 2030. Major emitting developing countries will also need to take action to reduce their emissions trajectories, with assistance from developed country Parties (…) all major emitting countries must, therefore, be engaged in global

The *AWG-KP* focused its debates again on mitigation potentials and "ranges of emissions reductions objectives for Annex I countries", going analytically into depth, while ultimately remaining inconclusive (ENB 2007c: 4–6). It did, however, discuss and clearly state options for a preliminary agreement on the indicative range of industrialized country emission reductions, which would gain further importance in the negotiations at and after Bali (ENB 2007b: 2; Spence *et al.* 2008: 146). In the debates on this issue, the EU successfully co-sponsored a proposal together with the majority of the G-77/China (e.g. ENB 2007c: 14), namely to include a reference to the IPCC science by mentioning a range of 25–40% GHG reductions for industrialized countries by 2020 (from 1990 levels) (ENB 2007c: 6). The Union also promoted its position on a long-term global emissions reduction target of 50% by 2050, reflecting the conclusions of the Heiligendamm G-8 summit (ENB 2007c: 4). Both references found their way into the Chair's conclusions: "The AWG recognized that achieving the lowest stabilization level assessed by the IPCC to date (…) would require Annex I Parties as a group to reduce emissions in a range of 25–40 per cent below 1990 levels by 2020", while also acknowledging that emissions should be "well below half of levels in 2000 by the middle of the twenty-first century" (UNFCCC 2007g: points 7, 6). Despite this small achievement, the EU and others voiced their disappointment about lacking progress in the closing plenary (ENB 2007c: 6).

The two tracks thus advanced discussions on future emission reductions targets for industrialized countries. The AWG-KP report made no explicit reference to the sharing of tasks between developing and developed countries in a future climate regime (UNFCCC 2007g). The Dialogue report however did (UNFCCC 2007h). Since the Umbrella Group, and here notably the US, but also the EU in its two written submissions and in the Dialogue meetings, emphasized that the participation of "all major emitters" was required in future emission reduction efforts, developing countries could not prevent this item from being reflected in the conclusions. The report even noted that the latter were actually prepared to be "part of the global response" in line with national circumstances (ENB 2007c: 8; UNFCCC 2007h: 7). This difference in approach between the two tracks *vis-à-vis* the interpretation of the CBDR principle would be set forth and become a characteristic feature of the talks that followed. With this preparatory work, the agenda was broadly predetermined for the Bali COP, which negotiators expected to move the debates "from dialogue to action" by extending the mandates for both groups, separately or jointly (ENB 2007c: 11).

---

efforts to mitigate emissions, according to their common but differentiated responsibilities and respective capabilities" (UNFCCC 2007e: 4–5).

From the Buenos Aires Action Plan to the Year 2007 (1998–2007)

During the months before the Bali summit, a number of additional **meetings** were held **outside the UN regime**. Among them featured the third ministerial meeting of the Gleneagles Climate Change Dialogue (9–11 September, Berlin), a UN high-level summit in New York (24 September), a US-initiated meeting of the 15 largest emitters and the EU in Washington ("Major Economies Meeting") (27–28 September) and a pre-COP ministerial meeting in Indonesia (25 October) (ENB 2007d: 2). If anything, they would underscore that – beyond the broad support for talks with a deadline for 2009 and the debate on building blocks – differences on the nature of a future agreement persisted among key countries: in contrast to the EU, other major players were either unprepared to discuss other than voluntary emissions reductions efforts (US, China, India) or remained rather silent on their approach (e.g. Japan) (Reuters 2007b; Spence *et al*. 2008: 146–148).

The Outcome: Small Advancements Towards a Regime Reform

In terms of regime development, the two conferences/meetings of the parties in 2005 and 2006 achieved mainly procedural results: the continuity of negotiations on the future of the UN climate regime was assured. The year 2007 witnessed then, also under the impulsion of the Fourth IPCC Report, a genuine intensification of concern about climate change across the different coalitions, with even the US administration showing openness for serious global talks. The immediate result was a proliferation of meetings within and beyond the UN arena, gradually leading to a growing impetus for COP 13/MOP 3 to deliver a mandate for a new regime reform process.

The EU's Influence on the First Two-track UN Regime Negotiation Sessions

The EU was certainly among the first and – together with AOSIS – most active parties at both COP/MOPs and in the year 2007 regarding issues related to climate regime development. It continued its trend from previous COPs, arguing forcefully for a comprehensive review of the regime with the intention of gaining leverage over developing countries emissions and, in this fashion, re-engaging the US in the negotiation process. Its major influence attempt was arguably the spring 2007 comprehensive proposal for a post-2012 regime, prominently featuring a 20% unilateral emissions reduction target by 2020 and a range of building blocks. As this attempt was partially already aimed at setting the agenda for a process that had not yet been started, it will be given closer consideration in the analysis of the post-2012 talks in Chapter 5. For the year 2007 itself, the Union's position contributed, together with other statements of intent by various players, to building the momentum for COP 13. The EU also engaged in broader foreign policy activities to promote its message more widely, placing climate change deliberately high on the agenda in many global fora (G-8(+5), UNGA, UNSC) and in its existing bilateral relations.

To assess the Union's influence during this period, it is first necessary to identify possible turning points regarding regime development and the topics of emission reduction targets and responsibilities during the global climate talks of the years 2005, 2006 and 2007. The only major turning point in this period was arguably the eleventh COP and the first MOP in 2005: "future action", a "taboo topic" before, was now systematically addressed in a two-track negotiations arrangement. At this turning point, the EU cannot be credited with any particular influence. Even if it had been the first and most fervent industrialized party supporter of a start of new negotiations ever since the early 2000s (*purposive behaviour, temporal sequence, interaction*), it *attained* its *goals* only to a very limited extent. The Union could be partially satisfied with the pursuit of multilateral talks on the future of the regime under the UN umbrella, and not, as apparently desired by the US, in multiple arenas (even if the latter did gain increasing importance from the mid-2000s on). However, it clearly did *not obtain* any substantial *advances* regarding its concrete demands for the future of the climate regime (discussion of developing country commitments, substantive re-engagement of the US, reference to 2°C target). All options remained on the table. Moreover, the EU was clearly not at the origin of the change of behaviour of the other parties. While it may be argued that it contributed to setting the agenda and to the build-up of momentum in the wake of the 2007 IPCC report, as a reputational analysis indicates (Interviews EU representatives 20, 27, Observers 3, 25), it was neither the only actor, nor can it – in counterfactual perspective, imagining its absence – be argued that the Union was decisive for bringing about the novel commitment to negotiating regime reform. Under the MOP, parties engaged to fulfil a treaty obligation imposed by Article 3.9 KP ("Commitments for subsequent periods for Parties included in Annex I"), while the COP decision to begin a "Dialogue" had originated from the Canadian COP Presidency (Depledge 2006: 17), and the EU had not had any leverage over its content (test on *absence of auto-causation: negative*).

The major constraint for greater EU impact during this period was arguably an external factor: the ever more protracted multilateral arena. The still obstructive behaviour of the US, the G-77/China's defence of its past acquis and the other Umbrella Group members' passiveness made it very difficult for a single player to gain leverage over key decisions. The diversification of the Union's influencing strategy can be interpreted as a partial and insufficient attempt to take this changing external environment into consideration by addressing other actors also outside the UN arena. It did not yield immediate success in this period.

From the Buenos Aires Action Plan to the Year 2007 (1998–2007)

Determining and Explaining the EU's Influence during the Period 1998 to 2007

Summing up the findings for this ten-year period of regime negotiations, key characteristics of the evolving EU foreign climate policy can be extracted. Its influencing strategy yielded overall limited impact, with two significant exceptions.

The Union's influence attempts were initially based on diplomatic tools again during this period: it argued with the help of substantive reform proposals in the beginning, and engaged in bargaining only towards the end of negotiation processes (e.g. at COP 6bis and 7). Just like during the Kyoto Protocol talks, the Union began the negotiations on the Protocol's operationalization with a very inflexible position, insisting on certain *conditions sine qua non* at COP 6, which ultimately forced it to give in to virtually all demands by the Umbrella Group in order to ensure its overarching aim (rapid enactment of the Protocol) at COP 6bis and COP 7. Following the US withdrawal from the Kyoto Protocol ratification process in 2001, however, the Union did slowly begin to alter its influencing strategy. It stepped up its diplomatic efforts to convince others to pursue the ratification of the treaty even without the United States. Later, to force ratification, the EU employed, for the first time in the history of its participation in the climate regime, explicitly and successfully economic tools and issue-linkage with topics outside the climate realm in order to convince Russia of joining the club. After this ratification, it broadened and diversified its strategy to adapt it to the proliferation of new fora (G-8+5, APP) and bilateral partnerships (US-India etc.). This approach provided not only the opportunity to forge hands-on technological cooperation, but also to develop a common understanding and institutionalize dialogues on climate change. In addressing developing countries, the EU thus explicitly employed more than "just" soft diplomatic tools, using economic incentives targeted towards what it perceived to be the demands of its interlocutors as a basis for the partnerships. It was hoped that such a pragmatic and wide-reaching (both in a thematic and geographical sense) outreach would also make up for the Union's damaged reputation in the developing countries after its very demanding attitude of the years 2002 and 2003. Even if no immediate impact could be detected during this period, the first contours of a much transformed foreign policy strategy helped position the EU for the post-2012 talks and in a long-term perspective (Schunz 2009). The major, once again more formal influence attempt during this period came, however, with the Commission proposal of early 2007, endorsed by the European Council. Its significance would show not so much at COP 13, but certainly during the talks thereafter.

EU influence could be discerned at two of the five key turning points during this period. During the debates on the operationalization of the

Protocol in the period 1998–2001, the Union exerted influence immediately after the US withdrawal with regard to the pursuit of talks under the UN, when it used a broad diplomacy approach to *argue* for the further continuation of the regime. During the period 2002–2004, the Union influenced the Russian decision to ratify the Protocol in order to guarantee the development of the UN regime. In this instance, *bargaining* through issue-linkage was the causal mechanism that helped the EU to exert influence.

All in all, when it comes to the extent and type(s) of EU influence during the analysed transitional period of the UN climate regime, the following observations can thus be made. While the Union had virtually no leverage over the substance of decisions during this period, it did – by contributing to the survival of the regime – manage to protect the emissions reduction target it had decisively helped to negotiate in Kyoto. As the rules around this target (supplementarity, sinks) were watered down in the Marrakech Accords and the US (with its -7% target) had withdrawn from the Protocol, the Union's overall influence on this item was, however, at best medium: the continuity regarding the target was assured for those who ratified the Kyoto Protocol (degree of legal bindingness), but the EU's degree of goal attainment was only partial. By contrast, it had no impact on the interpretation of the CBDR principle. Beyond assessing EU influence on the substance of the climate regime via these two central issues, the empirical analysis for this period strongly suggests a non-negligible EU influence over *where* (in which fora) and *how* (by what rules) the climate regime is developed – namely within the UN system, by the rules of multilateralism and with respect for what had previously been negotiated. This influence on the structural parameters of global climate negotiations can be characterized as high and rather enduring. Altogether, its influence on the operationalization of the Protocol and on the transition towards a new regime reform process must therefore be interpreted as medium, just like for the Kyoto Protocol negotiations.

Tentative explanations of EU influence were provided in the various sub-sections of this chapter and can be drawn together here to highlight some of the main changes since the Kyoto Protocol talks. While the external environment gradually became less favourable for an exercise of influence by the EU, internal changes in terms of foreign policy making and implementation constituted potential improvements bound to enhance its chances for influencing the climate regime. During this period, the immediate impact of the former tended to outplay the potential of the latter. As far as the *external environment* is concerned, the US withdrawal from the Kyoto Protocol ratification process in March 2001 altered the negotiation context in manifold ways. First, in a context of bargaining, it gave each of the major individual members of the Umbrella Group veto power on issues concerning the operationalization of the Kyoto Protocol for COPs 6bis and 7, since all of them were needed for a ratification of the Protocol. Second, as the US dropped out as *demandeur*, it became easier for the G-77/China to defend its achievements of the past and fight off any discussions about broadening commitments. Third, in a medium-term perspective, the US preferences for uni-, bi- and small multilateral action (e.g. partnerships with India, the APP) led to a fragmentation of the negotiations within (Convention vs. KP tracks) and outside (G-8, APP) the UN regime context. Under these conditions, the chances for EU influence decreased because (i) the heterogeneity of preferences further increased, (ii) the number of players (US, Australia) and fora increased; and (iii) some powerful players (US, G-77/China) refused to cooperate under the rules of the (reformed) regime. Partially in reaction to this changing environment, but also to its own weak performances at some of the first seven COPs and in anticipation of enlargement, the EU made *internal changes*, attempting to improve its actor capacity to enhance its chances for greater influence on the global scene. Both the inclusion of the Commission in the Troika and the introduction of the lead negotiator system were bound to have positive effects on its performance. Further, the EU adapted its influencing strategy by widening the scope of its influencing targets and employing a broader range of tools. Russian ratification of the Kyoto Protocol demonstrated that this change already bore fruit, suggesting that improved foreign policy implementation can enhance the chances of exerting influence. By contrast, within the UN framework, the EU's formal diplomatic strategy was not efficient, as it was unable to identify enough issue-linkages to allow for bargaining when it came to its desire to advance the regime.

CHAPTER 5

From the Bali Roadmap to the Copenhagen Accord (2007–2009)

EU Influence on the Post-2012 Global Climate Negotiations

This chapter provides an in-depth analysis of the EU's influence attempts and their effects during a time period that was supposed to lead to the substantial development of the climate regime, but resulted only in minor reforms. It traces the EU's influence on the "post-2012" climate negotiations during the period 2007 to 2009, starting with COP 13 in Bali and ending with COP 15/MOP 5 in Copenhagen, originally designated as the final point of this negotiation process (see Annex II).

The Context: Major Developments in Global Politics and Climate Science

The period 2007 to 2009 was marked by one important event that would impact on global politics. Moreover, long-term trends concerning global policy-making, the GHG emission trajectories of major countries and advances in climate science played a role in changing the overall framework in which global climate negotiations were conducted.

The major event occurred in mid-September 2008 in the United States, but immediately gained global significance. Following the file for bankruptcy of Lehman Brothers, one of the major global financial-services companies, US and global financial markets came under serious strains. This initial distress quickly turned into a major global financial and economic crisis when other financial institutes equally became insolvent. The crisis would have two major repercussions for the global politics of climate change from late 2008 on. Firstly, it made many governments across the world pay almost exclusive attention to the economic well-being of their populations, weakening the often already volatile interest for tackling climate change – despite voices from the UN, EU and US calling for solving the economic and climate crises together (Goldenberg 2009a; Dimas 2009; Ban Ki-moon 2009; Yuxia 2008). To stabilize the global economy (inter alia by rescuing major private financial institutions) and to attenuate other effects of the recession (such as growing unemployment), considerable amounts of public money were invested by governments all over the world. Although the stimulus packages initiated in major countries regularly dedicated a percentage to so-called "green investments" (Goldenberg 2009a; European Commission 2009h), the lion's share went into other measures, inter alia the stabilization of certain polluting industries (e.g. automotive). The sheer magnitude of public funding also implied that governments felt that fewer resources were available to support mitigation and adaptation measures in developing countries, constraining their own room for manoeuvre on this major cornerstone of the global climate negotiations. The second significant effect of the crisis was a tectonic shift in global (economic) politics more generally, with the sudden rise in importance of the G-20 (gathering the G-8 members, including the EU, as well as Argentina, Australia, Brazil, China, India, Indonesia, Mexico, Saudi Arabia, South Africa, South Korea and Turkey). After several meetings about new regulatory measures for the global financial markets, the forum proclaimed, in September 2009, that it would henceforth replace the G-8 as the key site for coordination among the most significant economies of the globe, including on such issues as climate change (CBS 2009).

Several gradual developments were equally to affect climate negotiations during this time period. Although an immediate product of the crisis, the rise of the G-20 also reflected the growing importance of emerging countries, such as China, India or Brazil (regularly referred to as BRIC, together with Russia, Keukeleire/Bruyninckx 2011), in world affairs. Their heightened significance was, in many cases, a direct result of long-term high economic growth rates, often coupled with equally exceptional demographic weight (e.g. China, with growth rates between 9 and 11.5% in 2004–2008, for 1.32 billion people; India, with growth rates between 7.3 and 9.8% in 2004–2008, for 1.15 billion people – Chinability 2009; CIA 2009). Economic growth had, however, also been accompanied by a stable trend of steeply rising GHG emissions in these countries. In 2007, China had overtaken the US as the world's number one emitter in absolute terms (PBL 2008). Moreover, studies suggested that the proportions between developed and developing countries were continuing to change dramatically not only in terms of annual absolute emissions, but also with regard to cumulative contributions to the problem of climate change. Under a business-as-usual assumption, China would overtake Western Europe as a cumulative CO_2 contributor during the 2020s and the US by mid-century, while India would arrive at an equal level of CO_2 emissions as Western Europe by 2080 (Botzen *et al.* 2008: 571). Taken together, these trends in economic growth, demography and the resulting effects on the global environment were bound to represent a gradual shift in the balance of powers in global (climate) politics. Where past debates on these issues had been dominated, oftentimes and on many agenda items, by the industrialized players, emerging economies were moving centre stage in the second half of the 2000s.

Additionally, the rise of the G-20 provided an emblematic example of the continued trend of an unprecedented increase in the number of fora, meetings and actors in global climate politics. With the partial de-localisation of talks into the G-8(+5), the Major Economies Meeting/ Forum and many regional or inter-regional gatherings (the APP, APEC, Asia-Europe meetings, EU/Latin America summits etc.), global climate policy debates became less restricted to the sole UN regime. In the late 2000s, extra-UN meetings would remain related to the UN regime talks, but would also take on a dynamic of their own. The change in quality and quantity of arenas also coincided with an increase in the number of meetings, enabling an almost continual exchange between the representatives of the major players (Interviews US, EU representatives 17, 6). This was further facilitated through new communication technologies that, at the time of the Kyoto talks, had only just begun to become more widely used (Oberthür/Ott 1999: 82–84). Finally, climate change became also increasingly subject of exchanges at high political levels and by different constituencies (development experts, finance experts etc.) in the various new fora. This increasing trend of "high-levelisation" would become particularly visible during the final days of the Copenhagen summit. Together with the further explosion of the number of non-party participants in and around global climate talks, be they from civil society, research institutes or the media, this contributed to the growing complexity of the climate politics arena.

Turning to the scientific knowledge about climate change, the successive release of the various parts of the Fourth IPCC Assessment Report (FAR) over the course of the year 2007 marked a major event during this time period. Without repeating details of the FAR already discussed in the Introduction to this work, some of its key messages shaped, together with the 2006 Stern Report that linked the science and the economics, the understanding of climate change and the perceptions of the issue among the interested public and politicians (Hasselmann/Barker 2008: 219). Those messages were above all: 1. "Warming of the climate system is unequivocal" (IPCC 2007b: 30); 2. Some impacts of climate change may be "abrupt or irreversible" (IPCC 2007a: 13); 3. Stringent early action is necessary to prevent its most worrisome consequences. With regard to the latter, it was especially various GHG stabilisation scenarios – linking, e.g., a 2°C global mean temperature increase above pre-industrial levels to a stabilisation at 450 ppm, which would necessitate a peak in global GHG emissions by 2000–2015 and a reduction of 50 to 85% by 2050 – that would acquire significance as points of reference in global debates (IPCC 2007b: 67). Finally, compared to its predecessors, the FAR had a much higher public resonance, underscoring the urgency with which climate change had to be tackled (Garber 2008).

Key Actors in the Global Climate Regime and their Positions

To further set the stage for the process-trace, the negotiation positions and their foundations as well as the strategies of the main coalitions/countries in the regime negotiations prior to the kick-off (and as far as relevant and not explicitly taken up during the discussion of the negotiation process also in the early stages) of the post-2012 negotiations require particular consideration.[1] The analysis focuses in the first instance on the key actors other than the EU, before explicitly discussing the Union's actor capacity and position.

Key Actors Other than the EU

The *Umbrella Group* continued to be split across UN negotiation fora in 2007: under the Framework Convention, the US was an active part of the Group, while it played (formally) a marginal role as observer in debates on the Kyoto Protocol.

In 2007, the *United States* possessed still the largest economy in the world, and had only just ceded the top spot as the world's biggest GHG emitter to China (World Bank 2008; UNSD 2009). As a result of its high dependency on fossil fuels (oil: 40%, coal: 23% of total energy production in 2006, US EPA 2008: ES-12), significant growth rates were still directly linked to rising emissions. As of 2006, these had increased by 14.7% compared to 1990 (US EPA 2008). In the face of these numbers, and despite observable ecological and socioeconomic impacts, the perception of vulnerability to climate change in the US had, for a long time, remained rather low (Romàn/Carson 2009: 41–42). It would only change in the course of the late 2000s under the more compelling evidence of climate science (IEEP/NRDC 2008: 60–61). One contributing factor to the fairly limited degree of concern had been the attitude of the Bush administration, which had downplayed or even opposed the findings of climate scientists ever since the 2001 withdrawal from the Kyoto Protocol ratification process.[2] In the particular US institutional context for climate policy-making (see Chapter 3), which grants a strong role to Congress (and here notably the Senate), the fact that the latter was Republican-dominated for much of the time during the Bush era meant that "no federal policies of significance" were enacted between 2001 and 2008 (Urpelainen 2009: 100). Yet, "while climate change policy appear[ed]

---

[1] The distinction between "prior to" and "during" the negotiations is fuzzy because reform talks had been ongoing ever since 2005, but it is fair to say that the official post-2012 negotiations commenced with the 2007 Bali COP.

[2] For one of the first times, George W. Bush mentioned climate change as a "serious challenge" in his State of the Union Address of 23 January 2007 (Bush 2007: 5).

hopelessly deadlocked in Washington, a set of state governments that cut across partisan and regional lines [was] demonstrating that it is possible to make some significant inroads on the issue" (Rabe 2004: 4).[3] These and other, civil society-based initiatives (IEEP/NRDC 2008) did, however, not have a major impact on the administration's overall position and strategy in the global climate negotiations. Until the end of 2007, it "rejected binding country-by-country limits on greenhouse gas emissions, focusing instead on a long list of voluntary bilateral and regional initiatives", such as special ties with India and China and the APP as well as, since September 2007, the "Major Economies Meeting" gathering countries that covered 80% of the world's emissions (Pataki/Vilsack 2008: 27–28). Although unique in its domestic inactivity, the Bush administration's negotiation position actually represented a continuation of long-standing US positions in global climate talks. For quite some time, the US had emphasized the importance of hands-on technology-based international cooperation to ensure cost-effectiveness of climate policies, notably through the use of flexible mechanisms, and had been concerned with the participation of major emerging economies, particularly China, in the global climate regime (Urpelainen 2009: 101–103; Biermann 2005: 276–277). It was not until late 2007 and 2008 that US climate policies would slightly alter, before undergoing significant transformations, paired with a change in public attitudes, in 2009. Without going into the details of the development of the US position, discussed where relevant in the process trace, the Bush administration did, in 2008, begin to acknowledge the necessity to act on climate change, with the President proposing to halt the growth of US emissions by 2025 (AFP 2008a). Further, during the presidential election campaign of 2007/2008, both Democrat Barack Obama and Republican John McCain displayed greater willingness to engage on climate policy domestically and globally (AFP 2008b). At the same time, a now Democratic majority in Congress attempted to have climate legislation passed: the most wide-reaching of several proposals, the Liebermann-Warner Climate Security Act, foresaw a stabilization of GHG emissions at 2005 levels by 2012, and a reduction by 70% until 2050. It was voted down in the Senate in June 2008 (Spiegel 2008b).[4]

---

[3] To highlight but two examples (Romàn/Carson 2009), at state level, the Regional Greenhouse Gas Initiative (RGGI) of 2005 joined ten states from the East Coast together in a cap-and-trade system covering utility sector emissions. Similar initiatives were created by five Western states and in the Midwest (six US states plus Manitoba). At the local level, the US Council of Mayors signed a Climate Protection Agreement in 2005: covering more than 900 cities in early 2009, the mayors committed their communities to individually complying with the Kyoto targets.

[4] A key feature of the bill was the proposal of a national cap-and-trade scheme covering electric utility, manufacturing and transportation industries. Emission trading would remain the policy tool of choice of US lawmakers in the period thereafter.

An almost complete reversal in terms of (i) the attitude towards climate change (both by political elites and growing parts of the public),[5] (ii) the necessity of global cooperation on the issue, and (iii) the means of achieving climate-related policy objectives could then be detected after Obama had taken office with the promise of reducing emissions by 80% by 2050 (Urpelainen 2009: 112, 100–117; Romàn/Carson 2009). Regarding his administration's approach to global climate politics, it quickly became clear that domestic policies would have to precede US commitment to a global climate agreement: "the US will have no international credibility until it acts decisively at home" (Stern/Antholis 2007/8: 177; Pataki/Vilsack 2008). By following this approach, the US government wanted to avoid a repeat of the frustrating experiences associated with the Kyoto Protocol (Interviews Observers 23, 4). This implied, as further developed in the story of the negotiations, that the Obama administration – and here in the first place the State Department negotiators – would be waiting for any type of legislation out of Congress as a basis for committing to climate policies negotiated globally.

As of 2007, *Japan's* stance in the global climate talks had not considerably changed since the Kyoto COP. The main economic and political conditions for policy-making on this issue had remained stable. As the second largest economy in the world, Japan remained the sixth largest emitter in the late 2000s (behind China, the US, the EU-27, Russia and India) (World Bank 2008; van Asselt *et al.* 2009: 321), with an important fossil fuel dependency (oil represented 46%, coal 21% of the energy production in 2006) (Korppoo 2009a: 71). Measures to curb emissions, mainly through the 2005 "Kyoto Protocol Target Achievement Plan", comprised policies and measures such as voluntary targets for large companies and energy efficiency standards for vehicles and appliances (Korppoo 2009a: 74–76). Advances in energy efficiency were, however, offset by rising car and household appliance sales. As a result, Japan's emissions had grown by 5% in 2006 compared to 1990 levels, and were further on the rise (Korppoo 2009a: 73). In 2007, its 6% Kyoto emissions reduction target seemed thus out of reach (Luta 2009: 4). Policy-making on climate measures had regularly been subject to intense struggles between the concerned Ministries of the Environment (MOE) and of Economy, Technology and Industry (METI, formerly called MITI). As "Japanese society and economy ha[d] traditionally been industry-oriented" and climate change had not yet become a major topic in public debates, business interests – defended by METI – played a decisive role in the definition of policies (Interview EU representative 24; Korppoo 2009a: 78–79). In the past, turf wars

---

[5] In mid-2009, this rise in US public support of domestic and international climate policies had already come to a halt again (Nordhaus/Shellenberger 2009).

between these two players had been settled through interventions of the Ministry of Foreign Affairs and/or the Prime Minister injecting foreign policy considerations into the debates (van Asselt et al. 2009: 322). Such considerations usually concerned the country's relationship with China and the US as well as its desire to sustain the global leadership it had provided regarding climate change through hosting COP 3 (van Asselt et al. 2009: 320–321; Korppoo 2009a: 79). This type of internal struggles based on tensions between ecological, economic and foreign policy considerations continued well into the analysed period. When defining the country's position for the post-2012 talks, successive Japanese governments had refused to quantify mid-term reduction targets in 2007, offering only a long-term aspirational aim of halving emissions by 2050 (van Asselt et al. 2009: 323). Further specification of this position became necessary in 2008 when Japan was to host the G-8 summit (Interview EU representative 24). During the debate preceding the summit, "the Ministry of the Environment (MOE), together with a network of environmentally-minded NGOs, pushed for a post-Kyoto commitment that would be in line with the EU's –20% target", but opposition was "organized, fierce and unapologetic" (Luta 2009: 4). The final position would only emerge in mid-2009, and be subject to an unprecedented reversal by a new government in September of that year, as further discussed in this chapter. A more stable element of Japan's position concerned the long-standing call for "meaningful participation" of all major emitters in global mitigation efforts (UNFCCC 2007j). A novelty in the expression of this interest was its proposal to distinguish among categories of non-Annex I countries on the basis of criteria such as emissions share or wealth indicators (e.g. GDP/capita) (van Asselt et al. 2009: 324). It proposed that different groups of countries should adopt varying measures, ranging from binding targets to voluntary commitments (Korppoo 2009a: 67; Interview EU representative 24). In the defence of these positions, Japan initially adopted a strategy best characterized as a "two-arena game", with participation in the UN regime as a major forum to discuss a global policy framework and in the APP as an arena to pursue interests regarding technological cooperation and to intensify relations with partners (US, China) (van Asselt et al. 2009: 326–332).

The *Russian Federation* had been the latecomer to the Kyoto Protocol, and joined it for reasons related to economic self-interest rather than climate mitigation as such. As the ninth largest economy on the planet in 2007, the country was the fourth largest emitter (behind China, the US, the EU-27) due to its high reliance on gas (53% in 2006), oil (21% in 2006) and coal (16% in 2006) (World Bank 2008; UNSD 2009; Korppoo 2009b: 88). Russia's role during the Kyoto Protocol's ratification process demonstrated that the country did not really possess a clear conception of the importance of climate change and its own vulnerability to its effects, with Russian elites and the public paying little attention to and, to some extent, denying the threats related to climate change (Andonova 2009: 38; Korppoo 2009c: 4). As Russian compliance with its Kyoto target of stabilizing its emissions over the period 2008–2012 (against 1990 levels) was never really endangered,[6] few domestic measures had been put into place. A key tool was the "Energy Efficient Economy" programme including a range of macroeconomic policies to reduce energy intensity by modernizing economic structures. The definition of both internal and external climate policies had long been led by the largely independent Federal Service for Hydrometeorology and Environmental Monitoring (Roshydromet), placed under the responsibility of the Ministry of National Resources during the period analysed here (Korppoo 2009b: 82). Gradually, the Ministry of Economic Development and Trade had also taken on a stronger role in the definition of the country's climate policies (Andonova 2009: 42). Together with the highest political leaders (President and Prime Minister) and the Ministry of Foreign Affairs, these actors would also be responsible for defining Russia's position for the post-2012 talks. Both economic interests and foreign policy concerns (improved relations with key partners such as the US and the EU) were therefore bound to loom large in its external strategy (Andonova 2009: 47; Korppoo 2009b: 97). In 2007, the Russian position for the post-2012 negotiations was yet fairly opaque. Certainly, the Federation wanted to defend the Kyoto Protocol framework and to "maintain some cushion of 'hot air'" in this new round of talks by refusing to accept much higher reduction targets for itself and promoting the continued use of the Kyoto flexible mechanisms (Andonova 2009: 47). At the same time, it "strongly support[ed] emission caps for developing countries" and thus a broader involvement in the emissions reduction efforts in a future climate regime, preferably through an indicator-based distinction between groups of countries, an approach similar to Japan's (Andonova 2009: 47). In this context, Russia called for taking national conditions into greater account and attempted to demonstrate that it would qualify as an emerging rather than a fully industrialized country (Korppoo 2009b: 82–83). More concrete components of the Russian position would, however, unfold only toward the end of the negotiations in 2009 (Korppoo 2009b: 81).

Among the *other players in the Umbrella Group*, two larger emitters require special mention. While *Canada* had not followed the US in the early 2000s by ratifying the Kyoto Protocol, but would at no point since

---

[6] Russian emissions had dropped steeply after the breakdown of the energy-intensive economic system of the Soviet Union. This meant that it possessed enough "hot air" to meet its target and engage in selling emissions rights.

1997 undertake sufficient measures to actually comply with its 6% reduction target, *Australia* moved in the opposite direction (Drexhage et al. 2008: 8; Cass 2009: 21). Where the Conservative Howard administration had opposed the Kyoto Protocol for a long time, following the US by withdrawing from the ratification process in 2002 – and this despite its allowance of increasing emissions by 8% during the first commitment period – a new Labor government voted into office in late 2007 reversed this tendency by immediately ratifying the Protocol (Cass 2009). Australia equally had difficulties in meeting its Kyoto objectives, however. Both countries would also define their negotiation positions with regard to the US stance. For Canada, this implied an ever closer alignment with its neighbour's position, based on own emission reduction goals of 20% by 2020 and 60–70% by 2050 (compared to 2006 levels) (Environment Canada 2007). For Australia, it meant more distance from the US and a (relatively) greater alignment with the EU. During his campaign, the new Prime Minister Rudd had called for GHG reductions by 60% until 2050 (from 2000 levels) (UNFCCC 2008n: 2).

The *G-77/China* had been acting as a UN negotiating bloc throughout most of the history of the climate regime, although this had never been self-evident, seeing its "internal heterogeneity along such key variables as prosperity, emissions and vulnerability to climate change" (Kasa et al. 2008: 114). The rise of the BASIC[7] countries (Brazil, South Africa, India, China) since the late 1990s raised the question of whether they were not gradually evolving into a distinct category of player, neither fully developed nor yet developing. Their more active role in global climate policy-making and greater bilateral engagements with major industrialized partners such as the EU and the US actually suggested that they might become "less dependent on group membership" in the G-77 (Kasa et al. 2008: 114, 119, 121–122). In addition to this emerging gap, the cleavage between AOSIS and OPEC persisted during the second half of the 2000s. What essentially held the G-77/China together was the shared feeling that only a united defence of common interests would guarantee enough resources to weigh in the negotiations (Kasa et al. 2008: 118). This rationale had proven quite reasonable during the Kyoto Protocol talks. The bloc's positions for the post-2012 negotiations remained thus similar to those promoted earlier, geared toward a defence of the main principles of the Convention and the Kyoto Protocol. Based on the CBDR principle, which reflected a shared concern about the right to economic development and the common idea that the main historic responsibility for dealing with climate change lay with the industrialized world, the bloc called on the

---

[7] BASIC did not emerge as a negotiation group before the autumn of 2009; it is used in this section to demonstrate the similarities between these emerging countries.

latter to undertake meaningful emission reductions and provide financial and technological support, while demanding to be exempt from own "further commitments" (Kasa *et al.* 2008: 116). The negotiation of concrete common positions flowing from these key concerns necessitated, however, a quasi-constant coordination, almost exclusively organised on the spot during UN climate negotiation sessions (Observation notes June, Nov., Dec. 2009). Despite all attempts to forge unitary positions, the coalition would frequently speak with many, diverging voices. Differences between key players within the bloc justify a closer look at their positions.

Due to its unprecedented economic growth (making it the third biggest economy in the world behind the US and Japan), demographic development and, as of 2007, top place in the global ranking of GHG emitters, *China* had made the perhaps most pronounced assent within the global climate policy arena since the 1990s (World Bank 2008; PBL 2008; UNSD 2009). This confronted the country with a major dilemma. To improve the living conditions of its more than 800 million poor, the Chinese government saw no alternative to a further continuation of its economic growth. Further growth of China's export-oriented economy, however, inevitably implied ever-increasing levels of GHG emissions, as the country's energy was generated – fairly inefficiently – primarily through the combustion of coal and oil (for an overview: Lewis 2007/8: 156–158; Jakobson 2009: 33, 37–39). At the same time, the Chinese government had begun to realize the importance of tackling climate change: immense forest loss and progressing desertification in Western regions were demonstrating the country's vulnerability, and it was expected that its large agricultural sector would further suffer from weather extremes in the future (Harris/Hongyuan 2009: 53–54). Signs of growing attention paid to climate change had been, internally, the adoption of laws and new institutions to improve the management of the problem. Measures foreseen in the 2007 Chinese National Action Plan on Climate Change complemented those already initiated with its 2005–2010 predecessor. They included, above all, energy intensity ("China will achieve the target of about 20% reduction of energy consumption per unit GDP by 2010") and diversification targets, but also measures on afforestation and vehicle standards (NDRC 2007: 26; Jakobson 2009: 39–41). Decision-making on these climate policies passed primarily by two bodies: the National Coordination Committee for Climate Change (NCCCC), set up in 1998 and regrouping 13 government departments coordinated by the National Development and Reform Commission (NDRC),[8] and a national leading group on climate change created in 2007 and headed by Premier Wen

---

[8] Among the departments were also the Chinese Academy of Science and the Ministry of Foreign Affairs.

Jiabao (Harris/Hongyuan 2009: 56; Lewis 2007/8: 159; Interviews EU representative 1, Observer 28). Decisions were prepared by lead ministries, notably the NDRC, a body concerned primarily with developmental as opposed to environmental concerns (Kasa *et al.* 2008: 120). This institutional set-up demonstrated that climate change had increasingly become a high-level affair in China. Given the overall rise of importance of the topic and the fact that climate change had long been considered a foreign policy issue in the country, this was not so surprising. Long-standing guidelines for its stance in global environmental and climate policies had been closely related to a set of central foreign policy concerns (Harris/Hongyuan 2009: 60–63): 1. The primacy of economic development over environmental protection (defended by the NDRC); 2. The concern for safeguarding its sovereignty, including the control over its natural resources; 3. Its preoccupation, shared with the G-77, about equity and "fairness" in global environmental affairs, embodied in the CBDR principle; 4. Its self-perception as a leader of the developing world and key player in global politics, promoted by the MOFA (Jakobson 2009: 44–45). In line with these considerations, China had also sought to strategically exploit its "dual status as a developing country (...) and its growing role as a top contributor to global environmental problems" (Harris/Hongyuan 2009: 57; Lewis 2007/8: 162): while representing the G-77/China position in the UN negotiations, it concluded bilateral partnerships around energy issues with the US, Australia or the EU, and participated in the APP (Kasa *et al.* 2008: 121). As a result of this set of premises, the 2007 Chinese negotiation position on key issues was characterized by a refusal of own binding emission reduction efforts, demands on developed countries to lead by reducing their emissions by 25–40% below 1990 levels by 2020 and by 80–95% by 2050 and by providing financial and technical assistance (Jakobson 2009: 24–25). China also favoured a second Kyoto Protocol commitment period.

Another significant player within the G-77/China, *India*, had also experienced years of steady high growth, elevating it to twelfth in the world in terms of the size of its economy, and to fifth regarding global GHG emissions in 2007/2008 (World Bank 2008; UNSD 2009). Despite high growth figures, India had been characterized as "a rich country with poor people", where more than 80% of the population lived on less than $2 per day (Imhasly 2008). At the same time, it was one of the most vulnerable countries to climate change: the low adaptive capacities of its poorest exposed them almost helplessly to, e.g., changed weather patterns. In spite of this, the topic was "not yet on the radar of everyday people, or even policymakers" before the second half of the 2000s (Bhandari 2006, cited in Korppoo/Luta 2009: 62, 64). Where the few domestic policy debates about climate change had focused on economic development issues until then, India slowly began to implement measures on renewable energy promotion, energy efficiency standards or reforestation (Rajamani 2008: 20; Korppoo/Luta 2009: 60). Despite these activities, it was estimated that its energy needs would double by the year 2020 under a business-as-usual scenario (Rajamani 2008: 19). As much of its energy was produced through coal (39%) and oil (25%) combustion, and energy intensity was high, emissions were thus further bound to grow (Korppoo/Luta 2009: 56–59). India's negotiation stance in the global climate policy arena had traditionally been based on the notion that climate change was above all a foreign policy issue. It had been conceived by a relatively small circle of experts from the Ministry of External Affairs and the Ministry of Environment and Forests (Korppoo/Luta 2009: 61–62). Their conviction was, ever since Indira Ghandi's appearance at the 1972 UNCHE, that India's economic development (poverty eradication, access to energy and electricity) should take precedence over environmental considerations (Korppoo/Luta 2009: 47). Insisting on its role as a developing country and the centrality of fairness in global climate talks, Indian negotiators regularly employed tough rhetoric to "shame and blame" the West by slashing at its alleged "luxury emissions" (Kennedy 2009; see also Michaelowa/Michaelowa 2011).[9] This moralizing argumentation strategy had not prevented the country from exploiting, similar to China, its emerging country status by concluding privileged bilateral accords on energy-related issues with the US (e.g., around nuclear energy) and the EU (on energy technologies more largely) and by joining the APP (Kasa et al. 2008: 122; Interview EU representative 31). In 2007, the concrete Indian negotiation position was fairly defensive: stylizing itself as a leader of the developing world, it refused to take on any binding emission reduction targets allegedly intended to "keep developing countries poor", demanded that the industrialized nations live up to their duty of leading in the global efforts (in line with the CBDR principle) by curbing their emissions and delivering financial and technological aid in the multilateral framework of the UN regime (Korppoo/Luta 2009: 47–52; Rajamani 2008). The country therefore also argued forcefully for a continuation of the Kyoto Protocol. Moreover, the Indian government promoted the concept of per capita entitlements to GHG emissions (Rajamani 2008: 21). In 2008, it declared that the country's per capita emissions would never exceed the OECD average (Shanka Jha 2009: 4).

Similar developments as in these two countries were noted for *Brazil*, and to some extent, *South Africa*: "fighting against commitments within

---

[9] This despite the fact that per capita emissions of the Indian middle class had been growing to about 5 t per capita, a level that was comparable to the global average in the mid-2000s, earning India the criticism that it was "hiding behind the poor" (Greenpeace 2007; Müller 2006: 29; Rajamani 2008: 23).

Kyoto through the G77 (...) and enjoying the fruits of their increasing global influence by participating in other agreements linking energy and climate outside the formal negotiations seems to be the chosen "opportunistic" strategy of China, India and Brazil at the moment" (Kasa *et al.* 2008: 122). This also implied, however, that the BASIC countries – as an increasingly informally organised group – still did share key positions with the large bulk of less and *Least Developed Countries* in the G-77/China. The latter group had, however, also a range of distinct interests. Many African nations, but also highly vulnerable Asian countries like Bangladesh were already – and even more so than during the 1990s – experiencing the negative consequences of climate change, although their contribution to this problem had been virtually nil. As a result, they not only demanded that mitigation efforts by industrialized countries be accelerated, but also that the latter provide urgent and sufficient funding for adaptation measures. Often insufficiently staffed, financed and organised, the delegations of these countries would experience difficulties in voicing their opinion in the negotiations, however (Observation notes June, Nov., Dec. 2009). A notable development as compared to previous rounds of negotiations was the attempt by the African Union – including many LDCs, but also South Africa – to forge a distinct common position.

With their demands, the LDCs and the African Group overlapped to a certain extent with *AOSIS*. As the most vulnerable players in climate talks, and in contrast to the LDCs, the AOSIS members had been able to strongly organise themselves. The coalition therefore also continued to be amongst the most vocal players well into the time period analysed here. It demanded more and faster mitigation efforts by developed – and (major) developing! – countries so as to limit mean temperature rise to a maximum of now 1.5°C (AOSIS 2009). Concretely, this meant that global emissions "should peak by 2015 at the latest, and decrease thereafter to at least 85% below 1990 levels by 2050" (AOSIS 2009). Like the LDCs, AOSIS called for increased Annex I funding efforts for adaptation activities (AOSIS 2009).

Finally, *OPEC* continued its attempts "to decelerate negotiation progress" by delaying discussions (Kasa *et al.* 2008: 114). What is more, according to observers, as "the OPEC country delegations – with Saudi-Arabia as its most powerful member – [were] resourceful enough to dominate the smaller and much poorer LDCs, they [were] able to influence G77 positions disproportionately in their own favour" (Kasa *et al.* 2008: 124). OPEC's positions on the substantial issues of interest in this analysis were clear: ideally, it desired no further development of the regime. If this could not be avoided, it wanted to ensure that no emissions reduction efforts were asked of developing countries, including its own members (Observation notes June, Nov., Dec. 2009).

The European Union: Actor Capacity, Negotiation Positions and their Foundations

With the enlargements from 15 to 27 members, the climate policy-relevant diversity within the EU regarding such indicators as prosperity, energy production systems and GHG emissions trends had further increased. As a result, the heterogeneity of interests and preferences on how and why the Union should be engaged in global climate policy had equally grown. Further, diverging beliefs about the EU's role in global (environmental) politics rendered climate policy-making more intricate.

The material differences leading to diverging interests within the EU could hardly be more pronounced. Regarding energy systems, while some countries were producing the bulk of their energy from a single fossil fuel like coal (Poland), others were already basing their energy production on a comparatively large share of renewables (Finland, Sweden); still others were relying to a larger extent on nuclear power (France, Sweden, Lithuania, Belgium) and a final group utilized energy mixes in which various fossil fuels (coal, gas) played important roles (e.g. Germany) (EEA 2008). In terms of GHG emissions, as of 2007, the largest contributors to the EU's overall emissions were Germany, the UK, Italy and France, with obviously much higher absolute emissions than smaller member states. Big differences persisted, however, in the per capita emissions between Luxembourg at the upper and Latvia at the lower end of the spectrum (UNSD 2009; EEA 2007). Some countries were over-complying (the UK) or largely in line (Germany, Greece) with their Kyoto target in the mid-2000s, whereas others, like Italy, Austria or Denmark, were substantially deviating from it (EEA 2009).[10] This non-negligible heterogeneity of climate-relevant national conditions found its expression in a cleavage between two extremes: countries with traditionally more progressive preferences on the issue (Denmark, the Netherlands, Germany, the UK) on the one hand and a majority of mostly smaller countries that was more concerned about the potential interference of progressive climate policies with their economic well-being (essentially the new, but also some old cohesion countries in the East and South of Europe) on the other hand (Lacasta 2008: 9). It made the EU appear – in its own understanding – as a "laboratory" for the international climate negotiations.[11]

---

[10] Nonetheless, the EU-15 was well on its way towards meeting its Kyoto target of 8% GHG reductions over the period 2008–2012: in 2008, it lay 6.2% below 1990 levels, while the EU-27 fared, with over 13% below 1990 levels and thanks to Eastern Europe's "hot air", even better (Phillips 2009b).

[11] From 2008 on, the economic crisis would blur this dividing line to some extent by reinforcing the concerns of more sceptical countries, while also impairing the commitment of some of the historically progressive players.

Despite this preference heterogeneity rooted in different interests and beliefs, the Union was able to define key contours of its position fairly early in the post-2012 negotiation process. Several reasons for this have been noted (on what follows: Oberthür 2009: 205–206; van Schaik/van Hecke 2008: 5–6). First, the beliefs of progressive EU members regarding the necessity to let the precautionary principle prevail and exploit economic opportunities of early action on climate change had been reinforced since the Kyoto Protocol negotiations, especially through the IPCC's Fourth Assessment Report and the 2006 Stern Report (Interviews EU representatives 7, 32; Scheipers/Sicurelli 2007: 445–450; Damro/ Mendez 2003: 79). This was notably the case in the UK, but also, for instance, in Germany or the Netherlands (Interviews EU representatives 7, 22, 8, 10). Their perceptions of climate change, largely shared also by the European Commission, would gradually become widely accepted within the Union and thus contribute to confirming a set of long-standing collective beliefs. In terms of environmental protection, the Union had – already since the mid-1990s – advocated to limit the increase of global mean temperature to 2°C. It now used the new IPCC findings to solidify this argument. Economically, the EU saw a further margin for decarbonising its energy systems, both in the context of intensified energy independence debates of the mid-2000s (Dehousse/Bekkhus 2007) and under international pressure through the Kyoto targets (Costa 2009). The wish to comply with the Kyoto targets also reflected long-standing beliefs of the majority of EU members in multilateral problem-solving and international law (Van Schaik/Schunz 2012; Scheipers/Sicurelli 2007: 448). Second, as of 2007/2008, the environmental and economic framing of climate change would be supplemented by a third vision of the problem: following a report by the EU's High Representative for CFSP and the Commission, the topic was increasingly perceived as a security threat (European Council 2008a). This raised the awareness of the EU's foreign policy community to this topic, contributing to its greater involvement in the Union's external climate policy, while further increasing the salience of the issue for EU policy-makers generally (Schunz et al. 2009). Third, institutional factors contributed to the Union's continued proactiveness on climate change despite preference heterogeneity. Differences among member states were attenuated through the unique institutional framework created under the Environment Council. Regularly gathering environmental experts from the member states and the Commission's DG Environment, it contributed to their socialization in a strongly pro-climate environment. This gradually led to the emergence of a fairly small group favourable of a strong role for the EU in climate policies at regional and global levels (Costa 2007; Interviews EU representatives 21, 6). In addition to these ideational, instrumental and institutional determinants of the Union's stance on climate change, the matter also acquired a high priority status due to public demand: a 2008 Eurobarometer report showed that 75% of the interrogated Europeans considered climate change to be a "very serious problem" (with responses ranging from 96% in Cyprus to 59% in the UK) and a majority thought that neither national governments nor the EU were doing enough about it (Eurobarometer 2008: 15–16; 46–50). At a time of constitutional paralysis – ratification attempts of the Constitutional (and later the Lisbon) Treaty would initially fail – these public demands allowed the Union's elites to use climate change as a driver for internal policy-making as well as to strengthen its profile as a global player.

The Union's actor capacity seemed further improved at the outset of the post-2012 negotiations in legal and institutional terms. To assess the legal bases for its involvement in the UN regime negotiations, it is first of all necessary to identify the international legal overture for its participation in UN bodies. A Regional Economic Integration Organisation clause in the treaty (Art. 24 KP) granted the EC and its member states the right to participate as full member in the Meetings of the Parties to the KP (Pallemaerts/Williams 2006: 39). Just like under the UNFCCC, the EC thus enjoyed the same rights as other parties regarding such matters as tabling, speaking and voting in debates on the Protocol. If it had to vote, it would do so on behalf of all its members, preventing them from exercising individual voting rights (and vice-versa) (Art. 22 KP). The provision would acquire crucial significance in a context of twin-track post-2012 talks under both the Convention and the Protocol. Internally, the legal bases and treaty objectives for its involvement in the UN regime in primary EU law had largely remained unchanged with the treaty reforms of Amsterdam (in force since 1999) and Nice (since 2001). This implied that the competences between the EC and the member states remained shared (Art. 174 TEC). The evolving internal climate policy acquis and modifications in the way decisions on internal and external climate policies were prepared, taken and interlinked meant, however, that "the complexity and character of EU climate decision-making ha[d] changed between 1997 and 2008" (Vogler 2008: 5). Where climate change had above all been an external EU policy during the Kyoto period, the gradual design of a European climate regime led to a closer interaction between internal and external policy-making (Jordan/Rayner 2010). This development came with some intricacies. On the one hand, internally and externally, the institutional structures created under the Environment Council had become the centre of decision-making, with the Working Party on Climate Change (WPIEI-CC) as the "engine" defining the Union's external position (Lenaerts 2009). On the other hand, internal and external climate policies were governed by different decision-making rules. Internally, significant measures were taken in co-decision between the Environment Council and the European Parliament, which played thus a major role in the continued set-up of the Union's regional climate regime. By contrast, the procedures for the definition of foreign policy positions largely excluded the Parliament[12] and required consensus decisions by the Environment Ministers.[13] Increasingly since the mid-2000s, it also required the confirmation through the European Council.

Once adopted, the EU's negotiation position would be represented in the global arena by the Climate Troika (consisting of the current and future Council presidencies and the Commission)[14] as well as the lead negotiators, together ensuring the outreach of the EU in bilateral and other informal meetings linked to the UN negotiations (Lenaerts 2009). In the UN talks, the lead negotiators would speak almost exclusively on behalf of the EU, with individual member states virtually never taking the floor (Observation notes June, Nov., Dec. 2009).[15] Under the impulsion of the Presidency, the expert groups under the WPIEI-CC would be responsible for the preparation of lines-to-take papers and the internal coordination on the spot. EU coordination meetings were held every morning during UN sessions, while expert groups met throughout the day, sometimes even several times (Observation notes Nov., Dec. 2009; Interviews EU representatives 20, 8, 29). EU coordination was to ensure constant exchanges between the 28 (member states and Commission), a flow of information between negotiators involved in different fora and – in the Expert Group on Further Action (EGFA) and the WPIEI-CC – the opportunity to discuss strategies (Observation notes June, Nov., Dec. 2009; Interviews EU representatives 13, 20; Vogler 2008: 3). Information, such as reports on informal meetings with third country parties, was also dispersed electronically so that each EU member state delegation would be up to date about the latest state of play (Observation notes Nov. 2009). During COPs,

---

[12] During the analysed period, the European Parliament demonstrated an increased interest in the EU's foreign climate policy. In April 2007, it created a Temporary Committee, whose task was inter alia to coordinate the Parliament's "position in relation to the negotiations for the post-2012 international climate policy" (European Parliament 2007). It would regularly adopt resolutions on global policies.

[13] The Environment Council, with input from the Economy and Finance Council (ECOFiN), would adopt the negotiating positions in the months preceding the COPs.

[14] The member states in the Troika during this period were: in 2007, Germany and Portugal (1st semester), Portugal and Slovenia (2nd semester); in 2008, Slovenia and France (1st semester), France and Czech Republic (2nd semester); in 2009, Czech Republic and Sweden (1st semester), Sweden and Spain (2nd semester).

[15] For the negotiations until 2009, representatives from the Commission's DG Environment took the lead in the discussions under the Kyoto Protocol track (AWG-KP), while the EU's lead negotiator under the second, "long-term cooperative action" track was Mars Goote, a civil servant from the Dutch Ministry of Environment, backed up by a team of issue leaders with representatives from the UK, Germany, France, Spain and the Commission (Interviews EU representatives 7, 9).

coordination meetings would take the form of informal WPIEI-CC meetings, Environment Councils and, at COP 15, even of informal European Councils, reflecting the pattern of increased high-level involvement in EU climate policy-making. Transcending the immediate framework of the UN negotiations, the EU's representation would increasingly involve actors from the Commission's DG Relex and member state diplomats (Schunz et al. 2009). Together with the Troika, these players ensured the foreign policy implementation of the EU in bilateral summits (EU-China, EU-India, EU-US etc.), through the promotion of bilateral partnerships with key countries or regions and in fora such as the G-8, G-20 and the Major Economies Meeting/Forum (Lenaerts 2009). Often, an implicit, but not officially validated task-sharing could be observed in exchanges with third countries, based on EU members' traditional relations with particular (groups of) countries (e.g. Spain with Latin American countries; the UK with members of the Commonwealth; Portugal with Brazil etc.) (Interviews EU representatives 6, 9, 20, 10). Further, diplomats from member state embassies and the Commission's delegations were increasingly being implicated in the EU's outreach in many third countries identified as key to the Union's interest, including the US, Japan, China, Brazil and India (Interviews EU representatives 1, 2, 30, 12, 24, 26; Schunz 2009). To cite but one example, representatives from (mostly the bigger and well-resourced) member states and the Commission would jointly promote the Union's position in the US capital. Coordination in Washington was ensured by the country holding the Presidency or the Commission. As seen earlier, the foremost activity of such "local Green Diplomacy Networks" was the carrying out of démarches before major UN sessions, on the basis of negotiation directives adopted by the Environment Ministers and targeted towards the conditions of the host country.

The Union's negotiation position and intended strategy for the post-2012 talks can best be understood when discussed in the context of the closely intertwined internal climate policy developments in 2007. The evolution of these parameters forms part of the process analysis. The overarching aim with which the European Union had entered the post-2012 negotiations after the entry into force of the Kyoto Protocol was laid out as early as February 2005 in the Commission communication "Winning the battle against climate change", which advocated the establishment of a post-2012 regime guided by a 2°C target and covering emission reduction actions by developed and major developing countries (European Commission 2005). After the European Council had already called for reduction pathways for developed countries in the range of 15–30% by 2020, discussions on the Union's own post-2012 reduction target were started in December 2006 (European Council 2005). In the Environment Council, the UK, Germany, Italy and Sweden, together with Environment

Commissioner Dimas, were reported as having pleaded for a 30% reduction by 2020 (compared to 1990 levels); Hungary, Slovakia, Poland, but also Spain and the Enterprise Commissioner Verheugen were more reluctant, arguing that the EU should wait to see what other major parties would propose before making "a hasty declaration of commitment" (European Commission 2005: 16; Vogler 2008: 22; Jordan/Rayner 2010: 73). At the same time, the ministers stressed "the need to significantly accelerate international negotiations (…) in 2007 with a view to their completion by the end of 2009" in an effort to avoid leaving a gap after the expiry of the first commitment period of the Kyoto Protocol in 2012 (Council 2006: 18). The EU's contribution to accelerating the negotiations would be specified in early 2007, settling the strategic divergences between the two groups of countries – as well as intra-Commission rifts between different DGs – through a compromise (Interviews EU representatives 6, 8). The January 2007 Commission communication "Limiting Global Climate Change to 2 degrees Celsius: The way ahead for 2020 and beyond" proposed that developed countries, including the EU, should ultimately reduce their emissions by 30% from 1990 levels until 2020 and that "until an international agreement is concluded (…) the EU should already now take on a firm independent commitment to achieve at least a 20% reduction of GHG emissions by 2020 (…) This approach will allow the EU to demonstrate international leadership on climate issues" (European Commission 2007b: 2). Based on this and the Environment Council conclusions of February (Council 2007a), the March 2007 European Council set out a full position on key elements of the "negotiations on a global and comprehensive post-2012 agreement, which should build upon and broaden the Kyoto Protocol architecture" (European Council 2007: 11). The heads of state and government called on industrialized countries to commit collectively to emissions reductions "in the order of 30% by 2020 compared to 1990. They should do so also with a view to collectively reducing their emissions by 60% to 80% by 2050 compared to 1990" (European Council 2007: 12). Provided "that other developed countries commit themselves to comparable emission reductions and economically more advanced developing countries to contributing adequately according to their responsibilities and respective capabilities", the EU showed preparedness to reduce its emissions by 30% (European Council 2007: 12). Publicly presented as leverage over the emissions of other major emitters, this conditional offer signified also a concession of the more progressive players within the EU, which allowed for maintaining the internationally well-known leading-by-example stance. Even if no global agreement was reached, the European Council adopted the Commission formula by making "a firm independent commitment to achieve at least a 20% reduction by 2020 compared to 1990" (2007: 12). Internal differentiation was called for to share the "effort" and the Commission was invited to propose comprehensive climate policies such as a reform of the ETS, which was to become the cornerstone of an international carbon market (European Council 2007: 12, 13, 11; Vogler 2008; Jordan/Rayner 2010: 74). With this, a clear link between the EU's external and internal policies was established. To provide the basic negotiation directives for COP 13, the Environment Council of October 2007 reiterated these positions, solidifying the Union's argumentation scientifically by including the findings of the Fourth IPCC Assessment Report (Council 2007b: 10–17).

To understand the further evolution of the EU's positions in the course of the post-2012 talks, a brief look needs to be taken at the range of legislative proposals introduced by the European Commission in 2008 in response to the request by the European Council. To achieve the 20% emission reduction pledged unilaterally, the Commission introduced a "climate and energy package" including a reform of the ETS, suggestions on effort-sharing for national emissions not covered by the ETS, directive proposals on carbon capture and storage and on the promotion of renewable energy sources as well as a draft regulation for car emissions (European Commission 2008c). The package was negotiated during the course of that year among the member states and between the Council and the European Parliament. It was politically adopted by the European Council and – in first reading – by the European Parliament in December 2008 (Council 2008d; European Parliament 2008; Agence Europe 2008m; Jordan/Rayner 2010: 74–76). Its formal adoption by the Council on 6 April 2009 marked a "momentous development", as it lifted climate policies to unprecedented levels of harmonisation within the EU (Council 2009d; Jordan/Rayner 2010: 76; van Schaik 2010: 270). Without going into any details on the genesis of the adopted measures, relevant aspects of which will be taken up in the process analysis, the importance of the regulatory package cannot be overstated (Morgera *et al.* 2011). Its significance for the EU's external climate policies was above all strategic: the Union was the first major actor in the international climate arena to come up with a set of ambitious legislative measures. In the eyes of the Union's decision-makers, this implied a huge credibility gain, which was to underpin – unlike during the Kyoto Protocol talks when the EU's ambitious target proposal had been characterized as "unrealistic" – its "leading-by-example" approach. It was expected to give the Union moral authority when making demands on other parties in the talks (Oberthür 2009: 200–202).

The Negotiation Process and the EU's Influence Attempts

The political developments of the early 2000s had transformed the post-2012 climate talks into a highly complex global process. To analytically deal with this complexity, this section provides a focused, phased narrative process-trace of the EU's activities and their effects in this process between late 2007 and late 2009, emphasizing particularly the talks on the core norm of the climate regime (emissions reduction targets) and its key principle (CBDR), but, where necessary, also on related issues (notably finance). To disentangle the web of interlinked negotiation arenas within and beyond the UN, several sub-plots are integrated into a single narrative, and the reader is provided a visual aid by highlighting the discussed level in bold.

Preparing for Copenhagen: the Years 2008 and 2009

The 13th conference of the parties to the UNFCCC and the third meeting of the parties to the Kyoto Protocol were held in parallel between 3 and 15 December 2007 in Bali, Indonesia (**COP 13/MOP 3**). Many parties had declared beforehand that they intended to initiate a comprehensive regime reform process (see Chapter 4). The EU had underpinned this willingness through its first major influence attempt in early 2007: the disclosure of its targets and overall position. Just before the COP/MOP, another Annex I party would underscore its seriousness about the talks: Australia's incoming government, led by Prime Minister Rudd, ratified the Kyoto Protocol, leaving the US isolated as the only major industrialized player who had not done so (Peake 2007).

In Bali, debates on the future of the Kyoto Protocol were mostly held in a contact group as part of the resumed **AWG-KP 4**. They focused mainly on the future work programme of the group, adopted relatively swiftly (Spence *et al.* 2008: 148). It was agreed that the AWG should pursue its work until MOP 5 in 2009 according to a clearly delimited timetable (ENB 2007j: 17; UNFCCC 2007b). Three conflicts on key issues emerged, however, all of which prominently involved the EU. First, and directly related to the issue of responsibilities, the EU had demanded in its written submission before the Bali meeting as well as in the contact group that the discussions on the future commitments of Annex I countries and the (second) review of the Protocol under Article 9 KP be coordinated (UNFCCC 2007i: 3; ENB 2007e: 2). Supported by Japan, this proposal was met with fierce resistance from the G-77/China, both in the AWG and in the group that was to prepare the second Article 9 review scheduled for MOP 4 in 2008. Anxious to avoid discussions about own binding emission reduction actions, the developing countries argued that the review should focus on the "implementation" of the Kyoto Protocol (ENB 2007f: 1). By contrast, the EU, Japan and other industrialized countries suggested debating the "effectiveness" of the Protocol, which would have cleared the path for reconsidering the utility of distinguishing between Annex I and non-Annex I countries (ENB

2007i: 2). As in past debates about this topic, it was the G-77/China position that would win out: the final MOP decision re-iterated that the review would not lead to new commitments for any party (UNFCCC 2007l; ENB 2007j: 17).[16] Second, some members of the Umbrella Group and the EU argued for a coordination between discussions in the AWG-KP and those to be held under the Convention on a post-2012 regime (ENB 2007e: 2.). Here again, the G-77/China was strictly opposed and managed to obtain its will in the parallel process under the Convention. Third, in its submission on the "Development of a timetable to guide the completion of the work of the AWG" of late October 2007, the EU had clearly stated its adherence to climate science as presented in the latest IPCC report, referenced – following EU and AOSIS proposals and insistence – in the AWG report adopted at the August 2007 Vienna meeting (AWG-KP 4, part 1) (UNFCCC 2007i: 2). During discussions in the relevant contact group, Canada, Russia, China and India argued against including the numbers (25–40% by 2020, 50% by 2050) in the final MOP decision, supporting a textual proposal that solely mentioned the report of AWG-KP 4, part 1 (ENB 2007j: 16). The EU, together with the majority of G-77/China and several industrialized countries, successfully defended its position in the closing plenary of the AWG against Canadian and Russian resistance (ENB 2007j: 17, 20).

Discussions on long-term cooperative action **under the Convention** were held in a contact group under the COP on the basis of the draft report on the Dialogue, which had met four times since COP 11 (UNFCCC 2007h). They focused on the precise shape of the future negotiation process as well as on the building blocks identified in the report, systematically addressing the major questions raised by the co-facilitators (above all: in which framework and "what to negotiate, when, and for how long"?) (ENB 2007e: 2). The EU had made its vision of regime development proceedings very clear: it desired the adoption of a "Bali Roadmap", i.e. an agreement covering all parties that would begin one comprehensive negotiating process to result in a post-2012 agreement by 2009, including components of the Kyoto Protocol and the results of broader Convention consultations (ENB 2007e: 2). After a first week of talks about the details of the various "building blocks" (ENB 2007g), a "non-paper" incorporating major proposals was circulated by the Indonesian Presidency on 8 December. Framed as a COP decision, it contained a "Bali Roadmap" encompassing two negotiation tracks and

---

[16] The review of the Protocol at MOP 4 in 2008 would be concluded "without any substantive outcome" (ENB 2008l: 18). As the procedures under Article 9 KP would thus not gain much significance in terms of regime development in this time period, this plot will not be further pursued in the remainder of this chapter.

the proposal to conclude negotiations by late 2009 (ENB 2007h: 1–2). Regarding the institutional set-up of negotiations, this "non-paper" limited the four options identified in the draft report on the Dialogue to three: a continuation of the existing Dialogue since Montreal; an open-ended ad hoc working group; and an open-ended ad hoc working group combined with the AWG-KP process (ENB 2007h: 2). The latter option best represented the EU's position, while most G-77/China members and the US preferred option 2 so as to keep discussions under the AWG-KP and under the Convention apart. This majority would also impose its view: a new ad hoc working group on long-term cooperative action was created (AWG-LCA) that would operate in parallel to the AWG-KP (ENB 2007j: 15). Two more substantial issues contained in the non-paper would be much more controversial, sparking considerable conflicts during the final days of talks. The first one concerned, in similar fashion as in the AWG-KP, the reference to IPCC science. It would thus have repercussions for the discussions on the level of ambition of the global emissions reduction target. The preamble of the non-paper remarked the "unequivocal scientific evidence" and pointed to the necessity that Annex I parties cut emissions by 25–40% below 1990 levels until 2020, that global emissions peak within 10 to 15 years, and that they need to be more than halved by 2050 (ENB 2007h: 1). These references, which had not been met with resistance during the August Dialogue in Vienna, and were supported by the EU and the developing countries, were now criticized by the US, Japan, Canada and Russia as "attempting to prejudge the outcome" (ENB 2007j: 15). The second issue touched upon one of the building blocks of the future talks that the non-paper elaborated on (mitigation, adaptation, finance, technology, shared vision): developed and developing country mitigation and the link between them (ENB 2007h: 1–2). Closely related to the CBDR principle, it generated a peculiar debate (ENB 2007i: 2).

Both issues would be at the heart of the **high-level segment** during the final days of the COP. During these talks, the EU attempted to increase its pressure on the US: if the latter refused to engage in a comprehensive negotiation process, EU representatives (such as the French and German Environment Ministers) threatened to boycott future Major Economies Meetings, i.e. the extra-UN processes initiated by the Bush administration to discuss climate and energy policy among the big emitters (Germanwatch 2007: 1; Interview EU representative 15, Video 12 Dec. 2007). Despite these European efforts, the final COP decision reflected US preferences regarding the references to the IPCC science. It only mentioned the IPCC's FAR in its preamble ("emphasizing the urgency to address climate change as indicated in the Fourth Assessment Report") and placed a footnote behind "urgency" (UNFCCC 2007n: 1). This footnote contained a reference to specific pages in the FAR, on which the different temperature scenarios assessed by the IPCC could be found. This allusion to the science represented a much less favourable outcome for the EU and the G-77/China under the Convention (with the US present as major antagonist) than under the Kyoto track (without the US) (Spence et al. 2008: 149). The second controversial issue essentially opposed members of the Umbrella Group (the US, Canada) and the G-77/China: while the former wanted the final document to contain strong references to developing country action/commitments, the developing countries refused this, even demanding clearer commitments by Annex I countries (ENB 2007j: 15).

Compromise language on the formulations in the non-paper was worked out on 14 December, the last official day of the COP, in a "small group of ministerial-level representatives" (Müller 2008: 2). This group reportedly forwarded the following text to the COP President: on developed country mitigation, it stated that a future agreement was to address "the consideration of enhanced national/international action on mitigation of climate change, including (...) measurable, reportable and verifiable nationally appropriate mitigation commitments or actions, including quantified emission limitation and reduction objectives, by all developed country Parties, while ensuring the comparability of efforts among them" (this was later to become part of the Bali Action Plan: UNFCCC 2007n: 2). On developing country mitigation, no agreement had been reached and so two options were presented (Müller 2008: 2–3). Option 1 read that "Nationally appropriate mitigation actions by developing country Parties (...) supported and enabled by technology, financing and capacity-building, *in a measurable, reportable and verifiable manner*" should be considered, while option 2 stated that "*Measurable, reportable and verifiable* nationally appropriate mitigation actions by developing country Parties in the context of sustainable development, supported by technology and enabled by financing and capacity-building" were to be undertaken. Both options clearly created a link between developing country action and industrialized country assistance, but differed with regard to one important item: option 2 was supported by the industrialized countries, including the EU, because it indicated that the nationally appropriate mitigation actions (NAMAs) of developing countries had to be "measurable, reportable and verifiable". By contrast, the formula preferred by the G-77/China was option 1, as "measurable, reportable and verifiable" referred to developing country actions *and* technology transfer and finance by industrialized countries (Spence et al. 2008: 149; Müller 2008: 3).

The "Proposal by the President" introduced in the COP plenary of 15 December did not reflect the differences on this issue, but simply stated option 2 (Müller 2008: 3). This prompted the Indian delegation to intervene and point out that the option at hand was not supported by the G-77/China, reiterating option 1 (Video 15 Dec. 2007, part 1). After two interruptions of the talks allowing for internal G-77/China coordination and consultations with the COP President, the meeting resumed with a speech by UN Secretary General Ban Ki-moon, "making an unscheduled return" to the meeting and telling the delegates that "everybody should be able to make compromises" (Reuters 2007c; Video 15 Dec. 2007, part 2). The COP President then suggested replacing his initial proposal (option 2) with the formula preferred by the G-77/China. The EU, which had kept a very low profile in this conflict until then – agreeing with the Umbrella Group demands, but wanting to ensure that an outcome would be reached – was the first party to support, "as a sign of the spirit of co-operation, compromise and trust, among us", the Indian proposal (Video 15 Dec. 2007, part 2). Other members of the Umbrella Group remained silent, until, a few minutes later, the US would request the floor and refuse the adoption of the Bali Action Plan. The willingness of developing country leaders to engage in emission reduction efforts had not been reflected in the text, according to the US delegate, and the new paragraph represented "a significant change in the balance" compared to what the COP had worked towards over the whole two weeks (Video 15 Dec. 2007, part 2). After intense further discussions in plenary, including interventions by South Africa assuring the US that "measurable, reportable and verifiable" applied to both developing and developed party actions, and immediately following an appeal by Papua New Guinea addressed to the US,[17] the Americans changed course (Video 15 Dec. 2007, part 2; ENB 2007j; Reuters 2007c).[18] Remarking that they had listened closely to the declarations of the developing countries, the US delegate affirmed to "join consensus with this today" (Video 15 Dec. 2007, part 2). With this, the path was cleared for adopting the Bali Action Plan (UNFCCC 2007n). Together with the MOP decisions, notably to continue the AWG-KP (UNFCCC 2007k), this formed what was informally referred to as the "Bali Roadmap" (for a summary: UNFCCC 2007m). A brief discussion of its major components sets the stage for analyzing the negotiations that followed until COP 15/MOP 5 in Copenhagen.

With the Bali Action Plan, the COP decided "to launch a comprehensive process to enable the full, effective and sustained implementation of the Convention through long-term cooperative action, now, up to and beyond 2012, in order to reach an *agreed outcome* and adopt a decision at" COP 15 (UNFCCC 2007n: Point 1, emphasis added). This process was to

---

[17] "We seek your leadership, but if, for some reasons, you are not willing to lead, leave it to the rest of us. Please, get out of the way."

[18] As this change of course was minor, concerning a few words in a document that started an open negotiation process – words that were, moreover, subject to different interpretations – what some called the US "u-turn" should not be overestimated (Müller 2008: 3–4). It did, however, signify US re-engagement in UN climate talks.

concentrate on the four previously identified building blocks as well as on discussions of a "shared vision for long-term cooperative action, including a long-term global goal for emission reductions, to achieve the ultimate objective of the Convention" (UNFCCC 2007n: Point 1(a)). To this end, the Action Plan contained an indicative schedule of four meetings in 2008 (UNFCCC 2007n: Point 6, Annex). Arguably central to the Plan were the paragraphs on developed and developing country mitigation and the link between the two. For the first time since the early 1990s, developing countries were prepared to discuss NAMAs, a sufficiently broad concept that could imply voluntary or binding actions, although the CBDR principle had been prominently retained (UNFCCC 2007n: point 1(b)(ii); ENB 2007j; Ochs 2008: 2). In return, the US joined the negotiation table again, prepared to equally consider emission reduction actions, including quantified emission reduction objectives (QELROs) (UNFCCC 2007n: point 1(b)(i)).

Commentators stressed the importance of the fact that the text did not mention the distinction between Annex I and non-Annex I countries that had characterized the climate regime for such a long time (Ochs 2008). UNFCCC Secretary General de Boer even referred to the "dismantling of the Berlin Wall" between groups of countries in the climate regime (ENB 2007j: 19). Industrialized country assistance would, however, be a precondition for developing country actions. The Bali Action Plan did not give any guidance on how to resolve the issue of who should go first (industrialized or developing countries), foreshadowing a possible catch-22 situation. Similarly, other major issues were completely left open (e.g. the nature of commitments or the legal form of the "agreed outcome" stipulated) (Spence *et al.* 2008: 151).

The first EU reactions to the Bali Roadmap were positive. The EU head of delegation, Portuguese Environment Secretary of State Rosa, was quoted in the press as saying: "It was exactly what we wanted. We are indeed very pleased" (Graham-Harrison 2007). This publicly displayed satisfaction about the type of outcome the EU had promoted (a "roadmap") could not mask the rather defensive role it had been forced to play during the second half of the meeting, when it tried to fight off attempts at watering down provisions on the science and the bindingness of commitments from both the Umbrella Group and the G-77/China (Ott *et al.* 2008: 93; Interview EU representative 27).

The **year 2008** witnessed a slow start into the post-2012 talks, with parties only beginning to unveil parts of their positions both within and beyond the UN framework. The EU used the year to re-design its internal climate regime in preparation of COP 15 as well as for outreach toward partners. Prior to the first sessions of the working groups created in Bali, a number of **fora outside the UN arena** allowed for loose exchanges of ideas on the building blocks of the Bali Action Plan. On 30–31 January

2008, the second "Major Economies Meeting on Energy Security and Climate Change" – following a US initiative the EU had originally threatened to boycott in Bali – brought together the 17 biggest global emitters (including the EU) in Hawaii to consult on the operationalization of the Roadmap (Xinhua 2008b). A similar debate was held between 11 and 13 February in the UN General Assembly, allowing for broader participation (ENB 2008a: 2). In mid-March, Japan hosted the fourth round of the G-8+5 Gleneagles Dialogue on Climate Change, Clean Energy and Sustainable Development, holding "open and meaningful discussions (…) on the 3 main issues: Technology, Finance, and Post-2012 International Framework" (MOE Japan 2008a). On the crucial question of mitigation objectives, the Chair's conclusions stated that parties "acknowledged the importance of sharing a long-term goal" and, regarding a mid-term objective, "reaffirmed the principle of common but differentiated responsibilities and respective capabilities as a premise of the discussion" (MOE Japan 2008b: 3). This wording underscored major parties' unwillingness to be more concrete on the crucial issues of targets at this early stage in the talks as well as the continued strong link between target debates and the question of "who will do what?" in the future regime. **Bilateral meetings involving the EU** included Troika exchanges with Latin American and Caribbean countries in February (Agence Europe 2008a: 5). Moreover, at a second Transatlantic High-Level Dialogue on Climate Change on 7 March in Washington, DC, the EU and the US stressed the importance of renewed cooperation and highlighted the significant role of extra-UN bodies in advancing the talks (US Mission 2008).

Within the EU, the Commission had, as early as January 2008, tabled its legislative proposals of the climate and energy package, providing for the Union's second major influence attempt in this still novel negotiation process. Notably the proposed legislation on the reform of the Emissions Trading Scheme contained a number of strong signals to the external world.[19] Especially *vis-à-vis* other developed countries (e.g. the US, Australia, Japan), the EU hoped to be able to use its ETS *positively* by making it the key reference for cap-and-trade systems in the world and providing incentives for these countries to design GHG emissions schemes compatible to its own (Benwell 2009: 100–101; Interviews EU representatives 26, 27).[20] *Negatively*, the EU attempted

---

[19] Already the original 2003 ETS Directive had comprised a call to conclude agreements with other Annex I parties (EP/Council 2003: Art. 25.1). The 2004 "Linking Directive" stated then that the Commission should pursue agreements with Kyoto Protocol parties to "provide for the recognition of allowances between the Community scheme and mandatory [GHG] trading schemes" (EP/Council 2004: para. 18).

[20] Following this approach, it had initiated, e.g., the International Carbon Action Partnership in 2007, involving US states, Canadian provinces, Norway and New

to employ its reformed ETS as a "club good" to which access could be restricted if others did not follow its preferences with regard to climate change mitigation (Benwell 2009). This approach was reflected in the explicit linkages between the ETS and other Kyoto mechanisms (notably the CDM).[21] As "carbon-market participants based in the EU [were] by far the main players in the CDM market, (and) China (...) by far the biggest host of CDM projects, followed by India, then Brazil", this was – potentially – a "strong piece of leverage" over emerging economies (Earthtimes 2008). Shortly before the first negotiation session under the Bali Roadmap, the Environment and European Councils of March 2008 would accept the package on the whole as a basis for negotiations in the Council and between the Council and the Parliament, to be finalized by the end of the year (Agence Europe 2008b, c).

In the UN climate regime, the first session of the Ad Hoc Working Group on Long-term Cooperative Action under the Convention and AWG-KP 5, part 1 were convened between 31 March and 4 April 2008 in Bangkok. Under the **AWG-LCA**, talks focused on designing a work programme (UNFCCC 2008a, e; ENB 2008b: 6–7). In line with the EU's position, it was agreed to focus on all items of the Plan in an equal manner in eight workshops organised throughout 2008 (UNFCCC 2008g: 3). Brief substantial discussions on key topics demonstrated differences: while many parties across all coalitions agreed on the need for a long-term goal and the necessity to differentiate among countries on the basis of "comparability criteria", key G-77/China members, like China or Brazil, argued that actions should be national and only nationally "measurable, reportable and verifiable" (MRV); by contrast, the EU and other industrialized countries called for further consideration of what "MRV" would have to mean for different groups of parties (ENB 2008b: 4–5). The **AWG-KP** focused on the analysis of the means at the disposal of Annex I countries to reach (yet to be specified) future emission reduction targets (UNFCCC 2008b). Delegates debated primarily the use of flexible mechanisms, LULUCF accounting rules and the possible introduction of sectoral mechanisms (ENB 2008b: 8–12). Their

---

Zealand, to lobby for the idea of an OECD carbon market, for which its own ETS could provide a blueprint (Benwell 2009: 102–103; van Schaik/van Hecke 2008: 17).

[21] The Commission proposed: "Once a future international agreement on climate change has been reached, CDM credits shall only be accepted in the EU ETS from third countries that have ratified the international agreement" and that therefore the EU "should only authorise project activities where all project participants have headquarters (...) in a country that has concluded the international agreement relating to such projects, so as to discourage 'free-riding' by companies in States which have not concluded an international agreement" (European Commission 2008d: point 5, para. 26; van Schaik/van Hecke 2008: 16). This formula found its way into the directive ultimately adopted in 2009 (EP/Council 2009: para. 33, 32).

conclusions indicated a general agreement on the need to keep LULUCF activities, "emissions trading and the project-based mechanisms under the Kyoto Protocol (...) available to Annex I Parties" (UNFCCC 2008h: 2). The EU greeted this result as "sending a strong signal to the private sector" (ENB 2008b: 12). "No formal link" between the AWGs "was extensively discussed" (ENB 2008b: 12). The session thus fulfilled its mostly procedural tasks, without achieving any substantial results.

Numerous **bi- and multilateral meetings involving the EU** were held in the spring of 2008. On 17–18 April 2008, a third Major Economies Meeting in Paris dealt primarily with the issues of technological cooperation and long-term emission reduction targets, but without "tangible agreement" (Xinhua 2008a). The EU's bilateral exchanges with some of the major players equally remained without results, but did facilitate mutual understanding of positions. Whereas the EU-Japan summit (23 April, Tokyo) helped to discover areas of convergence, especially with regard to the necessity of adopting *binding* targets, the Japanese expressed their unwillingness to specify a medium-term objective at this stage of the negotiations (Agence Europe 2008d: 3). From a meeting between Commission President Barroso and several Commissioners with high-level Chinese officials in Beijing (24–26 April), the EU retained: "*indications* of Chinese *readiness to include* its domestic emission reduction policies in an international agreement", provided that developed countries committed to 2020 reduction targets and promoted technology transfer (European Commission 2008b, emphasis added). China had thus clearly stated that it would not commit to anything but voluntary reduction measures. More fruitful exchanges were held with Latin American countries in May 2008, with both sides agreeing to strive for a legally binding agreement (Agence Europe 2008e: 4–5). Finally, the outcome of the US-EU summit of 10 June in Brdo, Slovenia, testified to continued differences. A confidential US strategy paper had leaked shortly before the meeting and revealed the Bush administration's preference for voluntary solutions in the framework of the Major Economies Meetings, pointing to a potential clash with the EU's multilateral approach (Spiegel 2008a). As a result, and besides a vague commitment to a UN-based agreement by late 2009, the parties' joint declaration contained no major points of convergence (Council 2008b).

While the next UN negotiation session was kicked off, the **EU's internal legislative process**, so crucial for its external activities on climate change, was picking up speed: taking stock of the climate and energy package debates, the Slovenian Presidency could report progress in the Environment Council of 5 June. At the same time, several member states were beginning to demand more flexibility in the approach to climate and energy policies, with only Germany, Finland, Denmark and the

Netherlands reported as strongly supporting the original Commission proposals (Council 2008a; Agence Europe 2008f: 8–9). The French Presidency would thus have to accomplish the bulk of the work during the second half of 2008 (Agence Europe 2008f: 8–9).

AWG-LCA 2 and AWG-KP 5.2 were held in parallel to the habitual **UNFCCC** Subsidiary Bodies meeting in Bonn between 2 and 13 June 2008. The **AWG-LCA** worked on each of the building blocks of the Bali Action Plan. On mitigation and differentiated responsibilities, debates illustrated the continued differences between the industrialized countries and the G-77/China: while the former considered it necessary to reform the climate regime in a way that would cover all major emitters, the latter insisted on the need to distinguish between developed country obligations and developing country "actions" (ENB 2008d: 3–4; ENB 2008c: 1–2). To overcome these bold oppositions, the Chair invited parties to submit written proposals on all elements of the Bali Action Plan (UNFCCC 2008i: 1). The work programme for 2009 foresaw four sessions of up to eight weeks to prepare COP 15 (ENB 2008d: 4–5; UNFCCC 2008j). In the **AWG-KP**, discussions of the means available to developed countries for reaching their future emission targets were pursued, focusing once again on flexible mechanisms, LULUCF rules and the possibility of adopting sectoral approaches (ENB 2008d: 5–8). No issues were settled. As in Bangkok, linkages between the two tracks (and between the AWG-KP and discussions of the Article 9 review of the Kyoto Protocol prepared by the SBI),[22] demanded by the Umbrella Group and the EU, met with resistance from major developing countries. Not surprisingly, the Union lamented that progress in both working groups was clearly too slow (ENB 2008d: 5, 8).

The **major non-UN climate-related meeting** during the summer of 2008 was held in Toyako, Japan. Between 7 and 9 July, both the G-8 and the G-5 (China, India, Brazil, Mexico, South Africa) leaders got together. The former reiterated their 2007 Heiligendamm agreement: "We seek to share with all Parties to the UNFCCC the (…) goal of achieving at least 50% reduction of global emissions by 2050 [through] contributions from all major economies, consistent with the principle of common but differentiated responsibilities and respective capabilities" (G-8 2008: point II). This formula linked a long-term emissions reductions goal to the recognition that "what the major developed economies do will differ from what major developing economies do" and added "in this respect,

---

[22] During the preparations of the Article 9 review of the Kyoto Protocol scheduled for the Poznan COP, the Umbrella Group demanded, as in previous sessions, a more comprehensive review so as to start a discussion on revising the differentiation between Annex I and non-Annex I parties – without success (ENB 2008d: 15–16).

we acknowledge our leadership role and each of us will implement *ambitious economy-wide mid-term goals*" (G-8 2008: point II, emphasis added). The G-5 summit, driven by China, by contrast, had a clear, quantified idea of the emission reductions to be adopted by industrialized countries: 25–40% of 1990 levels by 2020 and 80–95% until 2050, in line with the IPCC's 2°C scenario (G-5 2008; Jakobson 2008: 25). Only if the developed world committed to such targets, the G-5 countries would increase their NAMAs, if "supported and enabled by financing, technology and capacity-building with a view to achieving a deviation from business-as-usual" (G-5 2008: point 18). On 9 July, the two groups met to issue a "Declaration on Energy Security and Climate Change", which remained strikingly silent on mitigation. As the parties could not agree on a commitment to a quantified reduction goal, they only set out ideas on the other building blocks so as to "advance the work of the international community (…) to reach an agreed outcome by the end of 2009" (MEM 2008: point 3). The EU, as the driving force behind the Heiligendamm agreement, considered the G-8 and G-8+5 meetings as positive steps, especially with regard to the newly introduced notion of (yet unspecified) mid-term targets among the major industrialized players, but Commission President Barroso recognized that "much more" needed to be done, notably to get the emerging countries to take on emission reduction efforts (Agence Europe 2008g: 7).

The **next round of UN climate negotiations** was convened in Accra, Ghana (21–27 August). Ample party input to **AWG-LCA 3** was used to move from a position-stating to more engaged exchanges of views through the establishment of three contact groups: one on mitigation, one on adaptation and one on technology and finance (UNFCCC 2008c; ENB 2008g: 3–4). On mitigation, debates turned primarily around the issue of differentiation between parties and revealed the familiar developed/developing country cleavage (ENB 2008e: 2). The question of the legal bindingness of the outcome was equally raised: while many Umbrella Group parties called for a legally binding result of the AWG-LCA work, major forces in the G-77/China refused this, arguing that the Convention and the Kyoto Protocol had enshrined a distinction between the legal form of actions of different groups of countries (ENB 2008f: 1). Similarly, the Umbrella Group's repeated call for linking discussions under the LCA and KP tracks met again with resistance from developing countries (ENB 2008g: 4). In its conclusions, the AWG-LCA could therefore only request the UNFCCC secretariat to compile all proposals in one document (UNFCCC 2008k). It was hoped that this would provide a sufficient basis for "shift[ing] into full negotiating mode in 2009" by organizing "work accordingly", i.e. creating more time to exchange ideas (UNFCCC 2008l: 1; ENB 2008g: 5). The **first part of AWG-KP 6** pursued discussions on the methodological issues of the means for Annex I countries to reach future emission reductions targets (ENB 2008g: 6–9). With this and calls for further expertise, debates remained in an analysis and early positioning phase. None of the initiated debates were concluded at the end of the meeting, during which the EU had kept a rather low profile (ENB 2008g: 6–9).

A limited number of **important multilateral events** – overshadowed by the onset of the financial and economic crisis – was organized in the immediate run-up to COP 14/MOP 4. A preparatory meeting was organized in Warsaw in mid-October by the Polish COP Presidency: environmental ministers from over 30 countries exchanged views, but did not reach any type of agreement (AFP 2008c). They did, however, affirm their commitment to solving the problem of climate change despite the global crisis. A similar message emerged from the G-20, bringing together the world's biggest economies. At a special summit on the financial crisis (Washington, DC, 13–14 November 2008), leaders recognized the need to continue tackling climate change even in times of economic downturn (G-20 2008: point 15).

The **EU** also engaged in **bilateral outreach** before COP 14. Represented by the European Commission, it met with the African Union in Addis Ababa on 8 September to adopt an energy partnership intended to strengthen cooperation between the two continents (Agence Europe 2008h: 12). Three weeks later, the EU Troika got together with Indian representatives in Marseilles. The two parties agreed to reinforce their ties, notably through a "Joint work programme on energy, clean development and climate change", but could not synchronize their positions in any way, with India insisting on a strict interpretation of the CBDR principle (European Commission 2008e, f). In late October, the Asia-Europe meeting held in Beijing, though mostly concerned with the financial crisis, issued a declaration that expressed "the need to act with resolve and urgency" against climate change (ASEM 2008: point 21). This call could not mask substantial differences regarding the question whether emerging countries needed to make (binding) emission reduction efforts or not (Agence Europe 2008i: 11).

Within the EU, talks on the climate and energy package were just entering into their decisive stages in both the Council and the European Parliament when the Environment Council issued the negotiation directives for the Poznan COP in October. It re-affirmed positions outlined since the spring of 2007, essentially the call on industrialized countries to reduce emissions in the range of 30% and the need for major developing countries to deviate from business-as-usual by 15 to 30% by 2020 (at 1990 levels) (Council 2008c). It further "underline[d] the need to speed up preparation of the Copenhagen agreement" (overcoming the slow pace of the 2008 talks), "recall[ed] that the Copenhagen agreement must be reached within the UN process" (and not in the Major Economies format) and "underline[d] the importance of an ambitious mid-term target" (urging other industrialized countries to make theirs known) (Council 2008c: 7–11). Further, the Council highlighted "its intention to strengthen its partnership with Africa", the LDCs and AOSIS and the "need to build on the Kyoto Protocol" (Council 2008c: 7–11).

As key parties' positions had not evolved considerably in 2008, major players, including the EU, were hoping that **COP 14/MOP 4** in Poznan (2–13 December) would become a useful intermediate step towards Copenhagen. The European Commission estimated that the COP would be successful if the work programme for the negotiations in 2009 was specified, "consensus on a common vision of the future agreement with objectives between now and 2020 and 2050" was reached, and further talks on how to reinforce the Kyoto Protocol were conducted (Agence Europe 2008j: 11). Mirroring developments of the entire year 2008, the event turned out to become a "pit stop" of relative irrelevance, however (Santarius et al. 2009). While the story of the summit can thus be told swiftly, a second plot would become more intriguing for this analysis: in parallel to the COP, the EU was negotiating its climate and energy package.

In the opening plenary of the **COP**, organizational matters were dealt with, with the election of the former UNFCCC Secretary General Michael Zammit Cutajar (Malta) to the post of Chair of the AWG-LCA as the most noteworthy decision – he would preside over this crucial body well into COP 15 (ENB 2008l: 2). **AWG-LCA 4** then took up negotiations on the basis of an "assembly document" gathering all positions expressed up to that point (UNFCCC 2008m). In addition to the previously established contact groups (mitigation, adaptation, finance/technology), a fourth group was created on shared vision, reflecting the importance that many parties, including the EU, gave to this topic (ENB 2008l: 12–13). Neither this group nor an informal roundtable of ministers on shared vision would, however, deliver a concrete outcome on the key issue of a long-term target (ENB 2008l: 13; ENB 2008k: 2). Proposals were, in fact, quite divergent: the EU and some other parties spoke in favour of stabilizing global mean temperature rise at 2°C, whereas AOSIS argued that only 1.5°C would guarantee the survival of low-lying islands; still other countries were completely opposed to expressing their ambition in the form of a temperature limit (ENB 2008h: 2). Other familiar conflicts concerned the time horizon for mitigation actions. Industrialized countries, like Japan, were focusing on the long term, while developing countries were demanding emission reductions by industrialized countries in the range of 25–40% of 1990 levels by 2020 already (ENB 2008h: 2). When explaining its own proposals on this issue, the EU was confronted by South Africa:

stating that even the Union's upper target proposal of 30% reductions by 2020 was not ambitious enough, it questioned the scientific basis for the EU's demand that major emitting developing countries should deviate by 15–30% from business-as-usual by 2020 (ENB 2008h: 2). Two contentious points also arose on the issue of differentiation (Murphy 2009): rifts between Annex I and non-Annex I countries were set forth with the habitual arguments by industrialized ("the world has changed") and developing countries ("you have to lead") (ENB 2008l: 13), while, on the issue of MRV, the verification of NAMAs was once again taken up. The EU argued for a revision of the *outcome* of parties' actions, whereas India spoke for the emerging countries when it stated that it could not accept the international review of adequacy of developing country actions (ENB 2008j: 2). At the end of the session, progress on this and other issues under the Bali Roadmap (adaptation, finance) was virtually non-existent (Santarius *et al.* 2009: 20). The conclusions by the Chair remained therefore general, the detailed outline of a work programme for 2009 excepted (UNFCCC 2008n). Promising again to "shift into full negotiating mode", parties invited Zammit Cutajar to prepare "a document for consideration at its fifth session" and a negotiating text for the sixth session of the AWG-LCA in June 2009 (UNFCCC 2008p: 2; UNFCCC 2008n: 1). To prepare this document, the group requested parties to provide input on "the content and form of the agreed outcome at COP 15" (UNFCCC 2008p: 2). As basis for further talks, the "assembly document" prepared for this session would also be revised (UNFCCC 2008p: 2; revised document: UNFCCC 2008d). All ideas, including the EU's, remained thus on the table for 2009.

Discussions in **AWG-KP 6, part 2** were not more conclusive: although the EU and others had expressed their willingness to shift into "full negotiating mode" immediately, the debates were again mostly concerned with the means of reaching emissions reduction targets and fell short of a real confrontation of ideas (ENB 2008l: 14; Santarius *et al.* 2009: 4–5). For the first time since Bali, discussions were however held on the nature and magnitude of the actual emissions reductions to which the means that had been pondered over the entire year 2008 should contribute. Parties advanced concrete proposals on emission reduction ranges, with the developing countries and the EU advocating the 25–40% reduction range for industrialized countries by 2020 (from 1990 levels) of the IPCC's 2°C scenario (ENB 2008i: 2). Most industrialized countries preferred the bottom up approach of national pledges, to be made at a later stage in the process (ENB 2008l: 15). In its conclusions, the "AWG-KP agreed that further commitments for Annex I Parties under the Kyoto Protocol should, for the next commitment period, *principally* take the form of quantified emission limitation and reduction objectives"

(UNFCCC 2008q: 2, emphasis added). This obviously did not preclude any other outcome. The major agreement within the AWG-KP concerned therefore the concretization of its 2009 work programme (UNFCCC 2008o). Parties agreed to come to draft conclusions on "the scale of emission reductions to be achieved by Annex I Parties in aggregate" by AWG-KP 7 and on the "contribution of Annex I Parties, individually or jointly, (...) to the scale of emission reductions to be achieved by Annex I Parties in aggregate" by AWG-KP 8 (UNFCCC 2008o: 5). Other topics on the agenda until MOP 5 (duration of the commitment period(s); "how QELROs could be expressed, which includes how the base year is expressed"; coverage of GHGs) resembled those debated during the Kyoto Protocol talks (UNFCCC 2008o: 2). Parties were requested to provide input on these issues until mid-February 2009.

Altogether, the decisions validated by the COP and the MOP regarding the post-2012 regime reform were thus "meagre", amounting essentially to a copy-and-paste exercise of what had been agreed in Bali (Santarius *et al.* 2009: 20, 4–6).[23] Poznan, and with it the year 2008, had thus ended without "significant breakthroughs" on key issues (ENB 2008l: 1). It clearly fell short of what the EU had called and hoped for.

In parallel to the Poznan summit, **EU** member states had entered the final stage of **negotiations on the climate and energy package**. Several key conflicts had remained unresolved even after a special meeting between the French Presidency and reluctant Eastern European member states on 6 December in Gdansk, only a couple of hundred kilometres away from Poznan (Agence Europe 2008k: 7). Among them were the issues of wealthy EU members' aid for poorer states as well as the level of auctioning of permits under the reformed ETS. To overcome these problems, a financial solidarity clause was adopted and concessions made inter alia to Poland (Agence Europe 2008m: 4–6).[24] The agreement was endorsed by the European Council on 12 December, the official last day of the Poznan summit, and sent to the European Parliament for final approval.[25] Although this meant a major step forward in the Union's preparation

---

[23] Among the other issues that the COP dealt with the operationalization of the adaptation fund was regarded as one of the key outcomes (Santarius *et al.* 2009: 4).

[24] The solidarity clause foresaw that 10% of allowances would not be auctioned, but reserved for the poorest EU members; another 2% would go directly to the new members (Agence Europe 2008m: 5). Polish power plants were exempted from paying for allowances until 2020, whereas Germany, Finland and Italy, for instance, obtained that allowances would be completely free between 2013 and 2020 for industries with the greatest risk of carbon leakage (cement, chemicals, wood) if no adequate international agreement was reached (Agence Europe 2008m: 4).

[25] The European Parliament endorsed the package shortly after the COP, clearing the path for the final formal Council approval in April 2009 (Phillips 2008).

for the 2009 negotiations, the parallelism of talks had repercussions for its credibility in the global arena: the fact that it negotiated internal positions while UN talks were conducted produced misunderstanding among negotiation partners, notably from the developing countries (Interviews non-EU representative 18, EU representative 29). While Environment Commissioner Dimas used all press conferences to explain that the EU's internal negotiations would by no means jeopardize the magnitude of the Union's overall reduction target of 20% by 2020, the impression left by its internal rifts was quite the opposite (EU Press Conference 11 Dec. 2008; ENB 2008l: 17). An interesting influence attempt – the Commission's announcement, scheduled strategically for the final days of the COP, to invest € 22 million into clean and renewable energy projects in developing countries in Africa and Asia – could not offset that negative impression (Agence Europe 2008l: 9).

Shortly after Poznan, and foreshadowing events of 2009, the **Australian government** would be the first major Umbrella Group member to announce a mid-term target for the post-2012 talks. It pledged unilaterally reductions of 5% of 2000 levels by 2020 and a possibility to move to a 15% reduction in case of comparable efforts by other industrialized countries (Australian Government 2008: xxi). Emissions trading was to become the centrepiece of its climate policy (BBC 2008). With this proposal, the Australians would copy the EU's approach,[26] without, however, following its level of ambition (Australian Government 2008).

For **the first half of 2009**, parties had promised to shift into "full negotiating mode", but no industrialized party other than the EU and Australia was practically prepared to do so. Neither the Union's negotiation position nor its strategy were however fully worked out. Since Bali, the G-77/China had explicitly linked debates on developing country mitigation to the financing made available by industrialized countries for mitigation and adaptation actions in the developing world, but all Annex I countries still lacked a position on this crucial issue. To allow **the EU** to attempt to influence the global debates on this item, the European Commission started the year with another communication in late January, entitled "Towards a comprehensive climate change agreement in Copenhagen". The document comprised concrete suggestions on "innovative international funding sources" (industrialized country contributions on the basis of an "agreed formula" or funds generated through the flexible mechanisms, notably emissions trading) (European Commission

---

[26] In an op-ed, the Australian Minister for Climate Change, Penny Wong, explained the rationale behind this approach: "I hope the ambitious commitments of Australia and the European Union will encourage other developed countries to make comparable commitments, building momentum in international negotiations" (Wong 2008).

2009i: 9–10). It also contained a section on "funding early action" as "a bridging initiative in the transition period between 2010 and the full scale implementation of the new financial architecture to be agreed in Copenhagen" (European Commission 2009i: 10). What the communication did not contain was a specification of the European contribution to the outlined financial efforts. The Environment (2 March) and ECOFiN (10 March) Councils as well as the European Council of 19–20 March would, by and large, endorse the proposals and declare that the EU was prepared to assume its "fair share" of any future collective public funding effort (European Council 2009a: point 27).

Bilateral talks involving the EU resumed immediately in early 2009. On 16 January, a Ministerial EU Troika met with its counterparts in the framework of the EU-South Africa Strategic Partnership in Kleinmond to discuss inter alia climate change issues (Council 2009a). Two weeks later, the European Commission met with the Chinese leadership around Premier Wen Jiabao in Brussels in order to consult on the global financial crisis and climate change. After the meeting, Commission President Barroso "expressed confidence that there would be cooperation between the Union and China via 'constructive dialogue' with a view to the Copenhagen" COP (Agence Europe 2009a). Moreover, the Union's contacts with the US would be stepped up considerably after the presidential transition in January 2009, which had sparked many a hope for greater US engagement in global climate talks.[27] Before the EU would get to meet with the US, the new US Secretary of State Hillary Clinton would, however, visit both China and India, urging those countries not to "make the same mistakes we have" and instead grow on a low-carbon path (Landler 2009). The symbolic gesture of visiting the major emerging countries first set the tone for the rest of the year, as especially Sino-US exchanges were attaining central importance in the climate talks. It was not until early March that an EU delegation, including ministers from the UK, Denmark (COP host 2009), Poland (COP host 2008) and the Council Presidency Czech Republic, had the opportunity to meet with members of Congress and the US administration in Washington, DC, inter alia testifying on the European experience with climate change legislation (Santini 2009; Interview EU representative 26). A high-level EU-US summit, attended by President Obama, would then be held on 5 April in Prague. Its conclusions on climate change were quite general: "Together, the EU and the US will be in a stronger position to get on board key (…) emerg-

---

[27] This showed also in the new President's appointments: Obama filled positions that would be crucial for the climate dossier with the former administrator of the Environmental Protection Agency, Carol Browner, as White House coordinator on climate change, and Todd Stern, who had already negotiated for the US in Kyoto, as State Department "Special Envoy for Climate Change" (Romàn/Carson 2009: 21).

ing countries and achieve an ambitious outcome at (...) Copenhagen" (Council 2009c).

These and other exchanges prepared the stage for the next round of **UN climate negotiations**: between 29 March and 8 April 2009, parties came together in Bonn for AWG-LCA 5 and AWG-KP 7. The beginning of the talks was marked by a US declaration to fully re-engage in the multilateral arena, but only after a listening phase during which it would gradually refine its own position (Stern 2009a). Key contours of that position were, however, already visible. While the style with which the Obama administration approached developing countries had changed in comparison to its predecessor, the substantive claims had not: crucial for the US was a meaningful participation of major developing countries in global mitigation efforts. Moreover, as Jonathan Pershing, head of the US delegation, made perfectly clear, "it is not the point in time in 2020 that matters – it is a long-term trajectory against which the science measures cumulative emissions" (AFP 2009b). The US preference for the long term obviously conflicted with the EU's medium-term focus, a problem that, according to US Climate Envoy Stern, could be solved: "If [we] reduce relatively less between now and 2020, that will leave relatively more to do 2020–2050 (...) Our pathway accords with practical economics and political reality in a way that does not harm the environment" (Harvey 2009). Besides this pragmatic US re-engagement, the talks did not deliver much in the way of new proposals.

The **AWG-LCA** considered a "focus document", which represented an attempt at identifying areas of convergence in parties' proposals (UNFCCC 2009b). Without attributing the various propositions to particular parties anymore, it aptly reflected the state of the negotiations eight months before Copenhagen, and was considered by the Chair as a necessary transitional stage to narrow down options from party proposals before elaborating the negotiating text by June (Zammit Cutajar 2009: 5). On the subjects of interest in this analysis, it contained statements such as the following: "some Parties have proposed that developed countries as a group commit to emission reductions by 2020 in the (...) range indicated by the IPCC", i.e. 25–40% from 1990 levels; "another related proposal is that the overall deviation from a baseline for developing countries as a group by 2020 be quantified at 15–30 per cent" (UNFCCC 2009b: point 16). This specific example demonstrated that the EU's key positions were reflected in the text, an observation that could also be made of Zammit Cutajar's overall structural choice to focus the process on debates about medium and long-term targets. At the same time, the text provided, above all, a testimony of existing disagreements. In the opening plenary of the group, the document was criticized by the G-77/China for overemphasizing mitigation and neglecting elements such as finance and technology

(ENB 2009a: 1). With this, Zammit Cutajar's (and the EU's) approach was effectively discarded: the G-77 had clarified the bloc's interest in a broader, party-driven compilation exercise. Discussions held in the various contact groups throughout the week would confirm this observation. Debates on mitigation, moreover, testified to continued differences: the EU re-iterated its 30% aim for developed countries, AOSIS called for stabilization at 350 ppm and 1.5°C, India for developed country mitigation of 40% by 2020 (from 1990 levels), and the US detailed its long-term perspective, which entailed a mid-term goal of stabilizing emissions at 1990 levels by 2020 (ENB 2009d: 1). It further noted that "if only the EU and the US were to reduce emissions by 80% by 2050, such actions would still result in 630 ppm" (ENB 2009d: 1). This move was obviously aimed at shifting the focus of attention to the question of developing country mitigation: in a debate on the meaning of developing country NAMAs, the EU recalled its 15–30% deviation from business-as-usual proposition, while the G-77/China stated that its mitigation actions would be dependent on the support offered by industrialized countries. India and China further stressed that NAMAs were "voluntary" actions that should be considered in the context of development goals and "poverty eradication" (ENB 2009d: 1–2). Not surprisingly, no movement of key players' positions occurred during the remainder of the conference. Additional sessions were therefore agreed to for August and November. Besides this procedural outcome, parties were looking forward to the Chair's negotiating text, to be issued in June.

The **AWG-KP** equally continued its work towards a negotiating text for June 2009, focusing on discussions of the aggregate scale of emissions reductions by Annex I parties. To that end, a new contact group on "Annex I parties' further emission reduction commitments" (informally called "numbers group") was added to the existing ones (potential consequences, legal matters, flexibility mechanisms, LULUCF) (ENB 2009b: 2). Substantial debate in this group opposed Annex I and developing country parties. The EU called for a combined bottom up and top down approach: country pledges for determining the aggregate scale of emissions reductions (the preferred approach of industrialized countries) could be compared to the scientific demands (25–40% under the IPCC 2°C scenario, i.e. the preferred developing country approach) (ENB 2009c: 2). Other parties and coalitions made concrete proposals on emission reduction ranges for Annex I countries: AOSIS re-iterated calls for stabilizing GHG concentrations below 350 ppm, requiring reductions of at least 45% and 95% of 1990 levels by 2020 and 2050 respectively, whereas Australia (5 or 15% by 2020 from 2000 levels) and the EU presented their proposals, and Japan declared that it would announce a mid-term target in June (ENB 2009b: 1–2). Besides a growing range of propositions that often referred to different base years, rendering comparability difficult, the meeting did not result in specific outcomes. Parties therefore requested two documents for the next meeting: a proposal for amendments to the Kyoto Protocol, including suggestions on emission reduction targets, and a text on "other issues" (UNFCCC 2009d: point 74). There would also be the chance to submit Protocol amendments by 17 June, i.e. up to six months before the end of COP 15 (ENB 2009e: 2).

After "Bonn-1" (two more meetings in the former West German capital would follow), the **EU** pursued its engagement in **bi- and small-scale multilateral exchanges**. In April and May 2009, the Troika met, among others, with the African Union (Brussels, 28 April), Japan (Prague, 4 May), the Rio Group (Prague, 13 May), China (Prague, 20 May), Russia (21–22 May, Khabarovsk) and South Korea (23 May, Seoul) (Council 2009e, g, h, i, j; European Commission 2009j). The meetings consisted mainly of exchanges of positions on the global climate negotiations. **In the climate negotiations beyond the UN**, several meetings need to be highlighted. In their communiqué issued after the London G-20 summit on global recovery of 2 April, key world leaders found it necessary to "reaffirm [their] commitment to address the threat of irreversible climate change (…) and to reach agreement at (…) Copenhagen" (G-20 2009a: point 28). A few days later, the major emitters would follow a US invitation to meet in Washington for a "Major Economies Forum on Energy and Climate Change".[28] Attended by representatives from seventeen countries (including the EU Troika, Germany, France, the UK and Italy), COP 15 host Denmark and the UN, the Forum discussed its members' possible contributions to the negotiations under UN auspices in general (MEF 2009a). A second meeting was organized in Paris on 25–26 May, and focused much more specifically on key issues of the UN talks, including the notion of peak year, mid- and long-term mitigation targets for different countries and finance, notably the question of raising up to USD 100 billion per year for mitigation and adaptation actions, but without concrete results (MEF 2009b; Charlton 2009). Further in the global arena, activities in the US Congress – where climate bills had first been introduced as a "discussion draft" in late March and debated in April and May in the House of Representatives (US House 2009)[29] – provoked reactions in both China and India. While only timidly indicating areas of compromise, China maintained its official demand for developed countries to collectively reduce their emissions by 40% of 1990 levels until

---

[28] The MEF continued the "Major Economies Meetings" under a new name.
[29] The key proposal, the "American Clean Energy and Security Act", foresaw a cap-and-trade system which would reduce US GHG emissions by 17% by 2020 (from 2005 levels). This represented a 4% reduction from 1990 levels.

2020 (Buckley 2009a). This proposal was rebuffed by the US Climate Envoy Stern: "We are jumping as high as the political system will tolerate", but 40% "is unrealistic" (Hood 2009). India, while remaining firm – with its Climate Envoy stating that "what the US has offered is too little" – recalled what it sought: "our stand is not that we don't want to take on obligations. It is that any deviation from our position of not taking mandatory commitments for targeted reductions should be supported by financial commitments and technological aid" (Menon 2009). That an agreement between the US and the major emitters was to become crucial for a global deal became apparent when both a Congressional delegation and high-level US officials visited Beijing for several days of Sino-US bilateral consultations in parallel to the UN talks of June (Bodeen 2009; AFP 2009a). Reinforced Indo-American exchanges had to await late July, when Secretary of State Clinton visited New Delhi to prepare a new strategic dialogue around climate change (Lakshmanan 2009).

Between 2 and 13 June 2009, parties came to Bonn again for a subsidiary bodies meeting as well as **AWG-LCA 6 and AWG-KP 8**. Shortly before the session, many of them had submitted their visions of a future "agreed outcome" under the Convention as well as proposals for amendments to the Protocol. Further, the chairs of the two working groups had issued a first draft negotiating text (LCA) and new documents requested by parties (KP) (UNFCCC 2009f, g). The most significant party submission for the purpose of this analysis was the much-awaited US proposal, entitled "Copenhagen Decision Adopting the Implementing Agreement" under the Convention. On the key subjects of interest, the proposal foresaw "quantitative emissions reductions/removals in the 2020/[ ] timeframe, in conformity with domestic law" and "a low-carbon strategy for long-term net emissions reductions of at least [ ] by 2050" for developed countries (US 2009: 4). Regarding mitigation actions by non-Annex I parties, it differentiated between emerging economies and least developed countries. The former should undertake NAMAs "in the 2020/[ ] timeframe that are quantified (e.g., reduction from business-as-usual) and are consistent with the levels of ambition needed to contribute to meeting the objective of the Convention"; further, they "shall formulate and submit a low-carbon strategy for long-term net emissions reductions by 2050" (US 2009: 4). LDCs should implement NAMAs and "develop low-carbon strategies, consistent with their capacity" (US 2009: 4). On the other crucial interlinked issues of finance and MRV, the proposal remained very general, noting simply the need for "a dramatic increase in the flow of resources (…) from a wide variety of sources" (US 2009: Section 4). Finally, the notion of "implementing agreement" left the legal status of the "agreed outcome" open: it could be binding, but would not necessarily have to be – in any case, the Obama administration intended to have an international outcome of this type ratified by the Senate (Interview Observer 4). The US overarching approach was patently clear, however: rather than reproducing a Kyoto-type protocol, it essentially argued for a looser "pledge and review" approach focusing on the long term.

Other comprehensive protocol proposals under Article 17 UNFCCC were submitted by Australia, Costa Rica, Japan and Tuvalu. Australia's proposal called for a "single new instrument under the Convention", whose key feature would be "National Schedules" for emission reduction efforts by all parties (UNFCCC 2009i: 3–15). Japan's "draft protocol" had many features of the Kyoto Protocol, which it was to either replace or substantially amend. It distinguished between quantified emission reduction targets by Annex I countries and NAMAs in the form of intensity targets for all other countries, to be pledged into annexes (UNFCCC 2009e; ENB 2009i: 4). Tuvalu submitted a proposal for a new Kyoto-style "Copenhagen Protocol" to complement, not replace the Kyoto Protocol (UNFCCC 2009h: 9–22). Among many other submissions focusing on particular elements of the Bali Roadmap or amendments to the Kyoto Protocol, the EU's contribution under the LCA included the by then well-known positions on key issues, but contained also an urgent call for bringing the two negotiation tracks closer together: "in our deliberations in Bonn in April, it became obvious that a strict separation is not feasible" (Council 2009f: 1).

Deliberations under the **LCA track** focused then entirely on a first (and, on technology and adaptation, also second) reading of the Chair's negotiating text. On the issues of interest here (emission reduction aims, differentiation), the 53-page document stated:

> The long-term global goal for emission reductions {shall}{should} be set – Option 1 – as a stabilization of GHG concentrations in the atmosphere at {400}{450 or lower}{not more than 450}{450} ppm (…) and a temperature increase limited to 2°C above the pre-industrial level. For this purpose, the Parties {shall}{should} collectively reduce global emissions by at least 50 per cent {from 1990} levels by 2050. – Option 2 – as a stabilization of GHG concentrations in the atmosphere well below 350 ppm (…) and a temperature increase limited to below 1.5°C (…) Parties {shall}{should} collectively reduce global emissions by {81.71}{more than 85} per cent from 1990 levels by 2050 (UNFCCC 2009g: para. 12).

Similarly, all coalitions' positions were also reflected through options and/or brackets when it came to the question of differentiation, rendering a hardly legible ensemble (UNFCCC 2009g: para. 14, 15). Other paragraphs dealt at length with the different natures of targets for developed and developing countries. MRV was identified as requiring more elaboration, and differences persisted on finance about the sources of funding (public, private) and how to raise public resources, with eight different options mentioned in the text (UNFCCC 2009g: para. 67, 173). Despite the many brackets contained in the document, further consultations testified to parties' anxious – and successful – attempts at having specific formulations reflected in the text. The main consequence of transforming the Chair's proposal into a party-driven document was that it almost quadrupled in length. The final product, an over 200-page compilation of positions, would be issued as a revised version of the negotiating text forming the basis for further talks (UNFCCC 2009j). In this context, it is interesting to note that parties held informal consultations on the form of the outcome, evoking also manifold options (e.g. one or several COP decisions, protocol, single "legally binding instrument"), but taking no conclusions (ENB 2009g: 3, 2009h: 3).

Under the **AWG-KP**, parties continued considering further Annex I commitments, concentrating on various proposals for aggregate and individual emission reduction targets. The textual basis prepared by the Chair included mainly amendment proposals to Annex B of the Kyoto Protocol (UNFCCC 2009f). Most of these took the form of empty table shells with columns entitled, e.g. "quantified emission reductions (2013–2016)" and emissions reductions formulated in percentages against a base year, although some parties had proposed other values (e.g. per capita emissions) and/or varying base years (UNFCCC 2009f: 8–9). Interestingly, a limited number of concrete quantified emission reduction proposals for each Annex I party were also submitted by non-Annex I parties (such as South Africa and the Philippines) (UNFCCC 2009f: 12–13). The manifold options differed, however, with regard to almost all parameters (base year, length of commitment periods, criteria for calculating emissions etc.). Fundamentally, the main cleavage separating developed and developing countries still concerned the approach to setting emission reduction targets: whether bottom up or top down (Observation notes 11 June 2009). The EU's proposal of combining bottom up pledges with a top down approach that used IPCC science as a benchmark marked the middle ground. In the deliberations of the contact group on numbers, major developing countries wanted to focus, as China expressed it, on "numbers and not on text" (ENB 2009f: 4). Industrialized countries, including a very active EU, argued, by contrast, for first clarifying accounting rules (use of flexible mechanisms, LULUCF rules) to make pledges by Annex I countries comparable (ENB 2009f: 4; Observation notes 8–12 June 2009). They also repeatedly argued that pledges could not be raised as long as discussions under the AWG-KP and the AWG-LCA were kept apart. Any aggregate calculation of reductions that did not include the US would be incomplete (ENB 2009f: 4). Nonetheless, AOSIS, supported by the EU, requested that the Secretariat present a preliminary aggregation of Annex emission reduction pledges (ENB 2009g: 3). The result of this exercise was discussed during the second week of the meeting amidst a replay of the Kyoto debates about the length of commitment periods (with options ranging from four years to eight years, or twice four years) and base years (1990, 2000, 2005, 2006) (ENB 2009j: 2–3). Parties could not narrow down options within the contact group (ENB 2009k: 12–13; Observation notes 8–12 June 2009). For the AWG-KP on the whole, continued differences meant that the Chair was not given the mandate to prepare a negotiating text for the next session (ENB 2009k: 14). Talks would therefore have to be pursued, until MOP 5, on the basis of papers prepared by the Secretariat and party submissions.

In parallel to ongoing talks under both AWGs, two **major parties** would publicly discuss their positions during the session in Brussels. EU finance ministers debated elements of the Union's stance, repeating previous positions on the sources of climate finance (private, public, from all countries except the LDCs), without specifying the Union's "fair share" (Council 2009k: 4). On 10 June, Japan announced its long-awaited midterm target. Following months of internal debates, the country opted for a 15% reduction from 2005 levels by 2020, which amounted to a (compared to IPCC scenarios) modest 8% reduction from 1990 levels and was met with criticism from developing countries.

Over the summer of 2009, **bilateral and multilateral exchanges outside the UN framework** resumed. The EU met, for instance, with Asian partners at ministerial level on energy security (17–18 June, Brussels, ASEM 2009) and exchanged informally with the US. Reacting to cap-and-trade bill debates in the US House of Representatives, the Union estimated that the US targets (of roughly 4% below 1990 levels) were relatively weak, and indicated that a possible compromise was for the US to provide comparatively more funding to developing countries (Stearns/ Morales 2009). Multilateral meetings were organised inter alia under the so-called "Greenland Dialogue on Climate Change", an informal forum initiated by the Danish government as early as 2005 to feed into the UN talks on their way to COP 15 (MCE 2009a). The fifth gathering of this body (30 June–3 July, Greenland) brought together high-level representatives from 30 countries (including several EU members and the Commission) across all UN coalitions. Their conclusions listed some essential conditions for a future agreement, including that "Global warming must stay below 2°C" and "no money, no deal" (MCE 2009b). Further, a third and a fourth meeting of the Major Economies Forum took place in Mexico (22–23 June) and, at leaders' level, in L'Aquila, Italy (MEF 2009c; G-8 2009a). The latter meeting was embedded into a series of high-level exchanges, which would mark a high point of the 2009 extra-UN global climate talks. Between 8 and 10 July, G-8, G-8+5 and the

MEF were successively organized by the Italian G-8 Presidency, producing several declarations on climate change. The detailed conclusions of the G-8 stated, on the topics of interest here: "we recognise the broad scientific view that the increase in global average temperature above pre-industrial levels *ought not to exceed 2°C*" and "reiterate our willingness to share with all countries the goal of achieving at least a *50% reduction of global emissions by 2050* (...) we also support a goal of developed countries reducing emissions (...) in aggregate by *80% or more by 2050 compared to 1990 or more recent years* (...) we will undertake robust aggregate and individual mid-term reductions, taking into account that baselines may vary and that efforts need to be comparable" (G-8 2009b: point 65, emphasis added). Going beyond the 2007 Heiligendamm declaration, the language adopted by leaders would recur in the further course of the UN negotiations. It reflected a broad consensus between these major parties, while indicating some specific Umbrella Group demands with regard to mid-term targets (e.g. for different base years). The G-8+5 and MEF did not yield the same type of consensus declaration, reflecting earlier differences between the two groups of countries. The MEF did retain the 2°C declarative aim with the same phrasing and "resolve[d] to spare no effort to reach agreement in Copenhagen", but its declaration did not mention concrete targets and timetables (MEF 2009d: 1–2). EU reactions to the outcome of the meetings were therefore cautious: noting that key countries had now officially endorsed the Union's long-standing 2°C target (in a more compelling frame than the Greenland Dialogue), it also lamented the absence of commitment to concrete mid- and long-term targets, which in essence meant the rejection of central elements of the EU's post-2012 position (Agence Europe 2009b, c).

Between 10 and 14 August, the two **UN negotiation fora** came together for a fairly technical "Bonn-3" informal session without major debates on key issues (ENB 2009l). Under the **LCA track**, the second reading of the revised negotiating text was completed (UNFCCC 2009j; ENB 2009l: 7–14). Deliberations were now held in small groups, and based on newly formulated non-papers. They did not result in specific textual compromises on key issues. To "facilitate" future negotiations, the Chair promised to issue an information document as a guide to the revised negotiating text, allegedly still containing 2,000 brackets (ENB 2009l: 33)! Under the **AWG-KP**, debates on Annex I parties emission reductions continued, but did not reach any (even intermediate) conclusions either. The seven meetings of the "numbers group" were spent primarily with a consideration of "possible targets submitted by countries", with Annex I parties explaining the rationale behind their proposals (ENB 2009l: 14). The outcome of this work signified no substantial advances, but a reformulation of positions in the form of a non-paper that would supplement the documentation the AWG-KP Chair promised to prepare in advance of the next session (UNFCCC 2009k; ENB 2009l: 34). Altogether, the additional negotiation session of "Bonn-3" had thus delivered little added value, leaving problems unresolved and reinforcing cleavages from previous sessions (ENB 2009l: 34). In failing to deliver on its mandate of fixing an aggregate emissions reduction target for Annex I parties under the Kyoto Protocol, the AWG-KP was considerably delayed (UNFCCC Press Conference 14 Aug. 2009).

During the **final months before the Copenhagen summit**, parties would gradually disclose their positions on key elements of the Bali Roadmap. At the same time, the Danish Presidency would reinforce its efforts to forge a minimum agreement with key countries behind the scenes (Meilstrup 2010). For the EU, this would also be the moment to elaborate on its finance proposals, the key missing cornerstone in its position. **Multi- and bilateral exchanges involving the EU** were pursued right after Bonn-3. The Union would take repeated rounds of Sino-American exchanges as a starting shot to "kick-start talks at political level", beginning with another visit of the Troika to Washington (23–25 August, Agence Europe 2009d: 4). In an attempt to put greater pressure on the US and major emerging countries, it consistently communicated that it expected "greater ambition" from these players (Reuters 2009a; Brand 2009). At the same time, opportunities opened for the EU to build a stronger coalition with another industrialized country: Japan. Following a general election which had ended a 54-year reign of the (liberal-conservative) Liberal Democrats, the (social-liberal) Democratic Party promised to reduce the country's GHG emissions by 25% from 1990 levels by 2020, reversing less ambitious proposals of its predecessors (Spiegel 2009a; Tabuchi 2009). It later became clear, however, that the pledge was conditional upon comparable commitments by other Annex I parties (Eilperin/Lynch 2009).

Back **within the EU**, the European Commission introduced a new proposal on international climate finance in early September. Unlike its January communication, this version, entitled "Stepping up international climate finance: A European blueprint for the Copenhagen Deal", contained quantified proposals. Referring to debates about the magnitude of required financing, held notably in the MEF, it estimated that "finance requirements for adaptation and mitigation actions in developing countries could reach roughly € 100 billion per year by 2020" (as opposed to the 100bn USD discussed by the MEF); further, "domestic private and public finance could deliver between 20–40%, the carbon market up to around 40%, and international public finance could contribute to cover the remainder" (European Commission 2009k: 3). From 2013, the Union's contribution to this "remainder" "in the range of € 22 and 50 billion (…) would be from around 10% to around 30% depending on the weight given" to two criteria (ability to pay, responsibility for emissions) (European Commission 2009k: 3). "In case of an ambitious outcome in Copenhagen, the EU's fair contribution could therefore be between € 2 to 15 billion per year in 2020" (European Commission 2009k: 3). Moreover, the Commission proposed that "between 2010–2012, in the event of a successful agreement in Copenhagen, fast-start financing is likely to be needed (…) in developing countries in the range of € 5 to 7 billion per year" and "the EU should consider an immediate contribution of € 0.5 to 2.1 billion per year" (European Commission 2009k: 3). With these proposals, the Commission provided the ground for member states to decide on concrete numbers (Agence Europe 2009e: 12). Although the 27 initially reacted positively, debates on the precise amount of money were resolved only months later. However, two other strategic debates within the EU were receiving public attention at that time. Disunity reigned over if and when the Union should move to the 30% emissions reduction scenario (Ricard 2009). Officially for strategic reasons, but also because no agreement could be forged, the decision was held back for the expected hot phase of the global negotiations (Ricard 2009). Finally, EU members were divided over the (essentially French) proposal to introduce border adjustment taxes for goods coming from countries that had no ambitious climate legislation in place, in the event that COP 15 should fail (Spillmann 2009).

These internal rifts came at a time when an unprecedented amount of **multilateral meetings** lay ahead of the major parties. The month of September began with a G-20 summit of finance ministers, an opportunity parties wanted to use to discuss climate finance proposals. Emerging economies, anxious to keep finance discussions within the UN framework, resisted (Bergin 2009). It was set forth on 17 and 18 September with another meeting of the Major Economies Forum in Washington, DC, which debated notably adaptation and the (for the US central) question of MRV (MEF 2009e). On 22–23 September, parties then followed the UN Secretary General's invitation to meet in New York. Ten weeks before the Copenhagen summit, the UN headquarters provided an arena for stock-taking of parties' positions. Three leaders received specific attention: US President Obama, for his strong rhetoric in favour of ambitious climate action, the new Japanese Prime Minister for his government's improved emission reduction target proposal, and the Chinese President Hu Jintao for promising that China would reduce the "amount of carbon dioxide it emits to produce each dollar of gross domestic product" by a "notable margin" by 2020 from 2005 levels, increase its forest cover by 40 million hectares, and engage in fuel switches (BBC 2009a; UN 2009; Jintao 2009: 7). To conclude the month, a G-20 leaders' summit in Pittsburgh discussed climate change and energy security issues, adopting, however, only general conclusions on key issues, including finance (24–25 September, G-20 2009b).

Against this backdrop of – according to UNFCCC Secretary General de Boer – general, but "sincere commitment of leaders" to the UN process (UNFCCC Press Conference 28 Sept. 2009), parties met in Bangkok to accelerate **negotiations in the UN climate regime** (28 September–9 October). **AWG-LCA 7, part 1** continued to consolidate the revised negotiating text on the basis of documents prepared by the AWG Chair and the chairs of the various contact groups, altogether amounting to about 800 pages of text (UNFCCC 2009j; ENB 2009m: 4). In its opening statement, the Swedish EU Presidency called the text "unmanageable" and urged to finally bring the two AWGs together because "what we need from Copenhagen is one single agreement" (Video 28 Sept. 2009, AWG-LCA Opening Plenary). Repeating a formula the Union had already used in the AWG-KP plenary just hours before, this emphasis on "one single agreement" would cause a considerable uproar among the developing countries in both working groups (ENB 2009m: 2). Although it reflected a position the EU had defended for years,[30] this preference for one outcome had never been so clearly expressed, which had led the G-77/China to believe that the Union was in favour of a two-fold outcome (a new agreement under the LCA track, plus a second commitment period for the Kyoto Protocol). Many developing countries therefore voiced honest concerns that the EU – after all the strongest supporter of the treaty during the period 1998–2005 – now wanted "to kill the Kyoto Protocol" (Reuters 2009b).[31] When talks resumed in the relevant contact groups under the AWG-LCA, conflicts on key issues re-emerged: from the first day on, developed and developing countries clashed on the question of the nature of mitigation actions and their verification (ENB 2009m: 4). The US and the Umbrella Group proposed a *common* framework for reporting and verifying developed and developing countries' mitigation actions, an idea that was opposed by the G-77/China as going against the principles of the Convention (ENB 2009p: 19). While the EU sided with the Umbrella Group on this and many other issues, it also called for an aggregation of existing Annex I country target pledges so as to put pressure on those with either lower or no official proposals (such as the US) (ENB 2009n: 4). In this conflict-ridden context, industrialized countries asked during the mid-way stock-taking meeting that the Chair should take

---

[30] The Union's objective since at least 2007 had been a "comprehensive post-2012 agreement, which would *build upon* (…) the Kyoto Protocol architecture" (European Council 2007: 11, emphasis added).

[31] This was by no means the case, though, as the EU continued to be open for a two-track outcome (Interviews EU representatives 22, 8). In preparing its statement, however, it had apparently underestimated G-77/China reactions.

responsibility for reducing the scope of the negotiating text. This proposal was – successfully – opposed by the G-77/China, mindful of the "party-driven" nature of the negotiation process (ENB 2009n: 4). While technical advances were reported on certain building blocks (adaptation, technology), progress on key issues (shared vision, mitigation, finance) was thus very limited (ENB 2009p: 1). Existing documentation, including the non-papers that emerged from the contact groups, was forwarded to the final meeting before COP 15 (ENB 2009p: 3–13).

During **AWG-KP 9, part 1**, parties resumed consultations in the contact groups. As in the AWG-LCA, the EU's call for a "single legal instrument" would spark reactions in the "numbers group" in the afternoon of the first day. On Tuvalu's demand, the EU clarified that a single legal instrument would necessarily preserve key elements of the Protocol, but that it could be more easily ratified than two parallel outcomes (ENB 2009m: 2). Heated discussions followed on the legal form of the "agreed outcome", during which the EU received support from Japan (ENB 2009m: 2). In the further meetings of the group, emission reduction scales, but also the questions of base year and accounting rules were addressed (ENB 2009p: 14–15). The Union, at times supported by other industrialized countries, was again quite vocal in its argument that it was "not possible for Annex I countries to set targets without first knowing the rules" of accounting, a claim opposed by the G-77/China (ENB 2009o: 3). In her mid-term report to the plenary, co-Chair Wollansky thus had to observe that the numbers group was "facing problems with regard to how to proceed with its work" (Video 2 October 2009, AWG-KP Plenary). Despite numerous proposals and extensive discussions, the problematic situation would, however, remain unresolved. Developing countries wanted to ensure robust targets by the Annex I parties under a second commitment period (and financial aid under the AWG-LCA) *before* making commitments under the LCA track. Annex I parties to the Protocol wanted to know the rules of the game and the pledges of the US and emerging countries *before* committing any further. In this intricate context, the parties did what they usually did when they knew no way out: they asked the Chair to prepare yet further documentation.

Following the Bangkok talks, **major parties** began to openly recognize that the Copenhagen COP would only mark a step in a longer process toward a legally binding agreement. While the US had subtly dampened expectations beforehand, Chinese negotiators were quoted in late October as saying: "The real negotiations will be after Copenhagen [which] will be a starting, not an ending point", and even UNFCCC Secretary General de Boer spoke of COP 15 as simply laying the "groundwork" for further talks (Buckley 2009b). The question at this stage of the process seemed therefore how significant an intermediate advance it would represent. At the same time, key parties were beginning to solidify their ties with coalition partners. The BASIC countries increased their level of coordination, with especially China and India agreeing on practical cooperation on energy technologies, while also pledging to further coordinate their in many respects identical negotiation positions (Murray 2009). While attempting to preserve common interests, India in particular indicated, for the first time, flexibility under its new Environment Minister Ramesh, in office since late May 2009. Partially abandoning anti-Western rhetoric, he stated that India might be prepared to fulfil detailed reporting duties of its future NAMAs so as not to be a "deal-breaker" in Copenhagen (Goldenberg/Watts 2009).

Bi- and multilateral talks involving the EU equally resumed. A Major Economies Forum focusing on finance was held in London (17–18 October), but did not achieve any breakthrough in terms of quantified proposals (MEF 2009f). On 6–7 November, G-20 finance ministers equally came together in the UK to discuss climate finance proposals, but met with the resistance of emerging countries (G-20 2009c: 2). Bilaterally, the EU met with two major partners. During a summit on 3 November in Washington, the transatlantic consultations led to a new commitment to defending the 50% emission reductions aim by 2050 at COP 15 and the creation of a Joint Energy Council (European Commission 2009l: 1). At the same time, the US President reportedly informed the EU that, if it was for the US, COP 15 would not produce a legally binding outcome (Phillips 2009c). On 6 November, the EU-India summit in New Delhi expressed its hope that a "global goal of significantly reducing greenhouse gas emissions by 2050 compared to 1990 levels would be reached at Copenhagen" (European Commission 2009m: 2).

Within the EU, pressure had been mounting in the meantime: disappointed by the Bangkok meeting, and still facing criticism about its stance on the Kyoto Protocol, the Union sought to update its negotiation position so as to give new impetus to the global talks (Agence Europe 2009f: 8). A meeting of the Finance Ministers on 20 October brought, however, no breakthroughs on the issue of quantifying the EU's finance proposals, and the Environment Council of 21 October then effectively shifted the decision to the heads of state and government (Council 2009i, m).[32] On 29–30 October, the European Council reiterated key elements of the Union's negotiation position for Copenhagen, but its conclusions contained hardly any updates, except an endorsement of the Commission's

---

[32] Differences reportedly included substantial rifts over who would contribute how much to the EU's share, with the new member states trying to limit their expenses, and strategic discussions as to when to put money on the table, with above all Germany wanting to hold it back as a bargaining chip (Charter/Webster 2009).

September finance proposals that international finance "could amount to around EUR 100 billion annually by 2020", and that public financing was "estimated to lie in the range of EUR 22 to 50 billion per year by 2020, subject to a fair burden sharing at the global level" (European Council 2009b: points 12–14). Regarding fast-start finance for 2010–2012, "a figure will be determined in the light of the outcome of" COP 15 (European Council 2009b: point 17). Other than that, crucial decisions were shifted to an extraordinary European Council, to take place in Brussels during the first week of the Copenhagen summit. An element of the mandate worth mentioning was the continued EU attachment to a *"legally binding* agreement" for the period starting in 2013 (Council 2009m: point 59, emphasis added).

As parties were gathering for a **last preparatory UN session** in Barcelona (2–6 November), the hope for a legally binding outcome at COP 15 was further dampened by US Special Envoy Stern, for whom it did not "look like it's on the cards for December (...) We should make progress towards a *political agreement*" incorporating key elements of the Bali Roadmap (Goldenberg/Vidal 2009, emphasis added). **AWG-LCA 7, part 2** began with the Chair expressing hope that a single document could be prepared for Copenhagen (ENB 2009q: 1). This hope would be disappointed. Talks in the contact groups arguably advanced the consolidation of text on some issues (technology, adaptation) (ENBs: 8–15). On crunch items of the body's agenda (mitigation, finance), it would remain unchanged, though, just like it had ever since the first reading at Bonn-2 (ENB 2009s: 15). Parties decided therefore to forward the revised negotiating texts and the contact groups' non-papers as attachments to the Chair's report to COP 15 (UNFCCC 2009j; UNFCCC 2009l). **AWG-KP 9.2** moved debates quickly into contact groups to work on updated documentation. In the afternoon of the opening day, however, all deliberations were already suspended. Attempting to exert pressure on Annex I parties, the African Group threatened to boycott talks as long as the "numbers group" had not concluded its work, i.e. industrialized countries had not increased their emission reduction target pledges (ENB 2009q: 4). Just before this decision, the contact group had actually discussed a new compilation of Annex I parties' pledges, which amounted to only 16–23% reductions from 1990 levels until 2020, well short of the 25–40% range expressed in the IPCC's 2°C scenario (Observation notes 2 Nov. 2009). A compromise that would allow negotiations to resume was found late the next day: 60% of all time slots would be allocated to the numbers group (ENB 2009r: 1). Yet, availability of time did not seem to be the key obstacle. Quite obviously, it was not at the negotiators' (i.e. below the ministerial) level that Annex I Kyoto Protocol parties would move to increase their emission reductions targets, and certainly not in the face of uncertainty about US and major developing countries' efforts under the LCA track. The remaining talks touched thus on elements of a second commitment period of the Protocol (base year, commitment periods, etc.) without narrowing any options. Even the slight advance of identifying two instead of multiple alternatives on the commitment period (i.e. one eight-year period, supported by the EU, Japan, Russia vs. one five-year period, supported by the G-77 and Australia) was linked to so many conditions that it would not gain any significance in Copenhagen (Observation notes 5 Nov. 2009). Differences were reflected in non-papers, which would form part of the broad documentation the AWG-KP Chair would update for MOP 5 (ENB 2009s: 3–8). At the end of this final opportunity to prepare for Copenhagen, parties had thus made choices without taking explicit decisions. Behind the scenes, the Danish COP Presidency[33] and UNFCCC Secretary General de Boer were estimating – in the face of hardened positions – that only a continuation of the Kyoto Protocol and a political declaration on all building blocks of the Bali Action Plan, plus several annexes on developed country emission target and finance pledges and on developing country NAMAs, were feasible outcomes for COP 15/MOP 5 (Observation notes 6. Dec. 2009).

In the face of this lack of lack progress, the month between the Barcelona meeting and COP 15 would witness a further intensification of oftentimes simultaneous **bilateral and multilateral contacts**. Among the most significant exchanges were those involving the US (President) and various Asian countries, pursuing a trend started by the new US administration since early 2009. Obama began his Asia tour with a visit to Japan. On 14 November, he set the tone for subsequent meetings by referring to himself as "America's first Pacific President" (White House 2009a). On 14–15 November, he met with leaders of the Asia-Pacific Economic Cooperation (APEC, including the US, China, Japan, Russia, Australia) in Singapore. Draft conclusions on climate change reflected a compromise between G-8 formulas and major developing countries' concerns on the issue of long-term mitigation targets: "We believe that global emissions will need to peak over the next few years and be reduced to 50 percent below 1990 levels by 2050, recognising that the time frame for peaking will be longer in developing countries" (Coloma 2009). On the insistence of the latter, notably China, they were however left out of the final declaration (Spiegel 2009b; APEC 2009). A meeting in the margin of the APEC summit, led by COP 15 host, Danish Prime Minister Lokke Rasmussen,

---

[33] The Danish Presidency had been actively consulting with parties throughout 2009. A shift in its approach was remarked when the Prime Minister gradually engaged in the talks. Where Climate Minister Hedegaard had opted for a broad multilateral approach, Lokke Rasmussen preferred small-circle solutions among big emitters (Interview EU representative 22; de Boer 2010; Meilstrup 2010).

discussed more concretely possible outcomes of the Copenhagen summit. Rasmussen introduced his "one agreement, two steps" approach for further talks (Meilstrup 2010: 125): a "politically binding" agreement in December 2009, including pledges for mitigation targets and measures by developed and developing countries as well as finance pledges, followed by a legally binding treaty in 2010 (Adam *et al.* 2009). This met with wide-spread agreement, as it accommodated both the US and the emerging economies' interests (Eilperin 2009). In a press statement, a US representative was therefore also quick to publicly endorse the plan: "it was unrealistic to expect a full internationally legally binding agreement to be negotiated between now and when Copenhagen starts" (BBC 2009b). US President Obama had reportedly pleaded during the meeting not to let the "perfect be the enemy of the good", a pragmatism that did not go down well with EU leaders (Adam *et al.* 2009). Although the anticipation that the outcome of COP 15 would not be legally binding had been an open secret since the Barcelona meeting, the Union had wanted to keep up the pressure by not discarding this possibility already weeks before COP 15 (Interviews EU representatives 22, 8).[34] The US President's visit continued with a meeting with the ten smaller Asian countries reunited in the Association of Southeast Asian Nations (ASEAN), before culminating in a much-awaited US-China summit in Beijing (15–18 November). His exchange with Premier Wen Jiabao brought, however, no concrete advances on climate change. Their joint statement noted simply a "constructive dialogue", and that an agreed outcome was to include emission reduction targets of developed countries and NAMAs of developing countries (White House 2009b). A week later, the Indian Prime Minister met with Obama in Washington to conclude a "green" strategic partnership, based on energy technology cooperation (Goldenberg 2009b). The agreement included no specific common positions on the UN talks (White House 2009c). In parallel to these meetings, ministers from over forty countries, including several EU members, gathered in Copenhagen for a pre-COP meeting (16–17 November). On this occasion, textual proposals that would later re-appear during COP 15 were discussed, but parties' stances did not alter. At a last multilateral summit before the COP, the Commonwealth countries got together in Trinidad and Tobago (27–29 November). Leaders (representing all major UN coalitions) issued, together with the Danish COP Presidency, "The Commonwealth Climate Change Declaration", in which they called an "internationally legally binding agreement essential" (point 7) and proposed "a Copenhagen Launch Fund starting in 2010

---

[34] The new Danish strategy and Rasmussen's proposal at this meeting had apparently not been coordinated with the rest of the EU (Interviews EU representatives 22, 8), nor was it to the liking of Connie Hedegaard, who would have preferred pressuring the US until COP 15 (Meilstrup 2010: 125).

and building to a level of resources of $10 billion annually by 2012", which strikingly resembled EU proposals on fast-start finance (CHOGM 2009: points 7, 13).

Final **bilateral exchanges of the EU** included a summit with Russia in Stockholm (18 November). President Medvedev surprised his interlocutors when indicating that he was prepared to reduce emissions by 20 to 25% by 2020 (from 1990 levels), an improvement compared to previous offers (of -15%) (Rettmann 2009). While officially wanting to align itself with the EU's position, Russia would later clarify that the offer was not unconditional. On 30 November, the EU and China met in Nanjing, but only agreed to update their strategic "Partnership on Climate Change" (Council 2009n: point 10).

Following Russia's example, **other parties**, including the EU, used the last weeks before the COP for preparations and public disclosure of their positions on key issues: Brazil offered to reduce its emission by 38 to 42%, South Korea by 30% by 2020 compared to business-as-usual (Phillips 2009a). Japan openly thought about its contribution to fast-track finance (UNFCCC Press Conference 19 Nov. 2009). These developments were greeted by the EU and linked to an appeal to the US and China to follow suit (Agence Europe 2009g: 6–7). In a final stock-taking meeting before Copenhagen on 23 November, the Environment Council further displayed optimism about the prospects of the COP and satisfaction about its own degree of preparedness (Agence Europe 2009g: 6–7). Shortly before Copenhagen, the largest emitters would finally also clarify their positions. On 25 November, the White House announced that "the President is prepared to put on the table a U.S. emissions reduction target in the range of 17% below 2005 levels in 2020 and ultimately in line with final U.S. energy and climate legislation" (White House 2009d).[35] The country's position on finance remained unclear. Nonetheless, its proposal would incite China's government to publicly announce its target only a day later: concretizing the proposals made by President Hu Jintao at the September UN high-level summit, the State Council announced – as a "voluntary action" – that it would cut carbon emissions relative to economic growth by 40% to 45% by 2020 compared to 2005 levels (Watts 2009). Under pressure from the other major players, India, on 3 December, also suggested a "voluntary" and "non-binding" carbon intensity target of 20–25% by 2020 compared to 2005 levels, coupled to expectations about financial aids (Mohiuddin 2009).

---

[35] The US further pledged to reduce emissions by 30% by 2025, 42% by 2030 and 83% by 2050 (compared to 2005) to provide "a significant contribution to a problem that the U.S. has neglected for too long" (White House 2009d). The numbers reflected the discussions on climate and energy legislation held at that time in the Senate.

The Copenhagen Summit

COP 15/MOP 5, held between 7 and 19 December 2009 in Copenhagen, was supposed to deliver the "agreed outcome" stipulated in the Bali Action Plan. Expectations for the summit had gradually built up and found their expression in an unprecedented level of media and civil society attention. Civil society would not only attempt to make its voice heard outside the premises of the conference – through protests of hundred thousands of people on 12 December – but massively also at the COP. Altogether, about 45,000 registered participants sought access to the much too small Congress Centre during the second week of the COP.

The conference opened on **7 December 2009** with a short ceremony. Although the Danish Presidency did its utmost to create a spirit of optimism, negotiators were under serious pressure: according to the official planning, only six working days were left to prepare the negotiation texts for decisions by ministers. The last day of the summit was then foreseen for a high-level celebration involving the heads of state and government of 120 countries. During the organizational parts of the COP and MOP, delegates elected the Danish Minister for Climate and Energy, Connie Hedegaard, to preside the meetings (ENB 2009t: 1). Parties' opening statements reflected then well-known rifts: rather than displaying flexibility, they reiterated long-standing positions.[36]

In the first week, **talks in both AWGs** resumed where (and as) they had ended in Barcelona. **AWG-LCA 8** began with lengthy opening statements, before discussing the methodology of further proceedings. Work was pursued in the habitual contact groups – now referred to as "drafting groups" – plus an overarching contact group to be presided by the LCA Chair Zammit Cutajar (ENB 2009t: 2). The first textual discussions on key issues (mitigation, shared vision) did not result in major advances. On 9 December, the substantial difficulties in advancing talks on these issues were superseded by procedural gridlock in the COP. To discuss the protocol proposals that had been submitted (by Australia, Japan, the US, Costa Rica and Tuvalu), Tuvalu asked for the creation of a new contact group (ENB 2009v: 1). This triggered an intra-G-77/China debate opposing AOSIS and the LDCs to the emerging countries and OPEC. To resolve the conflict, the COP was suspended for informal consultations, stalling talks for more than a day. The AWG-LCA contact groups would not resume before the afternoon of 10 December, but did

---

[36] The only real novelty of the first day came therefore from the US, where the Environmental Protection Agency had formally declared CO_2 a public danger, sending an important signal to negotiation partners: even if no climate legislation was adopted in Congress, the US executive would be able to regulate emissions (Goldenberg 2009c).

not yield any tangible progress (ENB 2009w: 3).[37] **AWG-KP 10** started on the basis of broad documentation, without even a formal negotiating text (ENB 2009t: 3). Four contact groups were again formed, with the majority of eight time slots allocated to the "numbers group" (ENB 2009t: 4). Discussions in this group first focused on the left-overs from the Barcelona talks, i.e. essentially the level of ambition for aggregate and individual Annex I emission reductions, the use of flexible mechanisms and LULUCF in existing pledges, the length and number of commitment periods and the question of base year (ENB 2009u: 3). Work in small informal groups did not deliver any results during the first days. Despite the perceived pressure, parties were not displaying any willingness to compromise and continued repeating their by then well-known positions. On the issue of base year, for instance, parties spoke out in favour of one legally binding year per party, which implied that each party would be allowed to pledge whatever it felt most suitable to inscribe into the amended Annex B of the Protocol. On the central question of Annex I party pledges on emissions reductions (ENB 2009v: 3), debates opposed Annex I and non-Annex I countries: on 9 December, Japan and Russia stated that it was unrealistic to make and increase pledges if it continued to be uncertain how much and under what conditions other major emitters (i.e. the US and the emerging countries) would act under the parallel negotiation track (ENB 2009v: 3). Developing countries, headed by China, re-iterated their habitual legalistic argument that commitments under the Kyoto Protocol, also for a second commitment period, were legally binding and had to be honoured regardless of what happened elsewhere (ENB 2009v: 1). In this debate, the EU attempted to cut across lines by proposing means of increasing countries' pledges: (re-)introducing its four comparability criteria (population, GDP, early action, emissions, all weighted equally), it presented calculations comparing available party pledges against an assumed 30% overall emission reduction target for Annex I parties by 2020. If LULUCF rules and the current system of Assigned Amount Units, with its surpluses for countries that were subject to the "hot air" problem, were reformed, pledges could come close to the 25–40% range. Developing countries greeted the EU's presentation, but discussions during the following sessions of the group remained inconclusive.

Outside these two groups, the Danish Presidency had pursued its **talks behind the scenes** during the first week of the COP/MOP (Meilstrup 2010). Their intermediate result became known when a draft negotiation

---

[37] This scenario was repeated on 10 December in the MOP (ENB 2009w: 1–2). While Tuvalu wanted to ensure the adequate treatment of its proposal (built around a 1.5°C scenario), the larger G-77/China members refuted the notion of a new legally binding protocol outside the Kyoto Protocol. The MOP was suspended until further notice.

text was leaked on **9 December**.[38] Framed as a COP decision entitled the "Copenhagen Agreement", the draft contained seven main parts, largely reflecting the building blocks of the Bali Action Plan (UNFCCC 2009m). On the topics of interest here, it stated a commitment to keep mean temperature rise below 2°C, to strive for a peak by a specific year (2020 mentioned in brackets) and to halve global emissions by 2050 (at 1990 levels). Further, industrialized countries would pledge commitments into an attachment to the COP decision and reduce their emissions by 80% until 2050, while developing countries (LDCs excluded) would commit to NAMAs, "including actions supported and enabled by technology, financing and capacity building", which "could in aggregate yield a [Y percent] deviation in 2020 from business as usual" (UNFCCC 2009m: 4). To ensure developing country action, adaptation and finance were given a prominent place in the text. On the latter, it stated that "international public finance support [should/shall] reach the order of X billion USD in 2020" and, between 2010 and 2012, "[10] billion" USD per year (UNFCCC 2009m: 6). Finally, it was indicated that several subordinate COP and MOP decisions, among others on a "technology mechanism", a "Climate Fund" and "improvements of existing flexible mechanisms", would be annexed to this decision (UNFCCC 2009m: 13). The Agreement would be effective immediately, but negotiations also continued to reach a legally binding outcome by "COP XX" (UNFCCC 2009m: 1). The key structure and logic of this draft, based on voluntary pledges without any reference to the scientific benchmark of the IPCC's FAR, gave a flavour of what talks in parallel to the official negotiations would concentrate on during the remainder of the summit. Upon its leakage, the text sparked vehement reactions from various G-77/China spokespersons, notably the coalition's Chair Sudan, characterizing the Danish approach as "illegitimate" because it allegedly violated UN procedures (G-77/China Press Conference 9 Dec. 2009).[39] This started a series of attacks of G-77/China representatives on the Danish COP Presidency, which – through its attempts to forge an agreement in smaller circles behind the scenes – caused, in the eyes of observers, a growing distrust among parties during the further course of the talks (de Boer 2010; Meilstrup 2010: 128–129; Interviews EU representatives 22, 8).

As the summit was approaching the end of its first week, **EU leaders** were still bickering back **in Brussels** about precise financial proposals

---

[38] It would kick off a series of negotiating text leakages, published primarily through the website of "The Guardian" (for the story of those leakages: Vidal/Watts 2009).

[39] According to UNFCCC Secretary General de Boer, this criticism was, however, unwarranted: key members of the G-77/China, including Sudan, had been consulted on this draft, which apparently dated from 27 November (UNFCCC Press Conference 9 Dec. 2009; Meilstrup 2010: 128; Müller 2010: 10).

for fast-start finance between 2010 and 2012, but also about whether and under which specific conditions to move up its reduction commitment to 30% (DPA 2009a). The first problem would be dealt with swiftly: the EU's fast-start finance contribution agreed to on **11 December** would amount to € 2.4 billion per year between 2010 and 2012, with contributions from the EU's and all member states' budgets (European Council 2009c: point 37). With this, the Union launched a major influence attempt, hoping to provide new impetus to the talks. On the issue of targets, more reluctant member states (Germany, with the argument of holding back the 30% commitment as "leverage", and the Eastern European countries) won over those that favoured the more ambitious unilateral pledge (e.g. the UK, France, the Netherlands) (DPA 2009a).

On the morning of Friday, **11 December**, the chairs of the **two AWGs** attempted to streamline and accelerate talks by issuing new negotiating texts. In joint informal consultations, **LCA** Chair Zammit Cutajar was the first to present his draft (UNFCCC 2009n). Assuming the adoption of a second commitment period of the Kyoto Protocol, the seven-page document focused on key issues requiring political guidance, with placeholders on elements of the Bali Action Plan on which progress in the drafting groups was considered possible (e.g. technology transfer). Parties were generally to ensure that "the increase in global average temperature above pre-industrial levels ought not to exceed [2°C] [1.5°C]" (UNFCCC 2009n: para. 3a). To that end, all "Parties should collectively reduce global emissions by at least [50] [85] [95] per cent from 1990 levels by 2050", whereas "developed country Parties as a group should reduce their greenhouse gas emissions by [75–85] [at least 80–95] [more than 95] per cent from 1990 levels by 2050" (UNFCCC 2009n: para. 3b and 3c). On the issue of mid-term targets, the COP *"agrees"* that developed countries shall undertake

> individually or jointly, legally binding (...) commitments or actions, [including] [expressed as] quantified economy-wide emission reduction objectives with a view to reducing [them] by at least [25–40] [in the order of 30] [40] [45] per cent from 1990 levels by 2020,

while it only *"takes note"* that developing country Parties

> shall undertake nationally appropriate mitigation actions, enabled and supported by finance, technology and capacity-building provided by developed country Parties, and may undertake autonomous mitigation actions, together aimed at achieving a substantial deviation in emissions [in the order of 15–30 per cent by 2020] (UNFCCC 2009n: para. 11, 20).

Developing country reporting duties, another key issue, were to be met through national communications "and shall be [assessed at the national level] [considered in a [review] [consultative] process under the

Convention], in accordance with guidelines to be adopted by the" COP (UNFCCC 2009n: para. 24). Supported NAMAs "shall be subject to" MRV following COP guidelines (UNFCCC 2009n: para. 25). Finally, a distinction was made between fast-track finance, to be ensured through "individual pledges by developed country Parties to provide new and additional resources amounting to [XX] for the period 2010–2012" and mid-term finance (UNFCCC 2009n: para. 44 and 39). Like the Danish proposal that had leaked earlier that week, the draft was framed as a COP decision followed by other decisions (on LULUCF, mechanisms, etc.). While long-standing positions of all coalitions were reflected in the text, it leaned, with comparatively soft formulations on developing country actions ("takes note"), toward demands of emerging countries. AWG-KP Chair Ashe presented his text as content-wise "not new" and drafted on the basis that "nothing will be agreed until everything else is agreed" (UNFCCC 2009o: 1; Observation notes 11 Dec. 2009). It represented in essence a re-structuring of paragraphs long discussed in the AWG. Regarding specific targets for a second commitment period and their modalities (base year, commitment periods), it foresaw a MOP decision with amendments to Articles 3.1, 3.7 and an update of Annex B, into which Annex I parties would insert pledges "with a view to reducing their overall emissions of such gases within the range of [30 to 45] per cent below 1990 levels in the commitment period [2013 to 2018] [2013 to 2020]" (UNFCCC 2009o: Art. 3.1).

Reactions to both drafts were mixed: while the G-77/China was quite positive, industrialized countries remained more cautious. In informal consultations on the AWG-LCA draft, the EU called the text "a step into the right direction", but criticized the "enormous uncertainty on the steps to get to a legally binding outcome", and stated that the difference between the Kyoto rules and paragraph 15, cited above, was too pronounced (Observation notes 11 Dec. 2009).[40] A second shortcoming it identified concerned paragraphs 20–23 (developing country mitigation): the fact that the COP should simply "take note" of developing country action commitments was considered as "too loose" (Observation notes 11 Dec. 2009). Essentially for this latter reason, the US, which generally considered that the text "could be the basis for talks", refuted the entire mitigation section, which it regarded as "highly unbalanced" between what was expected from developed and developing countries (Observation notes

---

[40] The EU's concern was that the US would not be subject to the same accounting principles as the Kyoto Protocol parties. Within the EU, the texts stimulated also broader strategic debates: where the Commission and some states (UK, Sweden) did not yet want to accept the idea of two outcomes, others argued for accepting the texts as basis to proceed with discussions. This controversy also brought up the question of EU red lines on other issues (target pledges, finance).

11 Dec. 2009). It also spoke out against paragraph 15, which it found to resemble too much the Kyoto Protocol formulas. On the AWG-KP draft, developing countries like China noted "a very solid basis", while the EU stated that it should not prejudge the outcome, calling for a single legal instrument building on the Kyoto Protocol (Observation notes 11 Dec. 2009).

COP and MOP plenary sessions officially endorsed both drafts on Saturday **12 December** (ENB 2009x: 1–2). The first week of talks ended thus without any advances other than two artfully re-arranged compilations of parties' long-standing positions. For the EU, the week had equally been unsuccessful: it remained stuck with the refinement of its position regarding the crucial topics (legal outcome, red lines on mitigation) as well as many technical issues.

The second week of the summit began as the first had ended: slowly. With the official closure of the **two AWGs** only about 36 hours away, conflicts over procedure led to a temporary suspension of all contact groups on Monday, **14 December**. It followed renewed developing country criticism about the choice of the Danish Presidency to negotiate with a limited number of ministers that had arrived over the weekend (ENB 2009y: 2). After lengthy informal consultations and the assurance that the process would remain "party-driven", a new working method was agreed to, reflecting calls for stronger political steering. Additional drafting groups under the guidance of two ministers (one from the developed, another from the developing world) were to discuss crunch issues cutting across the two negotiation fora. They began their work on developed country targets under the Kyoto Protocol, in line with G-77/China priorities, and then moved on to AWG-LCA issues (developing country mitigation, long-term emission reductions and long-term financing).[41] When ministers would report back from their informal consultations in the afternoon of 15 December, it became obvious that this method had equally failed to produce but clarifications of seemingly insurmountable differences (ENB 2009z: 4).

Stock-taking meetings of the AWG-LCA and AWG-KP before the final plenaries would equally deliver little advances. The AWG-LCA formulated its report in the early morning hours of **16 December** (ENB 2009z: 1–2). Despite serious disagreements on key items (notably mitigation) reflected in the persistently high number of brackets, parties decided to forward the draft as "unfinished business" (UNFCCC 2009a; ENB 2009z: 2). The AWG-KP had completed its work a few hours before with an equally unfinished negotiating text (UNFCCC 2009p). Parties

---

[41] For three of these groups, EU ministers (from Spain, Germany and the UK) were appointed as co-facilitators.

requested the Chair to ask the MOP for another day of time to "clean out" the text to prepare clear options suitable for political decisions (ENB 2009z: 2–3). If parties wanted to avoid complete failure of the talks, already decried by the media, crucial issues (mitigation, finance) needed political scrutiny. This position was also expressed by the EU in the debates on the AWG-KP report in the **MOP plenary of 16 December**. The G-77/China, by contrast, requested further informal consultations. Similar discussions arose in the **COP**: after accepting the Chair's text as basis, parties requested the Presidency to clarify the further working method (ENB 2009aa: 3).

In the meantime, the **high-level segment** of the summit had already been opened. On **15 December**, the UN Secretary General and COP President Hedegaard reminded parties in their speeches that "failure is not an option" and that they should now choose between "fame and shame" when adjusting their behaviour for the final days of the talks (ENB 2009z: 1; Observation notes 15 Dec. 2009). On **16 December**, the ministerial part of the high-level segment started with a small surprise: Hedegaard resigned and parties elected Prime Minister Rasmussen to replace her (ENB 2009aa: 1).[42] The new President immediately had to deal with a point of order by Brazil, expressing concern that "a text" was being prepared "behind the scenes" to supersede the two AWGs texts (Observation notes 16 Dec. 2009). Brazil's concern was quite evidently linked to the person of Rasmussen himself, whose approach was not to the liking of the G-77/China, as it foresaw deal-making in smaller circles outside the AWGs (de Boer 2010: Meilstrup 2010: 130). Similar worries were voiced by China, India and the G-77/China Chair Sudan "and others, many of them among the parties whose leaders had shown [Rasmussen] support pre-COP" (Meilstrup 2010: 130; Müller 2010: 10–11). Rasmussen defended himself by stating that he had not yet presented any new texts (officially), and reiterated the Danish commitment to transparency. He also pointed out, however, that his duty was to "get things moving" (Observation notes 16 Dec. 2009). Although this did not satisfy the opponents of his approach, the chaotic discussion was (temporarily) interrupted to pursue the high-level meeting. Following speakers of all negotiation coalitions, many heads of state and government used the opportunity to present their world views, related or unrelated to climate change. An interesting development concerning the EU's role in the talks occurred when the African Group's representative, Ethiopian Prime Minister Zenawi, outlined his coalition's demands regarding finance: USD 10 billion per year between 2010–2012 (40% of which

---

[42] Officially, this move was to ensure that all participants during the high-level segment would be of equal levels of political seniority. Unofficially, it underscored the different approaches that had marked the Danish Presidency in 2009. Hedegaard continued her consultations as the President's "Special Representative" (Meilstrup 2010: 130).

should go to Africa) and up to USD 100 billion by 2020 (ENB 2009aa: 2). In contrast to the Africans' previous stance on this issue, he signalled openness to finance suggestions[43] – notably to the EU, the most fervent defender of financial proposals of such a magnitude. Speaking only shortly after Zenawi, Commission President Barroso stated – other than the EU's by then well-known positions and renewed appeals to China and the US – that he had listened "with great interest" to the Ethiopian comments (Observation notes 16 Dec. 2009).[44] A meeting scheduled for later that day between the EU Troika and Ethiopia was followed by a joint press conference in which the Swedish Prime Minister reported a very constructive exchange with the African Group that could provide "positive energy for this conference" (EU Press Conference 16 Dec. 2009). Zenawi himself explained: "we in Africa felt that if we could commence to resolve one of the issues, finance, other things could be solved as well" (EU Press Conference 16 Dec. 2009).

With the high-level segment and informal consultations about the working method ongoing, the next to last day of the negotiations, Thursday, **17 December**, witnessed the arrival of another important protagonist. At a press conference at midday, US Secretary of State Clinton introduced a new element of her country's position: responding to G-77/China finance demands, strongly voiced by the African Group the day before, the US envisaged up to 100 billion USD per year in finance for adaptation and mitigation in developing countries by 2020 (Clinton 2009). This money was to be generated from a variety of public and private sources, but Clinton failed to specify what the US contribution would be. Maximum transparency of emission reductions in emerging countries would be the precondition for benefiting from these funds. Replying to the media, she also stated that President Obama "was planning to come tomorrow (…) we hope there will be something to come for" (Clinton 2009). In reaction to her proposal, signs of openness were displayed in the evening when China's Vice-Foreign Minister indicated that his country would consider voluntary "international exchanges" of information on its climate actions (Broder/Rosenthal 2009). Further, Japan announced that it would be prepared to contribute USD 15 billion fast-start finance between 2010 and 2012 (WWF 2009).

Briefly after Clinton's statement, the **COP and MOP plenary sessions** resumed in order to debate the issue of procedure. President Rasmussen suggested pursuing on the basis of the texts delivered from the AWGs and in open-ended drafting groups under the guidance of Hedegaard

---

[43] Zenawi would come under criticism from the African Group for having given in on the firm position the coalition had defended on this item beforehand (Nazret 2010).

[44] The agreement had apparently been prepared in contacts between Ethiopia, France and the UK before the Copenhagen summit (Phillips 2010; Nazret 2010).

(ENB 2009bb: 1). Despite the procedural points raised by the G-77/China Chair Sudan, parties endorsed this proposal. It implied that a "Danish text" from behind the scenes was definitely discarded, and that the two "party-driven texts" would be further considered in the same type of arrangement that had dealt with them for two years. Unsurprisingly, as Hedegaard reported in the evening, it yielded the same meagre results on the crunch issues (ENB 2009bb: 2–4; Müller 2010: 12). At that point, she asked parties therefore for their procedural proposals. The EU was first to suggest setting up a smaller "Friends of the Chair" group to pursue negotiations.[45] Despite renewed developing countries' concerns about lacking transparency of such an arrangement, both COP and MOP officially charged about two dozen parties[46] with pursuing talks in parallel to the drafting groups under the two tracks.[47] The group debated at heads of state level until about three o'clock in the morning, working on a short, political draft agreement that made recourse to some of the formulas contained in the Danish leaked COP decision and was to provide a "chapeau" for the two negotiating texts – after which ministers took over until the early morning hours (DPA 2009b; Goldenberg/Stratton 2009; Goldenberg et al. 2009).

To conclude the conference, the dramaturgy of the summit had originally foreseen a ceremony at heads of state level, to begin at ten o'clock on Friday, **18 December**, its official last day. As nothing had been agreed, this timetable was necessarily altered. Upon his arrival in Copenhagen, US President Obama first invited leaders of key countries to talks in a hotel close to the airport (The Guardian 2009). While this meeting was held, the leader of the other major emitter, China's Premier Wen Jiabao was, to the surprise of many observers, awaiting the beginning of the proceedings in the half-empty plenary.[48]

---

[45] Behind the scenes, the original initiative had reportedly also come from the EU.

[46] Roughly 25 countries were permanently involved, i.e. most countries that had attended the Major Economies Forum (Australia, Brazil, China, France, Germany, India, Japan, Mexico, Russia, South Africa, South Korea, Spain, Sweden and the European Commission (these three as EU Troika), the UK, the US, and Denmark) plus several parties representing different coalitions. According to a statement by Grenada in the final COP plenary, these were: Ethiopia, Algeria (both for the African Group), Bangladesh, Lesotho (both for the LDCs), the Maldives, Grenada (both for AOSIS), Colombia, Norway, Saudi-Arabia (for OPEC) and Sudan (for the G-77/China).

[47] Unofficially, informal (bilateral) talks had been ongoing the entire time, as heads of state and government arrived in Copenhagen. From the EU side, the UK, Germany or France engaged in bilaterals, with, e.g., China and Brazil, while the EU Presidency itself was reportedly less and less involved in outreach activities.

[48] Wen Jiabao's absence from this meeting was explained in different ways: US sources said he had "refused" to meet informally, sending low-level officials instead; China claimed not to have received an official invitation (Müller 2010: 12). Reportedly, the country was represented at the meeting by Ethiopia.

When the heads of state meeting finally began, the first speakers would set the tone for the remainder of the day. Incidentally, the schedule foresaw a rapid succession of leaders from the world's major countries. To begin with, the Chinese Prime Minister used his speech to recall his country's climate policies and positions, stressing the fact that its 40–45% energy intensity reduction pledge by 2020 was to be incorporated into "Chinese mid- and long-term plans as a mandatory target" (Observation notes 18 Dec. 2009). He went on to state that his country would improve the internal monitoring and evaluation of its efforts to reach this target, and engage increasingly in "international exchange, dialogue and cooperation", thus responding positively yet cautiously to the US demand for verification (Observation notes 18 Dec. 2009). Showing no other signs of flexibility, Wen Jiabao concluded by underscoring that China envisaged its "voluntary action" independently of any other parties' targets/actions, and "will be fully committed to achieving and even exceeding" its target (Observation notes 18 Dec. 2009). Only minutes later, following a speech by Brazil's President Lula, Obama entered the plenary hall through a back door to deliver a short speech that largely resembled Wen Jiabao's, equally displaying a low degree of flexibility of the US position. Where many had hoped that he would bring another offer to the table, possibly in terms of a higher target proposal than the 4–5% by 2020 against 1990 levels, Obama had only the following message: according to the US, a "global accord", "in which we agree to take certain steps, and to hold each other accountable for our commitments" would be the ideal outcome at this stage of the talks, and "the pieces of that accord are now clear (…) mitigation, transparency, and finance" (Observation notes 18 Dec. 2009). He concluded with an appeal that was symptomatic of the US take-it-or-leave-it approach: "America has made our choice. We have charted our course, (…) made our commitments, and we will do what we say. Now it is time for the nations and people of the world to come together behind a common purpose" (Observation notes 18 Dec. 2009).[49] With this, he left the plenary again to engage in further informal consultations. For the EU, foreseen later in the programme, Sweden's Prime Minister reformulated earlier appeals to the two big emitters to go beyond what they had just proposed. Europe, he said, was serious about reaching an agreement and "not just talking about procedures", a clear side blow at the G-77/China (Observation notes 18 Dec. 2009).

Once the major speeches had been delivered, talks entered into their decisive stage: **informal consultations between members of the Friends of the Chair group** included an early afternoon exchange of

---

[49] That this was maybe not entirely the case was demonstrated by a press conference given by Republican members of the US House of Representatives on the same day in Copenhagen, in which several of them denied climate science and opposed its international regulation (Press Conference US House of Representatives 18 Dec. 2009).

Wen Jiabao and Obama (The Guardian 2009). In the course of the day, the political agreement sought by leaders would undergo significant transformations, reflected in several draft versions. To allow for a process-trace,[50] the most significant ones will be examined here with regard to the key topics under analysis.

While an overnight version of the political agreement had been unanimously refused as "too weak" by the EU (Becker/Nelles 2009), another, untitled version of the text became public around midday (UNFCCC 2009q). It resembled in some ways the Danish draft text rendered public the week before. "Affirming our firm resolve to adopt one or more legal instruments (…) as soon as possible and no later than COP 16/CMP 6", the parties formulated brief consensus language on key elements of the Bali Action Plan. The draft contained the recognition of climate science ("increase in global temperature ought not to exceed 2 degrees") (para. 1), but retained a marked differentiation between parties: Annex I parties would "commit to implement" quantified targets so that reductions would be in the order of "X per cent by 2020 compared to 1990 and Y per cent compared to 2005", laid down in an annex (para. 4).[51] Non-Annex I countries would "resolve to implement mitigation actions", which "shall be reflected through their national communications (…) every two years" (para. 5). Paragraph 5 on MRV contained an attempted compromise between the insistence on "voluntary" action by developing countries and international verification demands by the US and other Annex I countries.[52] On finance, paragraph 8 read: "Parties take note of the individual pledges by developed country Parties to provide (…) 30 billion dollars for the period 2010–2012" and "support a goal of mobilizing jointly 100 billion dollars a year by 2020". Money should go primarily to LDCs. Moreover, parties called for "a review of this decision and its implementation in 2016", indicating their desire to adopt the text as a COP decision. The final paragraph reaffirmed their commitment to extend the mandates

---

[50] As at the Kyoto COP, talks moved increasingly into the backrooms of the Congress Centre. Unlike in Kyoto, however, media presence and the wide use of social media (Blogs, Twitter) allowed for closely monitoring the evolving negotiations. Moreover, confidential audio tapings of the talks in the Friends of the Chair group – released in the spring of 2010 by the German news magazine "Der Spiegel" – confirm other reports about the difficult proceedings in this group. They also demonstrate that the debates among the world's leaders resembled much to those held at negotiators' level (Spiegel 2010). Although a few details of the high-level exchanges cannot be reconstructed, a clear overall message on the story and EU influence assessment emerges.

[51] This reflected the existing and the US-favoured base years.

[52] Supported NAMAs would be "registered in a registry (…) and shall be subject to international measurement, reporting and verification in accordance with guidelines elaborated by the COP (…) The Parties take note of the information on enhanced mitigation actions by non-Annex I Parties" (UNFCCC 2009q: para. 5).

of the two AWGs to arrive at "one or two legal instruments under the Convention", thusly shifting most issues to a later stage (para. 12, 13). Negotiations on the document, which lacked some crucial EU positions (mid-term 25–40% reduction range, long-term target), continued.

Another draft of the text negotiated by the Friends of the Chair reflected the status of the talks in the afternoon (ca. 4:30 pm, UNFCCC 2009r). Now entitled "Copenhagen Accord", in line with the wording chosen by Obama in his speech earlier that day ("global accord"), the draft had slightly changed on the key issues. The most marked difference concerned the parties' desire to adopt a legally binding outcome at a later stage, which had been completely dropped. Instead, the review of the accord had been predated to 2015 (para. 12). Language around the 2°C goal had been strengthened (para. 1 and 2), whereas a collective Annex I party target had been taken out. Countries could now pledge individual targets into an appendix by 1 February 2010 (para. 4), in line with US preferences. Non-Annex I parties would do the same with their "mitigation actions" (para. 5). The wording on MRV and finance had remained largely unchanged (para. 5, 8), but the idea of a "Copenhagen Green Climate Fund" was newly inserted into paragraph 10.

At about the same time, an **informal European Council** was consulting in situ on the draft and the question of whether the Union could give a last-minute impulse to the talks. As in the EU coordination meeting the night before, the UK and France proposed to move the unilateral emissions reductions pledge up to 30%, but met with opposition from, notably, Italy and Poland.

Talks on the 4:30 pm draft continued into the evening, with heads of states, assisted by high-ranking negotiators, engaging personally in concrete textual work (The Guardian 2009). The next intermediate outcome of these efforts was circulated at about 7:30 pm (UNFCCC 2009s). Although the prospect of a legally binding agreement had not been reinserted, the final paragraph stated that the review (in 2016) would consider strengthening the long-term goal in order to limit global temperature increase to "1.5 degrees", a strong AOSIS and LDC demand. Paragraph 2 mentioned, instead of the 2°C aim (now only in para. 1), the G-8 agreed long-term target of halving global emissions by 2050 (below 1990 levels). Further on mitigation, Annex I parties "commit to reducing their emissions (…) by at least 80 per cent by 2050" (without mentioning a base year), while the individual pledges immediately inserted into an appendix would be summed up into "X percent" in 2020 of 1990 levels and "Y percent" of 2005 levels, thus combining the formulas used in previous drafts (para. 4). Largely untouched, the section on mitigation actions by non-Annex I countries contained a placeholder "[Consideration to be inserted US and China]" on MRV (para. 5). Key changes concerned the targets and reflected industrialized countries' positions and formulas endorsed by the G-8 in previous years. This was not the end of the story, however (for an overview, see Table 4).

Table 4: The Negotiations of the Copenhagen Accord

DRAFT 1:00 pm	DRAFT 4:30 pm	DRAFT 7:30 pm	DRAFT 10:00 pm	COPENHAGEN ACCORD
(no title) (UNFCCC 2009q)	"Copenhagen Accord" (UNFCCC 2009r)	"Copenhagen Accord" (UNFCCC 2009s)	"Copenhagen Accord" (UNFCCC 2009t)	(UNFCCC 2009u)
Issue: Shared vision (regarding mitigation)				
"increase in global temperature ought not to exceed 2 degrees" (para. 1)	"We agree that deep cuts in global emissions are required (…) so as to hold the increase in global temperature below 2 degrees C" (paras. 1, 2)	"recognizing the scientific view that the increase in global temperature should be below 2 degrees Celsius" (para. 1); target of halving global emissions by 2050 (below 1990 levels) (para. 2); 1.5°C as aspirational target (para. 12)	"recognizing the scientific view that the increase in global temperature should be below 2 degrees Celsius" (paras. 1, 2)	"recognizing the scientific view that the increase in global temperature should be below 2 degrees Celsius" (paras. 1, 2); 1.5°C as aspirational target (para. 12)
Issue: Developed country mitigation				
"X per cent by 2020 compared to 1990 and Y per cent compared to 2005", laid down in an annex (para. 4)	Individual target pledged into appendix by 1 February 2010 (para. 4)	Individual pledges into appendix (at Copenhagen); Annex I parties "commit to reducing their emissions (…) by at least 80 per cent by 2050" (no base year) (para. 4)	Individual target pledged into appendix by 1 February 2010 (para. 4)	Individual target pledges into appendix by *31 January* 2010 (para. 4)

DRAFT 1:00 pm (no title) (UNFCCC 2009q)	DRAFT 4:30 pm "Copenhagen Accord" (UNFCCC 2009r)	DRAFT 7:30 pm "Copenhagen Accord" (UNFCCC 2009s)	DRAFT 10:00 pm "Copenhagen Accord" (UNFCCC 2009t)	COPENHAGEN ACCORD (UNFCCC 2009u)
Issue: Developing country mitigation				
"resolve to implement mitigation actions", pledged in an annex (para. 5)	Individual actions pledged into appendix by 1 February 2010 (para. 5)	Unchanged	Individual actions pledged into appendix by 1 February 2010 (para. 5)	Individual actions pledges into appendix by *31 January* 2010 (para. 5)
Issue: MRV for developing countries				
NAMAs "shall be reflected through their national communications (…) every two years"; Supported NAMAs would be "registered in a registry (…) and shall be subject to international measurement, reporting and verification in accordance with guidelines elaborated by the COP" (para. 5).	Unchanged	Mostly unchanged, but placeholder: "[Consideration to be inserted US and China]" (para. 5)	NAMAs "will be subject to their domestic MRV the result of which will be reported through their national communications every two years (…) *with provisions for international consultations and analysis under clearly defined guidelines that will ensure that national sovereignty is respected.*" Supported NAMAs "recorded in a registry along with relevant technology, finance and capacity building support. (….) *[Such] actions will be subject to international*	Unchanged

From the Bali Roadmap to the Copenhagen Accord (2007–2009)

DRAFT 1:00 pm (no title) (UNFCCC 2009q)	DRAFT 4:30 pm "Copenhagen Accord" (UNFCCC 2009r)	DRAFT 7:30 pm "Copenhagen Accord" (UNFCCC 2009s)	DRAFT 10:00 pm "Copenhagen Accord" (UNFCCC 2009t)	COPENHAGEN ACCORD (UNFCCC 2009u)
			measurement reporting and verification in accordance with guidelines adopted by the COP" (para. 5)	
Issue: Short and long-term finance				
"Parties take note of the individual pledges by developed country Parties to provide 30 billion dollars for the period 2010–2012", and "support a goal of mobilizing jointly 100 bn dollars a year by 2020" (para. 8)	Mostly unchanged; new: "Copenhagen Green Climate Fund" (para. 10)	Unchanged	Substantively unchanged, wording strengthened	Unchanged
"Affirming our firm resolve to adopt one or more legal instruments (…) as soon as possible and no later than COP 16/ CMP 6"; review of decision, implementation by 2016 (para. 13, 12)	No reference to legally binding outcome; review foreseen for 2015 (para. 12)	No reference to legally binding outcome; review foreseen for 2016 (para. 12)	No reference to legally binding outcome; review foreseen for 2015 (para. 12)	Unchanged

Note: The table concentrates on key analytical units of this study (mitigation by different groups of countries) and closely linked topics (MRV, finance and legal bindingness). It covers key draft versions and indicates the approximate time at which each draft was circulated.

Obama and Wen Jiabao were still supposed to meet in order to clarify the issue of verification of developing country emission reduction actions and thus lift one of the final brackets in the draft text (Melstrup 2010: 132; The Guardian 2009). Upon Obama's arrival at the venue where he was to encounter the Chinese Premier, the latter was in a meeting of the BASIC countries (Broder 2009). According to US and Brazilian sources, Obama (and Secretary of State Clinton) interrupted that meeting and (were)[53] asked to join in (Broder 2009; Müller 2010: 13). In this exchange, obviously very difficult to reconstruct here, (essentially) China and the US reportedly worked out the terminology regarding MRV. In return for Chinese acceptance of formulas going into the direction of US preferences, the US (and the Friends of the Chair group) would have to accept Chinese (and BASIC) demands of deleting passages containing the G-8 formula of "halving global emissions by 2050", 80% reductions by Annex I countries and the 1.5°C target reference dear to AOSIS (Merkel 2009; Lynas 2009; Spiegel 2010).[54] None of those were found back in an otherwise largely unmodified draft of the accord of about 10:00 pm (UNFCCC 2009t). When the Maldives, supported by the majority of the Friends of the Chair, later defended the 1.5°C reference, China had to accept its re-insertion, but not without considerably weakening the language around that target (Lynas 2009). For the rest, the final "Copenhagen Accord" resembled very much the previous version of the text, with the exception of the date by when parties would inscribe their pledges into the appendices (31 January) (para. 4, 5) and a stronger emphasis on the Green Climate Fund (para. 8, 10). Finally, the text endorsed the outcomes of the AWGs, without specifying what to do with these or how their work would be continued.

While parties were still discussing final details, and those that had not been involved in the Friends of the Chair group had not actually seen the final draft, one of the key players of the day reported already to the press. Calling the Accord "a meaningful and unprecedented breakthrough", US President Obama emphasised that for the first time "all major economies have come together to accept their responsibility to take action to confront" climate change (US Press Conference 18 Dec. 2009).[55] After

---

[53] The sources diverge on whether Obama forced his entry into the meeting or was asked to join in.

[54] China's rationale in demanding the scratching of those Articles was arguably based on a fear of having moved to the rank of Annex I countries and having to comply with this rule by mid-century (Lynas 2009; Müller 2010).

[55] As one commentator noted, the "White House mounted a surgical strike of astounding effectiveness (and cynicism) that saw the president announcing a deal live on TV" to make "anyone (...) not particularly interested (...) believe that a deal (...) had been done, with the US providing leadership to the global community" (Black 2009).

acknowledging that progress was "not enough", he went on to express his hope for the "beginning of a new era of international action" and concluded on a pragmatic note ("it's important for us, instead of setting up a bunch of goals that end up just being words on a page and are not met, that we get moving"), before returning to Washington (US Press Conference 18 Dec. 2009).

EU leaders, who had last met in an informal European Council to discuss the draft of 7:30 pm, took a bit more time before coming in front of the press. Both UK Prime Minister Brown and German chancellor Merkel stated that the Accord marked a "first step", which was far from perfect (Brown 2009; Merkel 2009). It would take another few hours before the EU Troika would give insights into the final hours of the negotiations as well as into the delegation's feelings at the end of this summit. Swedish Prime Minister Reinfeldt admitted that this was "not a perfect agreement", and that it would not limit global temperature increase to 2°C (EU Press Conference 19 Dec. 2009). On the EU's role, he commented that "we were very well prepared, but we saw that there was not the same level of preparation on other parts" (EU Press Conference 19 Dec. 2009). Commission President Barroso did not want to "hide his disappointment"[56] and already turned to the future: "we need to take this process into a new phase and learn the lessons from here" (EU Press Conference 19 Dec. 2009). Asked about the EU's implication in the process during the final hours, Barroso stated that "in ambition, we were always leading, but we were not leading when it was the point of lowering the ambition (…) it is true that others were much more influential when it was about reducing the ambitions" (EU Press Conference 19 Dec. 2009). Reinfeldt completed: "We have not been chased by others to go to 30%", and concluded: "It seemed sometimes that we were not in a climate change negotiation" (EU Press Conference 19 Dec. 2009).

Despite the (reluctant) endorsements by major leaders, the conference was not yet over: the Accord negotiated by the Friends of the Chair still had to be agreed – by consensus – in the formal decision-making body under the Convention, **the COP**. This proved extremely difficult (Müller 2010: 13–17). Resuming shortly after 3 am, the overtired delegates engaged again in a battle of words about the nature of the process and its outcome. The Danish Prime Minister tried to get the Accord, elaborated by a "representative group of leaders", quickly accepted, giving parties only limited time to consider their reactions. To this end, he opened the **final MOP plenary** with the intention to suspend it and pursue discussions on

---

[56] This was certainly an understatement of how many EU negotiators felt about this outcome. Some, certainly completely overtired after nights with little sleep, were seen crying after learning of the final "Copenhagen Accord" (Minten 2009).

the Accord an hour later (ENB 2009cc: 7). Several parties raised points of order, however, in which they expressed disagreement with the working method and the Accord itself. For Tuvalu's representative, this method was "disrespectful of the UN", which is why "Tuvalu cannot accept this document" (Observation notes 19 Dec. 2009). Venezuela, Bolivia, Cuba and Nicaragua equally resisted. As there was no consensus, Costa Rica suggested having the Accord issued as an "information document" (INF), while Nicaragua proposed to issue it as a "miscellaneous document" (MISC), i.e. a party submission (ENB 2009cc: 8). After a first interruption, Rasmussen proposed to consider the text as MISC document, which was refused by, inter alia, India on grounds that such a document was too informal given the fact that the Accord had been negotiated by its head of state (ENB 2009cc: 8). After many other parties, including the EU, the LDCs, the African Union, Japan, the US, Grenada and the Maldives, had supported the Accord, UK Environment Minister Miliband – reportedly alerted by his staff to regain the conference premises in the face of the problematic development of the talks (Pearce 2009) – suggested that it be adopted as a COP decision, after which Slovenia proposed a COP decision with a footnote stating the dissenting parties. This was further opposed by five countries (Venezuela, Bolivia, Cuba, Nicaragua, Tuvalu). When Rasmussen was at the point of concluding the session without results at 5:30 am, the UK moved for an adjournment (ENB 2009cc: 8). Two and a half hours later, Rasmussen had been replaced by a COP Vice-President who proposed that the conference "takes note" of the Copenhagen Accord and that those parties supporting the Accord could associate themselves with it. He then quickly gavelled this decision through. Following another debate on the operationalization of the agreement, a decision was taken to continue work under the AWG-LCA on the basis of agreements reached in Copenhagen (UNFCCC 2009v). The COP closed shortly after two o'clock in the afternoon. The MOP then adopted a decision on the further work of the AWG-KP (UNFCCC 2009w; ENB 2009cc: 11). With this, the Copenhagen summit closed at 3:30 pm on Saturday, 19 December 2009, almost 24 hours after its scheduled ending.

The Outcome: the Copenhagen Accord

The main product of the two-year negotiation process culminating in COP 15 was the "Copenhagen Accord" (CA), a document best characterized as a political declaration. Unlike the originally envisaged legally binding agreement (i.e. an international treaty that would, at least in principle, be enforceable), this two and a half-page document had only moral value: for those that had negotiated it – as well as for whoever would associate with it –, it would be politically inopportune to not strive for its implementation. This implementation was, however, seriously threatened.

By framing the texts discussed by the Friends of the Chair as COP decisions, the Danish Presidency had hoped to obtain an outcome with soft law status (Rajamani 2010a, 2009). It came differently: due to the resistance of several countries (Tuvalu, Nicaragua, Bolivia, Venezuela, Cuba) in the final plenary, the COP only "took note" of the Accord. This, in the words of UNFCCC Secretary General de Boer, represented a "way of recognizing that something is there, but not going so far as to directly associating yourself with it" (UNFCCC Press Conference 19 Dec. 2009). It also meant that the Accord was not an official UN document and certainly not the "*agreed* outcome" stipulated in the Bali Action Plan, with two main consequences (Müller 2010: 1; Rajamani 2010a): 1. Certain provisions requiring a reliance on UN structures and procedures could not be "immediately operational"; 2. Negotiations would, as the two associated COP decisions on the work of the AWGs suggested, have to continue (UNFCCC 2009v, w).

A brief discussion of the key provisions of the Accord relevant for this study concentrates on the sections on *mitigation commitments and actions* by developed and developing countries and their verification as well as on finance. Recalling Article 2 UNFCCC, the document first concretized the objective of the climate regime: to stabilize GHG "at a level that would prevent dangerous anthropogenic interference with the climate system, we shall, recognizing the scientific view that the increase in global temperature should be below 2 degrees Celsius (…) enhance our long-term cooperative action to combat climate change" (para. 1). This formula was repeated and specified in the subsequent paragraph with a reference to the IPCC's Fourth Assessment Report. Providing for an implicit link between the 2°C target and the mitigation scenarios developed in that report, it amounted to an indirect recognition of the relevant emissions reduction trajectories (of 25–40% by 2020 and 50–85% by 2050 compared to 1990 by industrialized countries), promoted especially by the EU and the majority of the G-77/China ever since the Bali COP. The essential approach to mitigation, a "pledge and review" of the most voluntary kind, reflecting US and emerging countries' preferences, was, however, in no ways capable of ensuring that this weakly formulated aim would be attained (Rajamani 2010a). Moreover, the Accord stipulated that Annex I parties should submit commitments "to implement individually or jointly the quantified economy-wide emissions targets for 2020" and non-Annex I parties NAMAs, both "in the format given in" two appendices "to the secretariat by 31 January 2010 for compilation in an INF document" (para. 4, 5 CA). LDCs and AOSIS members "may undertake actions voluntarily and on the basis of support" (para. 5 CA). In addition to the bottom up approach inherent in the pledges, shifting the information on commitments/actions (i) to the future and (ii) into an INF (informal) document of no other than an informational value underscored the weak engagement parties made with this agreement. Nonetheless, these provisions, notably with regard to emerging country actions, have been interpreted as the major novelty and "breakthrough" of the post-2012 negotiations (Purvis/Stevenson 2010). For the first time, non-Annex I parties would bind themselves, albeit politically, to NAMAs. Yet, the *differentiation* between groups of countries, also reflected in the language of the Accord (Annex I vs. non-Annex I), was retained. Verification of target fulfilment and actions, a third important item in the Accord, was equally differentiated. A lengthy paragraph, reflecting US concerns and compromises with BASIC, spelled out the details for developing countries (see Table 4): actions would be reported every two years and "subject to (...) domestic [MRV]". Reports on their implementation would be subjected to "international consultations and analysis under clearly defined guidelines that will ensure (...) national sovereignty" (para. 5 CA). NAMAs "seeking international support will be recorded in a registry along with relevant technology, finance and capacity building support"; "supported (...) actions will be subject to international" MRV in accordance with COP guidelines (para. 5 CA). By contrast, developed country commitments "will be measured, reported and verified in accordance with existing and any further guidelines adopted by the" COP (para. 4 CA). This type of MRV applied not only to mitigation, but also to finance, the final key component of the agreement. On this issue, the deal foresaw the provision of fast-track financing via the "collective commitment by developed countries (...) to provide new and additional resources (...) approaching USD 30 billion for the period 2010–2012" as well as long-term financial resources. "Developed countries commit to a goal of mobilizing jointly USD 100 billion dollars a year by 2020 to address the needs of developing countries", which would come from public and private, bilateral and multilateral sources (para. 8 CA). Besides a "Technology Mechanism", the Accord stipulated the creation of a "Green Climate Fund (...) as an operating entity of the financial mechanism of the Convention" (para. 11, 10 CA). Due to the uncertain legal status of the Accord, the operationalization of both as well as of the financial proposals in general was, however, unclear (Rajamani 2010a).

Besides these – for the purpose of this study – central elements of the Accord, paragraph 12 called for the assessment of its implementation in 2015, at the mid-point between its adoption and the delivery of the 2020 targets and after the next IPCC report expected for 2014. At the insistence of AOSIS, this paragraph also contained a loose reference to their preferred temperature limit: the 2015 assessment would include "considerations of strengthening the long-term goal (...), including in relation to temperature rises of 1.5 degrees Celsius". Finally, the CA also endorsed two related decisions to continue the work of the two AWGs, but without any deadline or indications as to the further working procedures of these bodies. The two decisions themselves were a bit more concrete about these proceedings. The COP decision stipulated "to extend the mandate of the [AWG-LCA] to enable it to continue its work with a view to" finalizing it for COP 16 (UNFCCC 2009v: point 1). Talks would be continued on the basis of the negotiating text as it stood at the end of COP 15. The respective MOP decision stipulated the same for the AWG-KP and its negotiating text (UNFCCC 2009w). Although "cleaned" on many technical items, both negotiating texts still resembled compilations of party positions on a range of key issues.

In the final analysis, the Accord appeared weak on its targets and form, and unspecific regarding its operationalization and implementation. To allow for a full assessment of the document, but also of EU influence in the next section of this chapter, it is instructive to briefly highlight what it did not achieve besides legal certainty. Several issues discussed at length between Bali and Copenhagen were simply not reflected in it. Most prominently, this concerned the reference to emission reduction goals for 2020 and 2050 for different groups of countries. Closely related to this was its silence on a (common) base year. Finally, amid many other technical points, accounting rules (regarding LULUCF and flexible mechanisms) were not specified. All this would render the calculation and comparison of targets extremely difficult. It would form the basis of continued negotiations from 2010 on.

The EU's Influence Attempts: Extracting Patterns

The process-trace of the negotiations between COP 13 and 15 allows for an extraction of patterns of the Union's external activities during the post-2012 talks. To begin with, a brief consideration of global climate politics seems, however, in order so as to place the EU's influence attempts into a broader context. The story testifies to the gradual transformation of global climate policy-making into a complex multi-site process, with two main tracks under the UN umbrella and a partial delocalization of talks into restricted arenas outside the UN regime (G-8+5, Major Economies Forum, G-20, Greenland Dialogue). Closely linked to this was a gradual "high-levelization" of talks. As seen, the finale of the post-2012 talks was characterized by the involvement of the highest political level in a process of combining selected bits of texts prepared within the UN regime and in fora outside the UN realm.

The EU was prominently represented in all these arenas and at multiple levels, mostly through the Climate Troika, but also through key countries. Although a logistical challenge – given the number of meetings held notably in 2009 – the Union thus made adequate use of its ample diplomatic resources. The internal coordination between different actors involved in the various fora, however, did not always function smoothly (Interviews EU representatives 8, 10). In fora outside of the UN, the EU attempted to influence talks through the promotion of its overall position without, however, taking any specific initiatives. Notable exceptions existed only when EU members held the G-8 Presidency (in parallel to the EU Presidency), as in 2005 (the UK) and 2007 (Germany). In both cases, the presidencies undertook – in close coordination with the rest of the EU – significant attempts to engage other major emitters: the US and the emerging powers through the G8+5 Gleneagles Dialogue from 2005 on, and the US specifically at the summit in Heiligendamm in 2007 with regard to the 50% reduction goal by 2050. Mostly, however, the EU's participation in fora outside the UN was a reaction to invitations from other parties (e.g. the MEF as a US initiative). More important than the participation in these fora was arguably its increased bilateral diplomatic activity. Although the process-trace could highlight only the most significant examples of these efforts, the EU appeared as more energetic in the promotion of exchanges with third countries or other world regions than in the past. While many of these took primarily the form of exchanges of positions aimed at trust-building and promoting mutual understanding, some bilaterals also went further (Interviews EU representatives 8, 10). As briefly discussed for the period 2005 to 2007 (see Chapter 4), the Union applied specific strategies with regard to a limited number of countries. *Vis-à-vis* the emerging powers China and India (and later: Brazil and South Africa), identified as crucial for global (climate) politics, concrete cooperation projects were initiated.[57] In its relations with the LDCs, examples of practical cooperation based on economic and other aid were alluded to in the process analysis (e.g. during COP 14). These concrete approaches all served one overarching aim: promoting the Union's objective of concluding a comprehensive global climate treaty (Interviews EU representatives 12, 22, 10). Although the EU thus defended its overarching positions in many arenas of global climate policy and through many channels, key influence attempts were targeted, as in the past, at the UN negotiation process. They would come in the form of positions expressed through Environment and European

---

[57] Though not specifically traced here, it has to be noted that, especially *vis-à-vis* these larger trading partners, a certain parallelism between genuine EU outreach and activities of (big) EU member states like the UK, Germany or France existed. While these countries' activities did not necessarily go against EU objectives in the UN talks, they certainly operated with different approaches, giving at times the impression of fragmentation of the Union's approach (Interviews EU representatives 1, 12).

Council conclusions, internal legislation, written submissions to the UN and the oral defence of positions in the AWGs, COPs or MOPs and through the media. To further analyse the logic of the EU's approach, several key influence attempts linked to the UN talks can be identified:

1. **MARCH 2007**: Following the January 2007 Commission proposal, the European Council determines the EU's position for the post-2012 process. The EU attempts to affirm its leadership ambition even before the official start of reform talks through a unilateral 20% reduction offer, linked to the conditional 30% offer that it hoped to employ as leverage.

2. **AFTER COP 13** (January/March 2008): In January, the Commission publishes its proposal for a climate and energy package, which the Spring European Council endorses. Right after the Bali COP, the EU thus reconfirms and strengthens its commitment for the post-2012 process. Even if it would take the entire year to adopt legislation, the proposals sent clear signals to the outside world about the Union's seriousness and approach. They would form the backbone of its communication strategy and be dispersed via submissions to the UN.

3. **COP 14** (December 2008): Although not a major influence attempt, the content and timing of the EU's commitment to invest into renewable energy projects in Africa announced in Poznan stands emblematically for its attempts to rally parties behind its position.

4. **AFTER COP 14** (January 2009): Through the Commission communication "Towards a comprehensive climate agreement in Copenhagen", the EU attempts to set the agenda on the crucial topic of finance by identifying crunch issues and possible solutions (fast-start finance, quantification of overall amount of finance required, sources). The proposals would be concretized in the autumn of 2009. Although member states would not finalize the concrete finance position until COP 15, the Union used its general proposals as basis for appeals to other countries, notably the US and China, to reveal and/or increase their ambitions.

5. **COP 15** (December 2009): As first Annex I party, the EU reveals its proposal for fast-start finance and recalls its commitment to contribute a "fair share" to an overall amount of € 22–50bn funding by 2020.

Although each of these influence attempts merits recognition by itself, emphasis will be placed on the central proposal, the 20/30% reduction demand and offer, before moving on to discuss the overall picture that emerges in light of the story, which allows for the further extraction of patterns concerning the EU's broader foreign policy approach.

Regarding the Union's conditional target offer, the approach chosen for this negotiation round displayed a certain parallel to the 1997 target proposal: at that time, 15% reductions were publicly promised as the target the EU would adopt (and that it demanded from other industrialized countries), while the internal burden-sharing covered only 10% of these efforts, which signalled to a certain extent the bargaining character of the position. In 2007, the EU applied the more cautious version of the very same construction, proposing only what it had agreed to internally (20%), but demanding nonetheless 30% from others. From a foreign policy (and here notably a bargaining) perspective, the two approaches had different virtues: while the former could (and did) help the EU to pull other major emitters (the US, Japan) towards the higher end of the proposed targets, the conditional offer for the post-2012 talks did not give the EU any leverage over other parties because, as the Swedish Prime Minister had to realize right after the conclusion of the Copenhagen Accord: "We have not been chased (...) to go to 30%". In other words, parties completely neglected the Union's 30% offer.

Diverging from its strategy in past talks, the Union relied not only on formal diplomatic tools, but combined these with economic foreign policy instruments, partially in the UN regime (as finance became such a key issue in the talks), but also outside of it. From an analytical perspective, the Union complemented thus its still predominant problem-solving approach (in the language of the WEIS coding scheme: "to make proposals") by "offering and or granting economic rewards", coupled to concrete demands (Smith 2003: 52–68; Wilkenfeld et al. 1980). This also meant that its influence attempts were not only, as mostly in the past, structure-focused (targeting regime structures), but also actor-focused (targeting other actors' behaviour, preferences). Furthermore, the central strategy of the past, the positing of politically strong and (with regard to the targets) ambitious positions to "lead by example", was explicitly strengthened and expanded through a legal and economic approach. Internal legislation was developed not only to underscore the EU's seriousness with regard to mitigation, but also with the express purpose of utilizing them to exert influence at the global level, since many of its new policies comprised specific external dimensions.

Most prominent in this regard was certainly the notion of a global carbon market, for which the ETS could be the "prototype" (van Schaik/van Hecke 2008: 17), and "linkage" to which the EU intended to employ "as political leverage", essentially *vis-à-vis* other developed countries (Benwell 2009: 105). The US was approached with a strategy that highlighted that climate change was manageable without economic losses, possibly even with gains (using the ETS and own legislation as references) (Interviews EU representatives 26, 12). Other industrialized countries were asked to follow the EU's model internally to increase their mitigation ambitions and to cooperate toward setting up joint carbon emissions schemes.[58] Vis-à-vis LDCs, the market also played a key role in the Union's finance proposals regarding aids for adaptation and mitigation measures. Numerous other concepts that it introduced into the international negotiations also originated from internal policies: the EU's "effort-sharing" was essentially based on Commission proposals around the principles of "fairness and solidarity", taking into account, inter alia, GDP/capita (Vogler 2008). Even if derogations were allowed in the end for some member states in the EU's internal deal, it advanced its criteria-based approach as an example of how common but differentiated responsibilities could be practically implemented internationally (E3G 2009: 4). Similarly, other proposals introduced during the UN talks would include combined indicator-based approaches that intended to balance out different interests by taking into account varying national circumstances (Swedish Presidency 2009). For instance, to assess the comparability of other parties' efforts that would have allowed the EU to move to a 30% reduction under a future global climate agreement, the criteria it proposed were: ability to pay (GDP/capita), reduction potential (GHG emissions/GDP), population trends and domestic early action (European Commission 2009i: section 3.1; Council 2009b). Beyond ideas stemming from internal legislation, the EU would also make other, often technical proposals in an effort to promote problem-solving in the global talks. In AWG debates, it made a range of advances to rationalize negotiations or to occupy the middle-ground between the G-77/China and the Umbrella Group. A key example of this would be its attempt to combine bottom up pledges for mitigation targets (preferred by Annex I parties) and compare them to top-down numbers proposed by IPCC science (favoured by the G-77/China).

What the analysis of the Union's influence attempts during the analysed time period ultimately boils down to is a reproduction of previous patterns, subjected to a few significant adaptations:

1. *an update of its traditional proactive leadership approach*: ambitious, early target proposals exploiting, in the Commission's view, "The Power of Example" (Runge-Metzger 2008) were now backed up by internal legislation and linked to conditionality, with an attempt to employ the 30% proposal as leverage over other industrialized countries and to appeal to others on the basis of scientific benchmarks (25–40% for developed countries, 15–30% deviation from BAU for major developing countries);
2. *continued rational argumentation aimed at problem-solving* through

---

[58] This strategy arguably bore fruit *vis-à-vis* countries of the Pacific (Australia, but also Japan), and – to a more limited extent – Russia. It requires further consideration in the context of the assessment of EU influence.

concrete technical and policy solutions, often targeted at reconciling other parties' diverging positions, and now based on a cost-effective, managerial rather than a policy and measures approach, which tried to use the ETS as an incentivizing (model) and/or coercive (club good) foreign policy tool; 3. *new forms of alliance-building* with bilateral partnerships on the basis of economic and technological incentives; 4. *broader diplomatic outreach* to disperse its messages more widely, corroborating its arguing strategy.

With this combination of a diplomatic and a managerial approach, the Union was arguably able to ensure a role in agenda-setting in the UN process. It did not, however, render itself independent of its long-standing overreliance on the quality and timing of proposals. As in the past, the leadership approach was based on the premise that the EU could posit its (ideal) position and that other major players were either already on the same wavelength regarding the desirability of a legally binding post-2012 agreement centred on market instruments, or prepared to follow the EU at a later stage. This belief was largely unwarranted. The story also reveals, however, other "blind spots" and trade-offs of the EU's influencing strategy. Given their potential significance when it comes to explaining its actual impact on these talks, they require closer inspection. The most evident downside of "frontloading" detailed propositions was a limited room for manoeuvre towards the end of the negotiations. Inflexibility during the final stages in the process was further increased through the high-levelization of talks. Their move from the negotiators' to the heads of state level meant that the EU had to coordinate ad hoc among non-experts, which impaired decision-making and outreach. This was especially the case because the coordination between negotiators and the foreign policy community and cabinets of heads of state was unprecedented and, thus, regularly insufficient (Interviews EU representatives 8, 10). The same problem occurred also with the increased overall EU outreach: information was not always shared coherently so that negotiators in the UN process and EU representatives in other processes did not consistently possess the same type and/or degree of insights into the ongoing talks at different levels (Interviews EU representatives 22, 8, 10).[59] Partially as a result of this, the EU also had difficulties in adapting its negotiation position and strategy to the evolving negotiation processes. Shortly before Copenhagen, issues that had been pending for months remained unresolved. On finance, the absence of a common position was disguised as a strategic move of "backloading" for the final bargaining session. This led to incoherence in its approach that arguably weakened its posi-

---

[59] The most obvious example of this was the Union's involvement in the Friends of the Chair meetings at COP 15, during which its key negotiators in the AWGs were more or less sidelined by members of the cabinets of the heads of state.

tion in Copenhagen (Interview EU representative 22; Observation notes Dec. 2009). The question of when and under what conditions to move to 30% reductions equally remained unanswered. Further, the EU continued to negotiate on the assumption that the outcome of COP 15 would be legally binding, although this had de facto been discarded by APEC.

These three examples also form a pattern: choosing to live with its own ambiguities, the EU left crucial decisions more than once to ad hoc coordination on the spot (e.g. the European Council informals at COP 15) or, oftentimes, to parallel meetings in Brussels (e.g. during the Poznan COP). Where an absence of fall-back positions and preparation of alternative scenarios to the one it envisaged further impaired its adaptive capacity, the parallelism of talks in the UN climate regime and in Brussels confused negotiation partners and EU negotiators alike.

The EU's Influence in the Post-2012 Climate Negotiations until 2009

To determine the EU's influence in the talks leading to the Copenhagen Accord, and thus answer research question 2 for this time period, it is now necessary to zoom in on the main turning points on the two analysed issues (mitigation targets, differentiation). The process-trace provided not only a comprehensive overview of the EU's influence attempts during this time period, but also a detailed reconstruction of the talks on the selected core issues. In this regard, it represents also the documentation of a failure: the climate negotiations between 2007 and 2009 inside and outside the UN had remained, for an exorbitantly long time, stuck in a positioning and loose formula-building phase. Within the UN, virtually no options were eliminated from draft texts before the final COP. A few trend-setting decisions were, however, taken by smaller groups of major players outside the UN and would later feed into the AWG talks. Nonetheless, crucial decisions were "backloaded" so that real detailing began only in Copenhagen. This implies (i) that the story of the post-2012 negotiations did not contain many evident turning points, and (ii) that the ones that can be identified primarily occurred outside UN fora. On the two analysed issues together, which – as core pillars of the climate regime – would become crunch issues of the talks, the narrative reveals a total of four major turning points. As during the Kyoto Protocol negotiations, a first turning point on both analysed issues could be observed early in the process at the agenda-setting COP 13/MOP 3 in Bali. A second turning point with regard to the issue of targets occurred then outside the UN process in July 2009 when the G-8 and the MEF met in L'Aquila and major emitters would rally behind the 2°C stabilization target. Thirdly, when negotiations steered towards failure because central players (US,

China, India) hesitated to disclose elements of their position (notably on mitigation), the APEC summit of mid-November 2009 clarified that the outcome of COP 15/MOP 5 would by no means be legally binding, clearing the way for last-minute target pledges and moving negotiations into the detailing phase. Finally, talks held at the highest political level in a small group of parties on the last official day of COP 15 represented arguably the major turning point regarding ultimate decisions on both key issues analysed here. Focusing on these four turning points allows for determining if, how and to what extent the EU was influential in the post-2012 talks until 2009. Central for the final outcome was certainly the last turning point, as it superseded previous decisions. Nonetheless, as the post-2012 talks did not end with COP 15, all turning points would potentially remain of importance.

Where the 1995 Berlin Mandate had set the agenda for talks leading to the Kyoto Protocol, the Bali Roadmap provided steering for the post-2012 negotiations. Yet, while the former had de facto eliminated already some options from the negotiations (notably regarding differentiation between groups of parties), the latter identified mostly broad topics ("building blocks") that talks were to focus on. Nonetheless, crucial concepts were pre-defined in Bali, which, in retrospect, set the post-2012 negotiations on certain rails which ran right into the Copenhagen Accord. For that reason, COP 13 marked a *first* significant turning point on both key issues. On the issue of differentiation, the Bali Roadmap arguably provided a first step toward overcoming the "wall" separating Annex I and (major) non-Annex I parties in terms of their respective obligations in the regime. The notion of "NAMAs" provided the necessary, sufficiently broad concept for major developing countries to accept further talks about own mitigation efforts and also brought the US, effectively disengaged since 2001, back to the negotiation table. While the EU arguably attained its (intermediate) aims of engaging both the US and the BASIC in systematic negotiations on their future contributions to the climate regime with the Roadmap (*goal attainment*), it had not been the first and most vocal player to demand the reform of the differentiation enshrined in the CBDR principle and in the Kyoto Protocol. The revision of the Annex I/non-Annex I divide had been a long-standing US and Umbrella Group request, which the EU had only – certainly purposively and very actively – articulated from the 2000s on (*purposive behaviour, interaction*). This observation already excludes the fulfilment of one other condition (absence of *temporal sequence*). A check of the negative pole of the concept (test on *absence of auto-causation*) then reveals that the EU did indeed not exert any significant influence on the item of differentiation as the central component of the Roadmap. Asked in counterfactual manner, did the EU effectively alter other key players' behaviour on this issue and at this point in time, notably that of the major emitters that had resisted the kick-off of a new negotiation process in the years before, i.e. essentially the US, China and India? And had the EU not been as active in calling for a post-2012 regime reform, including the overcoming of the Annex I/ non-Annex I divide, would these players have behaved differently? On the basis of the analysis of the general global context and domestic developments of the year 2007, the answer to these questions must be negative. In the US, the George W. Bush administration's previous positions on climate change had become domestically untenable under the impression of the fourth IPCC report. It was therefore prepared to subscribe to a new negotiation process, provided it could ensure the fulfilment of its long-standing *sine qua non* condition of engaging major emerging countries in serious talks about emission reduction efforts (Ochs 2008: 4). This interpretation is supported by the change of course the US delegation made during the final COP plenary which adopted the Bali Action Plan. After having refused to accept a formula that could have meant an unbalanced approach in favour of the developing countries at first, it later joined consensus when obtaining further assurances from the emerging economies. The latter, essentially China and India, as the discussion of their domestic circumstances revealed, were already experiencing problems resulting from climate change and beginning to undertake domestic actions *anyway*, i.e. whether there was an international regime reform process or not (Ochs 2008: 3). For them, it was therefore possible to bow to external pressure – from, ever since 1995, the US and the Umbrella Group, but later also from the EU – and agree to talks on something as diffuse as NAMAs. In conclusion, as the most vocal and pushy agenda-setter, the Union certainly contributed to an existing overall "wind of change" during the year 2007, helping to build political momentum for discussing the start of a new negotiation process in Bali, as argued in Chapter 4. Yet, EU influence on this crucial, but intermediate outcome regarding a possible re-interpretation of the CBDR principle was not discernible.

As far as the talks on the concrete issue of the magnitude of mitigation targets were concerned, a clear link was established under both tracks between future target discussions and/or NAMAs on the one hand and climate science as reported by the IPCC on the other. As seen, the reference to the IPCC scenarios (with the 25–40% emissions reduction range for industrialized countries associated with temperature rise of max. 2°C) was directly (AWG-KP) or indirectly (AWG-LCA) referred to in the documents composing the Bali Roadmap. On this item, the EU did exert influence. Firstly, it largely, albeit not completely (given the weakened outcome under the LCA), achieved its aims (*goal attainment*) (Interviews EU representatives 20, 27, 22; Observers 3, 19, 25). It had been the first major industrialized party to embrace the scientific findings of the IPCC in its positions, ever since the 1990s (*temporal sequence*). Its science-inspired positions were repeatedly defended through often proactive outreach activities in 2007, which had prepared grounds for what would later happen in Bali (*purposive behaviour, interaction*). On this basis, and in coalition with many developing countries (notably AOSIS), the Union successfully managed to alter the behaviour of previously very reluctant industrialized countries (Canada, Russia) under the Kyoto Protocol track (test on *absence of auto-causation*). In the debates under what would become the AWG-LCA, resistance to the EU and G-77/China proposals in Bali was more fierce, however, including also the US and Japan. The Union therefore had to concede a substantial weakening of the reference to the science in the final decision. Even as a footnote, the reference would, however, remain on the negotiation table for future talks. In sum, by exerting influence on the link between IPCC science and the mitigation target negotiations in both arenas, the EU contributed to heightening the chances that the post-2012 reform process would be guided by climate science, and impacted thus on the talks leading to the Copenhagen Accord. The diverse contexts under the two tracks illustrate, however, the limits to the EU's impact: in the face of US-led opposition, its influence was considerably restrained. Beyond these two issues, the overall approach of the Bali Roadmap, although it did not preclude any outcomes, largely reflected earlier EU ideas on the process expressed in the second half of 2007. In a similar vein, the Union would become an important agenda-setter on various items beyond those focused on here through its proactive, problem-solving approach in the post-2012 negotiations. Its positions would, however, not necessarily find their way into final decisions.

A *second* key turning point, exclusively with regard to the issue of emission reduction targets, occurred in mid-2009 outside the UN climate regime. While the Bali Roadmap had only (weakly) indicated reduction ranges for industrialized countries, talks on shared vision and mitigation under the AWG-LCA would remain inconclusive regarding clear emissions reductions paths and targets until the very end of the negotiation process. In July 2009, industrialized countries gathered for the G-8 summit in L'Acquila would then "recognise the broad scientific view that the increase in global average temperature above pre-industrial levels ought not to exceed 2°C" (G-8 2009b: point 65). They would be joined by other key global economies when the Major Economies Forum endorsed the same formula a day later. This outcome represented a major achievement for the EU (*goal attainment*), which had actively built its position and outreach around the 2°C target from the mid-1990s on, and further used it to justify its submissions on emission reduction ranges after the IPCC's FAR (*temporal sequence, purposive behaviour, interaction*). Would the other players have adopted this objective without the EU pushing for it (test on *absence of auto-causation*)? The answer is two-fold. At that stage in the process, after lengthy debates about a shared vision of the future regime in the AWG-LCA, many options had been pondered and other industrialized parties seemed – judged by their submissions to the UN – convinced *that* a numerical target would be beneficial to advance negotiations. The fact that the choice made by key Annex I parties in the G-8 fell on the 2°C formula (and not on another temperature, an objective expressed in ppm or in percentage cuts) can be attributed to the EU's repeated lobbying for this target. In the past, e.g. at the 2007 G-8 summit in Heiligendamm, the Union had already – via key member states – successfully convinced major industrialized players of similar targets (at that time, the US subscribed to 50% cuts by 2050, arguably also under the impression of the fourth IPCC report). To also persuade the emerging countries in L'Aquila, the dramaturgy of the two-part meeting certainly helped: the MEF met after the industrialized countries had endorsed the 2°C target. Together, the G-8 members could convince the emerging economies to also accept this aim. Indirectly, the EU was thus also the key influence-wielder when it came to changing the latter countries' previous preferences for not stating such a goal. If the adoption of the 2°C goal, which would later become part of the Copenhagen Accord, represented an instance of EU influence, the question regarding the significance of this achievement needs nonetheless to be raised. A temperature range by itself does not have much effect if it is not associated to either a stabilization target expressed in ppm or a reduction aim expressed in percentages. The IPCC's FAR provides for linkages between temperature scenarios and reduction prescriptions, and references to it were made under the Bali Roadmap. In that sense, the affirmation of the target could (have) become significant under the UN negotiation process (Mrusek 2009). Its importance would, ultimately, largely depend on a shared understanding of its nature, however. Here, UN talks testified to deep differences between those parties who interpreted the temperature goal as "aspirational" (e.g. India, Russia) and other parties, like the EU, who wanted to translate it into concrete targets and measures. Similar problems of interpretation would arise after the conclusion of the Copenhagen Accord, and are taken up in the discussion of turning point 4.

The *third* turning point in the post-2012 negotiations could be observed only a few weeks before the Copenhagen summit. At the final AWG preparatory meeting in Barcelona, major parties had refused to disclose key elements of their positions, let alone to engage in compromising. As a result, although not everyone would publicly acknowledge it, key players began to realize that COP 15 could not lead to a legally binding, comprehensive agreement. Nonetheless, the most vulnerable countries (AOSIS, LDCs), but also the EU, still appeared to desire keeping up the pressure for a more ambitious outcome. Their efforts would be rendered futile when negotiations were catapulted into a new phase at the APEC summit in mid-November 2009. The summit marked a major turning point in two regards: it not only pre-determined the form of the Copenhagen outcome (and thus of both the form of the core norm of a target and of the type of differentiation adopted with this outcome), but also triggered a process during which parties would finally release missing components of their positions, notably relating to the issue of emission reduction targets.

When the APEC country leaders gathered in Singapore, they jointly called for the adoption of a "politically binding" agreement at the Copenhagen summit, followed by a legally binding outcome at a later stage. The process analysis highlighted the role of the Danish Presidency on this occasion. Following Prime Minister Lokke Rasmussen's preference for searching for pragmatic solutions in small-scale meetings of major parties, the Danish proposals had been geared toward the preferences of, notably, the US, which was domestically unprepared for concluding a legally binding international agreement. Not surprisingly, US representatives immediately embraced the Danish suggestions, creating – together with the other major Pacific countries – a fait accompli that other players involved in the UN negotiations, including the EU, simply had to accept. From this point on, legally binding emissions reduction targets were thus de facto excluded for COP 15. It goes without saying that the Union could not and did not exert influence on this crucial decision. Before having the major Asia-Pacific players decide on its proposal, the Danish Presidency (after all an EU member state) had apparently not even consulted with the rest of the Union (Interviews EU representatives 22, 10). Although the formal proposal came from the Danish, US influence over all other parties to the UNFCCC at this turning point was undeniable. Since the beginning of the year 2009 (and in actual fact ever since the Bali COP), many US actors, from the administration over Congress to the ENGOs and think tanks (such as the renowned Pew Center for Climate Change) had argued that the US would domestically not be "ready" by COP 15. This message turned into a self-fulfilling prophecy of sorts. Yet, the US did not have to exert much pressure, finding natural coalition partners in the emerging economies, notably China and India, which were equally reluctant to agreeing to anything legally binding at that stage. While the APEC decision thus buried hopes for a legally binding outcome, it also had a positive effect on the negotiations: liberated from the "burden of bindingness" and extremely high public expectations, key parties began to reveal missing parts of their positions, especially on mitigation targets and finance. In this process of gradual disclosure of positions between mid-November and the Copenhagen summit, limited EU influence could be detected. Already beforehand, the Union's active promotion of its long-standing unilateral mitigation pledge and its clearly formulated expectations to other groups of countries (30% from developed countries, 15–30% deviation from BAU from developing countries) had provoked reactions. As seen in the story, the EU's model, including the management approach centred around emissions trading, had deliberately been followed by Australia in late 2008 (Wong 2008). Arguably, the improved –25% target proposal by Japan of August 2009 – which implied that the Union was not the most ambitious industrialized actor any more – was also facilitated by EU target promises. After the APEC summit, many other countries would disclose their targets: Brazil, South Korea, Russia and, finally, the US, China and India. While the APEC summit, but also traditional negotiation tactics of "backloading" had determined the timing of those pledges, the EU arguably influenced the magnitude of some of the proposals. On all occasions, the three conditions *purposive behaviour, temporal sequence and interaction* were fulfilled (the EU was the first to make proposals, used its "leading-by-example" approach explicitly as a foreign policy strategy), and its *goals partially attained* (the emerging countries, Japan and Russia lay within the emission reduction ranges prescribed by the EU). The question needs, however, again to be posed whether any of these countries would have acted in the same way without the Union's proactive, demanding approach? The answer needs nuancing: it is unlikely that Australia would have chosen for the exact same modalities (conditional offer, emissions trading) (Wong 2008; Interview EU representative 22). Moreover, Japan or Russia would most probably *not* have proposed targets of this magnitude in the absence of EU activity, with Russia even publicly acknowledging to having aligned itself with the Union (*absence of auto-causation*) (Spencer et al. 2010: 3). By contrast, it is difficult to assume any EU leverage on major developing country pledges: the EU's proposals for emerging country emissions reductions in the 15–30% range compared to business-as-usual may have served as a benchmark, but this cannot convincingly be affirmed. The fact that the EU had some limited impact on the decisions of major Annex I parties is not to insinuate that it was the only cause of these countries' decisions. Yet, it was certainly a key contributing factor in the overall mitigation debate. This reasoning could even be extended to the US, where more progressive forces used the Union (its approach, rather than level of ambition) as an example, at least in early 2009, as the brief discussion of the EU's outreach in Washington, DC, demonstrated (Egenhofer 2010: 167; Interviews EU representatives 30, 26, Observer 23). Nonetheless, like in the case of China and India, where the former's pledge had strongly informed the latter's, domestic factors and the reception of scientific knowledge were certainly predominant in the determination of the US position. In essence, the EU therefore exerted (albeit limited) influence over the important debate on the magnitude of the target pledges – with repercussions in some Annex I countries, but not so much in the US and the BASIC countries – which would later become unaltered pledges in the Copenhagen Accord. One cannot help but notice parallels between the post-2012 and the Kyoto Protocol negotiations in this regard.[60] While the EU had leverage over the magnitude of the emission targets adopted in 1997, it failed to influence the key structures of the Protocol. Twelve years later, limited influence over the target ambitions, facilitated by other factors (e.g. IPCC report), was observable, but the modalities of the agreement (e.g. regarding the binding character of those targets) seemed even more out of the Union's control, due in part to an increasing delocalisation of the negotiations out of the UN regime and into smaller bodies in the Asia-Pacific.

The *fourth*, and for the final outcome of the analysed negotiation process most significant turning point was detected at the end of COP 15/ MOP 5. No key issues had been resolved in the official meetings of the AWGs so decisions had to be taken at heads of state level. In the Friends of the Chair group, the text developed by the Danish Presidency since the autumn of 2009 was combined with formulas that had first emerged in the G-8 or MEF meetings and items from the AWGs' negotiating texts. The turning point concerned both analysed key issues, coupled to two other major topics, which were all inter-linked in a package deal involving differentiated targets/actions for Annex I and non-Annex I parties, MRV provisions and finance. Due to their intertwinement, EU influence on each of those items needs to be assessed.

On *mitigation*, the key overarching provision of the Copenhagen Accord was the recognition of the view that stabilization at 2°C should be pursued, which was coupled to a reference to the IPCC's FAR (para. 1, 2 CA).[61] For the rest, major indicative target ranges expressed in percentages (e.g. 50% for all countries by 2050, 80% for industrialized countries by 2050), dear to the EU, were dropped at the last minute on the insistence of China and the other BASIC countries. On this key pillar of the climate regime, the EU exerted thus only limited influence. The inclusion of the 2°C aim in the Accord represented a confirmation – and thus an indirect result – of the EU's successful influence-wielding at the G-8 and MEF in L'Aquila in July 2009 (turning point 2). Moreover, the Union had been responsible for the reference to the IPCC science, which had first

---

[60] Another striking parallel is that the pledges and/or policies adopted in the late 2000s were abandoned or scaled down at later stages by new governments in several countries (e.g. Australia, Japan).

[61] The Maldives, for AOSIS, and supported by developed countries, managed to retain a reference to the 1.5°C target in the paragraph of the CA that dealt with its review in 2015. This reference was retained in later negotiation documents at COPs 16 and 17.

been mentioned in the Bali Roadmap and was integrated in all negotiating texts ever since COP 13/MOP 3, on the EU's and the G-77/China's joint insistence (turning point 1). Yet, the weak formulations of the reference implied that the 2°C aim continued to be essentially unrelated to any emission reduction target ranges, which considerably relativized this achievement, at least in the short term. Finally, as observed for turning point 3, the EU also exerted limited influence on the magnitude of some other countries' mitigation pledges, which would be confirmed by those countries when they made their submissions under the Accord in January 2010. Obviously, the fact that non-delivery on those pledges could never be sanctioned under the Accord considerably weakened the significance of this successful instance of influence-wielding. In essence, if the Accord reflected some, albeit limited degree of EU influence over the key norm of the regime (emission reduction target), the Union did not achieve this outcome at Copenhagen, but through previous decisions taken in non-UN bodies that the world's leaders had referred back to when making the deal in the Danish capital.

Differentiation between Annex I and non-Annex I countries de facto remained part of the climate regime with the Copenhagen Accord, both in the types of mitigation commitments and in their verification. Nonetheless, all major emitters were reunited under one single agreement. With this, the EU partially reached one of its aims, but not as a result of its own influence. As argued for turning point 1, it had been clear since its failed Kyoto Protocol ratification that the US would not have committed to any, even voluntary measures without the major developing countries doing their share, and vice-versa. The final outcome represented thus a compromise between the US and the BASIC countries, notably China. It was this compromise that also led to the elimination of the target references for mid-century, on the insistence of the emerging countries, and against the explicit will of the EU.

Closely linked to this, the issue of *finance*, which had provided the enabling condition for many developing countries to accept own mitigation actions, marked probably the strongest instance of EU influence at this final turning point. The notion of "fast-start finance" had emerged from EU proposals put forward by the Commission in early 2009, as highlighted in the narrative, and later taken over in the Danish proposals and their consultations, for instance with the Commonwealth countries right before Copenhagen. Also, the magnitude of these short-term financial provisions adopted in the Copenhagen Accord had been influenced by EU positions, as the Union had been the first to make a public, quantified pledge during the first week of COP 15, with the explicit aim of convincing the developing countries to join a larger agreement (*purposive behaviour, temporal sequence, interaction*). On the magnitude of long-term finance, the EU had taken up and increased earlier proposals of 100bn USD, discussed in the first half of 2009 in the MEF, by suggesting to raise 100bn € by 2020 (*purposive behaviour, temporal sequence*). The positive response of the African Group to the proposals of 10bn USD per year in 2010–2012 and 100bn USD by 2020 during the last days of COP 15, enabled through a close coordination between France, the UK and the African Group Chair Ethiopia (*interaction*), would clear the way for an agreement of the majority of smaller and poorer G-77/China members to the final agreement. With its proposals and proactive behaviour on both short- and long-term finance, the EU could have thus exerted influence over these items. The fact that the numbers it proposed would also find their way into the Accord regarding both time horizons further implies that the Union reached its aims (*goal attainment*). The question needs to be posed, however, whether other actors would have behaved differently without the EU's interventions on this item (test on *absence of auto-causation*). The fact *that* money would become an important factor for reaching agreement was certainly acknowledged by all major players. Yet, the precise amount of the finance provisions ultimately adopted with the Copenhagen Accord must at least in part be attributed to the Union's activities: it changed the previous behaviour of the US and other Annex I countries (who had remained mostly silent on this agenda item until the last days of COP 15) as well as of the African Group (as the most vocal solicitor of funds among the G-77/China, with previous demands far above those finally agreed to). All in all, the EU certainly did exert influence over the concept of short-term financing and the magnitude of both short- and long-term finance. Yet, finance was not an end in itself for the Union. It wanted to use it as leverage and "sweetener" for developing countries to support its desired comprehensive, legally binding outcome. This did not work out: while the EU attained its aim with regard to the finance issue in itself, it received no meaningful commitments in return. Ironically, it actually strongly contributed to "buying" developing countries into an agreement that it had itself difficulties accepting.

In conclusion, the EU's overall influence on the key pillars of the Copenhagen Accord was low. While a leverage over the 2°C target is non-negligible, the immediate "'success' of having a 2°C target referenced in the Accord seems somewhat irrelevant" considering that with the reduction pledges in early 2010, "the world is headed for a global warming of 3.5°C by 2100" (Curtin 2010: 6). In this perspective, the EU's limited leverage over the magnitude of some countries' targets equally appears as insufficient. Moreover, even if its influence over the finance provisions was probably indispensable for getting to an agreement, it did not yield impact on the main components of the regime. This latter observation leads to an important consideration, however, namely the question of whether the EU was actually crucial to getting an agreement *at all*. The number of instances on which the Union had set the agenda during this negotiation process might point in this direction. The EU was the most proactive and demanding industrialized actor. It invested considerable resources into this negotiation process and, from a very broad perspective, also attained its aims: an agreement which involves the US and the major developing countries into (some form of) global mitigation efforts (*purposive behaviour, interaction, temporal sequence, goal attainment*) (Interviews EU representatives 8, 10, 16). In this sense, the EU could have indeed exerted influence over the overall outcome, albeit not its specifics. But even this is uncertain. A counterfactual analysis reveals that many other factors may have played a role: climate science and media attention, the change in US government, with an administration that had become more interested in a global agreement on this issue, an overture of the major developing countries responding to the US. While the EU contributed to the fact that there was "something to agree to" in Copenhagen, claiming that without the Union there would have been no such agreement would be overstretching the interpretation of the evidence (*test on absence of auto-causation: negative*).

Synthesizing the findings of the preceding section allows for determining the Union's *overall influence* on the development of the global climate regime during the analysed period. Regarding *emission reduction targets*, the analysis revealed that the EU had limited leverage over the magnitude of targets in a few countries, but not in others. Further, it had very little influence on the approach taken to the targets: its own preferences of science-oriented, relatively high legally binding targets arguably provided a benchmark for many parties in the negotiations, but was not retained in the Copenhagen Accord. The only substantial item of importance in terms of a key pillar of the regime on which the EU did exert influence was the 2°C target, mentioned in the Accord, but with the discussed limits. Concerning the issue of *differentiation*, no influence was discerned. Finally, one can claim that the EU contributed to the *overall result* of concluding an(y) agreement, not so much through its general approach, but through rallying developing countries behind the aim of sealing a deal through its finance proposals. In that sense, the EU co-prepared the soil for an agreement. This would never have seen the light of day without many factors being reunited, however, above all new scientific findings, but also internal developments, notably in the US and some other key countries. As a result, and despite instances of EU influence on one key pillar of the regime and indications of impact on agenda-setting during the early stages of the talks (Purvis/Stevenson 2010), the Union's overall influence has to be evaluated as very low because it did not attain its aims on the assessed items (legally binding QELROs and

NAMAs) and in general (one single legal instrument based on the KP) (*very limited extent of goal-attainment*). Moreover, the outcome was as voluntary as could be imagined (*very low degree of durability*). The assessment of EU influence as very low at the critical juncture of COP 15 is confirmed not only by many observers' commentaries on the post-2012 negotiations (Egenhofer/Georgiev 2009; Curtin 2010; Purvis/Stevenson 2010; Spencer *et al.* 2010), but also by reputation analysis with EU and non-EU negotiators at different levels. Commission President Barroso's view of the talks was that "it is true that others were much more influential when it was about reducing the ambitions" (EU Press Conference 19 Dec. 2009). The Swedish Environment Minister called the outcome a "disaster" for the EU (Pawlak 2009) and the AWG-KP lead negotiator from the Commission referred to the EU as the "unpopular class goody-goody" of the climate talks (Agence Europe 2010: 24). Other negotiators highlighted EU influence on agenda-setting and the fact that it did attain its minimum objectives (a basic, potentially science-guided agreement with all major emitters on board), but acknowledged that the outcome fell way short of its broader expectations, indicating a clear lack of influence (Interviews EU representatives 22, 8, 10). The same tenor came from outside of Europe (Reuters 2010; Interview US representative 2).

Having determined the extent of the EU's influence, further analytical operations can help to specify the *type of influence* exerted by the Union regarding both the time horizon and the underlying logic of its impact. Other than during the Kyoto Protocol negotiations, the EU's influence attempts were targeted at the medium to long term. Relying on a science-based argumentation strategy coupled to economic incentives (ETS, partnerships, finance proposals) and timid allusions to coercive instruments (ETS, border adjustment taxes), the Union expressed a clear vision of the future shape of the climate regime, specifically with regard to its key components: ambitious mid- and long-term targets and pathways for getting to those target (a regulatory approach centred on market instruments). Although its position contained more potential "give and take" through its economic instruments than in previous negotiation rounds, the continued absence of bottom lines rendered bargaining very difficult. It was further complicated by the fact that clashes of positions necessitating bargaining were held back collectively by the parties until the very last moment. At this crucial moment in Copenhagen, as the narrative indicated, the Union was effectively sidelined by other major parties (US, the BASIC group). Its offers lined up for bargaining (essentially the finance proposals and the 30% reduction offer) were gratefully accepted (finance) or neglected (30% target) by other parties, but stood in no real relation to a concrete service in return to the EU. By consequence, one cannot conclude that the Union's influence was bargaining-based.

Rather, on the few items on which the EU did exert influence (the link between the objective of Art. 2 UNFCCC and the 2°C target, magnitude of targets in some countries, finance), this influence was, if anything, argumentation-based. Influence attempts involving arguing based on the science or economic rationales had convinced other parties to adhere to what originally were ideas promoted by the EU. Further, although its influence was low, it was, contrary to the Kyoto process, potentially enduring *and* surely incidental. The EU had the described limited leverage over the potentially – in the long run – significant 2°C objective, which might form the nucleus of a future science-based regime. On the finance proposals, its influence was only incidental, contributing to start-up finance and the conclusion of an "immediately operational" deal. Altogether, it did not, however, as it had desired, achieve an enduring reform of the climate regime on the whole.

By way of comparison, and to set the EU's influence into a broader context, it is interesting to briefly reflect on the other major players' influence in this negotiation process, focusing on the US and the emerging countries. The US had already determined the regulatory approach of the Kyoto Protocol, which dominates debates until the present. Reproducing past patterns, it also exerted influence over the voluntary "pledge and review" approach of the Copenhagen Accord. Right down to the title of this document ("Accord"), it effectively created structures (or prevented the creation of alternative structures) by uploading its own preferred policy approach to the global level. This approach had already been inherent in the proposal for an "implementing agreement" of May 2009 and reflected exclusively domestic institutional necessities. Besides attaining the objective of having loose medium-term commitments anchored only in domestic rather than in international law, it also achieved its long-standing aim of engaging major non-Annex I countries in (yet very general) mitigation efforts. At the same time, its influence was limited – by the opposition of the BASIC group – when it had to sacrifice its desired long-term targets for 2050 in the Copenhagen Accord. The major developing countries, especially China and India, exerted influence on the issue of differentiated responsibilities. Using a legalistic strategy based on arguments centred on Article 3 UNFCCC and the Kyoto Protocol, and successfully engaging partners from the G-77/China in their foreign policy strategy, they managed to employ structures they had created in the 1990s to retain a substantial differentiation between Annex I and non-Annex I parties regarding mitigation efforts and their measurement. On issues that opposed them to the US, other Annex I countries and parts of their own coalition, these countries further demonstrated their much-increased clout, rooted both in their growing standing in world politics and in their current and future GHG emissions profiles. This was most evident for the decisions on the mid-term targets. Fearing to be bound by such goals a few decades later, (especially) China attempted to influence talks by demanding the deletion of the relevant paragraph from the final draft of the Accord. In so doing, it effectively altered the behaviour of industrialized countries and betrayed the interests of many players in the G-77.

Explaining the EU's Influence during the Period 2007 to 2009

To facilitate the explanation of EU influence, two sets of assumptions were made when designing the analytical framework for this study: causal mechanisms underlying the exercise of influence (bargaining, arguing), and potential variables that may qualify as conditions enabling or restraining EU influence were identified. Two analytical strategies can be employed that capitalize on these assumptions: pattern-matching and explanation-building (see Chapter 1).

When it comes to pattern-matching, a first observation is that during the decisive final day at COP 15, the post-2012 negotiations conducted in the Friends of the Chair group undoubtedly involved bargaining. The trade-offs in the agreement between the US and the BASIC countries or the finance proposals to get developing countries to approve the deal were indicators of a collective search for compromises to "seal a deal" involving concessions by all major parties. The narrative had shown that the EU was physically present during these instances, but not effectively engaged as a player in this bargaining session. There was no real give-and-take involving the Union: while it certainly gave – it gave up on a legally binding agreement, it gave up on its target proposals, and it gave money – it received nothing in return. The causal mechanism "bargaining" was thus clearly not triggered by the EU's foreign policy performance during the period 2007 to 2009.

The rare occasions on which the EU was actually influential resulted from "arguing", i.e. "non-manipulative reason-giving" (Keohane 2001: 10; Kleine/Risse 2005: 9), on the basis of rational, often science-based proposals. On the issue of targets, the EU's 2°C, previously not part of other major parties' preferences or positions, was taken over by these latter – even if they received nothing in return for it. Umbrella Group and BASIC countries alike seemed simply convinced, in light of the science and the overall negotiation context, that the adoption of such an aim was useful as a yardstick. On finance, notably in the short term, the rationale of EU proposals was – given the overall negotiation context – equally accepted as reasonable. Finally, the Union's leverage over countries' targets in these negotiations was also the result of successful arguing for the feasibility of fairly ambitious reduction targets paired to a management approach. It incited, for instance, Australia to change its preferences in the direction of what it perceived as the best argument.

These accounts of the logic of EU influence raise a number of questions as to the scope conditions that did or did not trigger the causal mechanisms. On the one hand, the central question is certainly why the Union could never successfully bargain, especially during the decisive final stages of the talks. On the other hand, the scope conditions for arguing need to be explored, notably on the issue of the 2°C target. To do so, both pre-specified explanatory factors of EU influence and new factors that emerged during the process analysis are considered.

As highlighted in the narrative analysis, significant *external events and trends* during this time period comprised above all the emergence of the financial-economic crisis in 2008, an overall trend in global (climate) politics toward the creation of smaller fora (e.g. G-20), and the growing importance of the emerging economies. While the financial crisis became a topic in the climate negotiations of late 2008 and early 2009, no indications were found that it permanently impaired talks beyond a non-negligible loss of momentum at that time. It did not modify the EU's positions or strategy in any major way, despite the fact that more reluctant member states invoked it as a justification for limiting EU climate measures both during the debates of the climate and energy package and at later stages. The shift into smaller fora and the increased role of major emerging countries would, by contrast, have major consequences for the Union. Changing GHG emission profiles of the major emitters meant that the EU was already – and more obviously in a medium-term perspective – becoming less significant to overall global mitigation efforts. At the same time, it now had to face – in addition to the US and some Umbrella Group members – even more players whose principal interest it was to sustain economic growth and who perceived this objective to stand in contradiction to ambitious mitigation policies. Especially in smaller circles outside the UN, it would become more difficult for the Union to promote its more progressive positions. All in all, these events and trends determined EU influence moderately to strongly negatively, as further discussed below.

Climate science, by contrast, had the expected positive impact on the Union's influence. As it frequently used the science in its argumentation strategy, the fact that the findings of the fourth IPCC report had been more compelling and widely accepted facilitated arguing and contributed to successful influence attempts regarding, above all, the 2°C target. The importance of climate science notwithstanding, differences in the reception of new findings remained. The debates on science references under the two tracks during the Bali COP/MOP illustrated the continued weariness of notably the US to let science and the precautionary principle prevail over conflicting economic interests. When the US government in 2009 embraced the science to a larger extent, it still interpreted it differently than the EU, stressing mitigation in the long run (2050), despite the IPCC's emphasis on the importance of medium-term peak years. Nonetheless, advances in climate science constituted an undeniable enabling factor of EU influence during the period up to 2009.

Before tackling the multilateral regime dynamics, the *domestic contexts in some of the key countries* need to be recapitulated. Most importantly, the story revealed the central role of the US domestic legislative process, which had evident repercussions on global climate politics: the delay of the UN negotiations was in large part the result of internally diverging preferences about climate change between a more progressive administration and the opposition in Congress, unresolved due to an unfavourable political system. In Japan, the situation was quite different: less geared toward the US than in past negotiations, the new social-liberal government demonstrated in August 2009 that it was possible to cut through internal preference divergences and adopt ambitious mitigation targets. China and India had not only similar positions centred on equity and sovereignty concerns, but these had also resulted from similar internal institutional contexts and debates. In both countries, many agencies and/or ministries were involved in the definition of the position, making it difficult to deviate from long-lived defensive, often anti-Western stances. The involvement of the highest political level in talks at COP 15 reinforced this inflexibility. The uncertainty about positions of key countries during the final stages of the talks rendered debates very complicated and also had a restraining effect for the EU's capacity to exert influence during the endgame in Copenhagen.

These individual domestic contexts obviously had repercussions at the aggregate level of *regime dynamics*. The story aptly illustrated the differences in preferences among key parties/coalitions. With regard to the issue of emission targets, there was a clear cleavage between the EU, Japan (at later stages in the process) and many developing countries on the one hand, and the US, other Umbrella Group members, and the emerging countries on the other hand. While the former insisted on legally binding, ambitious, science-based targets, the latter preferred by and large a bottom up approach. On differentiation, the cleavage opposed developed and (major) developing countries. On other important items, like the form of the outcome, matters were even more complex due to the two UN negotiation strands: the EU, with its preference for a single outcome incorporating elements of the Kyoto Protocol occupied the middle ground between the developing countries, who argued for the continuation of the Kyoto Protocol and another agreement under the LCA track, and the US, who preferred a single LCA outcome. In all cases, the *interest and belief constellations* were heterogeneous and the EU defended either a minority or an outlier position. Regarding the predominant *logic of action* of these talks, a potential explanatory factor that emerged from the analyses of the time periods 1995–1997 and 2007–2009, parties' strategies and initial absence of clear stances resulted in the long persistence in a positioning mode. Many were playing the "waiting game" until the autumn of 2009, when pragmatism and lowered ambitions – steeped in distrust among the large emitters – determined the nature of the talks. During the long period of assessment, position-building and limited engagement by the big players, opportunities emerged for the EU to set the agenda and exert influence. On mitigation, this seemingly worked fairly well *vis-à-vis* Australia, and it had repercussions for the later stages of the talks when other countries pledged their targets. When the overall logic of action turned into bargaining during the decisive stages of the talks, the Union was, however, marginalized, since no pivotal player was interested in what it had to offer. This absence of interest in the EU's offers during the last stages of talks was certainly in part the result of own miscalculations, but it must also be explained by the *relative power of other actors* in the global climate negotiations. Since the 1990s, when developing country mitigation was effectively excluded from the regime and the Union was still among the key emitters, the relative power of others has gradually grown. While the story demonstrated the continued power, and capacity to convert it into influence, of the US, it also showed the increased potential of the BASIC countries. As a result, essentially the US and China could behave strategically on the basis of take-it-or-leave-it offers in the negotiations. The EU, by contrast, with ever lower emissions, had to struggle to remain among the most powerful players since the 2000s, and clearly did not succeed in doing so during the analysed period. Altogether, the regime dynamics, especially the power and difficult bargaining positions of other actors, added to the awkward and somewhat inappropriate mediation attempts by the Danish COP Presidency, provide partial explanations for the EU's absence of influence during the decisive stages of the talks. They can also convincingly account for the outcome of the post-2012 talks. The few examples of EU influence can be explained by the negotiation context during early stages of the talks and the reliance on science, both opening up windows of opportunity.

To complement this more structuralist view, the EU's own activities need to be injected into the overall picture. Apparently, the Union was itself not well-prepared for the final bargain. By contrast, its leadership and problem-solving strategy predisposed it for exerting influence through arguing. Regarding the EU's *actor capacity*, the internal conditions for position-building had become even more complex over time. Heterogeneous beliefs and preferences of the 27 member states and the Commission led to proposals such as the 20/30% reduction offers, essentially a compromise agreement, and to quasi-permanent strategic discussions on the more political issues, resulting in some significant instances of incoherence (e.g. the backloading of finance proposals). A potential institutional remedy to these problems had been introduced in 2004 with the system of lead negotiators and issue leaders under the Environment Council. While this did seem to improve the Union's internal coordination and external representation to some extent, producing much more refined positions on the more technical issues and an overall broader outreach strategy, the evolution of the external arena constantly provided new challenges. The diversification of fora and of bilateral exchanges led to new transaction costs: information had to be shared among EU negotiators, assessed and fed back into the Union's decision-making machinery. Yet, with only a preparatory and representative function, the new lead negotiators system could not by itself alter the negotiation positions of the EU. This implied that it was still necessary for member states to coordinate ad hoc and at length on the spot during many UN sessions. As in the past, a trade-off between internal coordination and foreign policy implementation could thus be detected. Despite continued problems of this type, both EU and non-EU negotiators and observers perceived the Union, by and large, as a unified actor for most of the talks (Interviews EU representatives 9, 20, 25, non-EU representative 18, Observers 3, 19, 25). This unity was, however, hampered by the implication of heads of state during final informal talks at COP 15, when the large members dominated the scene alongside the Presidency. Although they reportedly also tried to distribute tasks in the Friends of the Chair group, the task-sharing system that functioned at the level of negotiators would not work well at the highest political level. Despite its undeniable actor capacity during this time period, the EU's adaptive capacities came thus under strain toward the end of 2009, especially when there was a rapid succession of negotiations functioning according to a bargaining mode. The limits to its adaptive capacity reflected and reinforced a lack of coherence and strategic thinking, producing thus less positive preconditions for exerting influence.

As a final explanatory factor to consider, the EU conceived its *foreign policy implementation* much broader and, at first sight, more coherent than in the past. It combined a proactive problem-solving approach relying on diplomatic tools with economic instruments and broader outreach activities. Yet, while it was attempting to build support for its preferences through this form of outreach, its positions remained very static, qualitatively seemingly unaltered by the information it assembled in all its exchanges with third countries. This suggests that it was not so much the actually fairly developed efficiency of its foreign policy implementation (the EU did the things it wanted to in a correct manner) or a lack of outreach that limited its influence, but rather the lower quality and effectiveness of its overall approach. As seen from the narratives of this and previous negotiation periods, the EU's position regularly strongly determined its strategy, to a point that "making proactive proposals", i.e. having a front-runner position, was almost considered equivalent to possessing a strategy. Yet, a strategy for a dynamic negotiation process has to be – per definition – flexible, something that can hardly be said of the EU's positions. The Union's (absence of) influence during this period must thus partially be explained by the strategy underlying its foreign policy implementation. Proactive positioning did yield limited influence in early stages of the process when other parties were still building their own positions, as discussed above. Yet, several examples from the narrative point to an inadequate conception of this strategy when it came to anticipating the final stages of the talks. As the post-2012 negotiations evolved, and although clashes of ideas failed to appear, the Union's strategy remained as if this inactivity did not challenge its leading-by-example approach based on mobilizing others' support for its own policies. What is more, it negotiated the final months on the assumption that the outcome of talks would be legally binding even after this had effectively been discarded by the November 2009 APEC summit. In the endgame, it had nothing to contribute to the bargain: no one was interested in its offer to move to 30% reductions; no party gave anything in return for its finance offers; and it had no coalition partner to put pressure on major players during the final hours of COP 15. All these examples can only be explained by the fact that the Union must have assumed that other parties would be prepared to follow its example and step up their efforts at the very end of the negotiations. In retrospect, this proved wrong: the EU's strategy and foreign policy implementation seemed to be built on incorrect premises, insufficiently geared toward the arguably very intricate and evolving negotiation context. Together with the latter, a deficient strategy can thus partially also explain the lack of EU influence in Copenhagen. Obviously, foreign policy strategy, position and actor capacity are closely linked. One could therefore also assume that an actor capable of defining a proactive position is able to define and adapt a strategy for defending that position. If this assumption is accepted as valid, the Union's underperformance in the post-2012 negotiations is not so much the result of problems related to its actor capacity and an inability to forge common positions, but of wrong choices the 27+1 took regarding the positions they did adopt. The EU's strategy had been updated in the period 2005–2007 on the basis of *past* experiences, especially with the Kyoto Protocol negotiations, and *past* global transformations, but had clearly not anticipated potential future constellations.

CHAPTER 6

Gradually "Back on Track" (2010–2012)
EU Influence on the Resumed Post-2012 Global Climate Negotiations

The aftermath of one of the most high-level and media-covered global summits in history resembled a hangover of sorts for many: not only had "Hopenhagen" not delivered on what it had been mandated to bring about, it had also led to a great deal of distrust among major parties – both between industrialized players themselves and between actors from the developed and the developing world – undermining parties' and peoples' confidence in the multilateral process. It would take some time to adjust to this new situation, re-build trust and re-ignite the far from finished global negotiations. Also within the EU, the "Copenhagen disaster", as the Swedish Environment Minister had called it, sparked debates about changes to its foreign climate policy. In both cases, however, the controversies had limited effects on actual positions, actors' behaviour and final outcomes. For that reason, and since the period 2010 to 2012 was more about "saving" the multilateral negotiation process by transitioning toward a novel negotiation phase than about any substantial developments of the climate regime regarding the two issues of targets and inclusiveness, this chapter highlights the major developments in the EU and in the global climate negotiations after the Copenhagen Accord. They resulted in the incorporation of that Accord into the UN process via the "Cancun Agreements" at COP 16/MOP 6, the adoption of the "Durban Package" on a new negotiation process at COP 17/MOP 7 and the endorsement of a general work plan for this process at COP 18/MOP 8 in Doha ("Doha Gateway").

The Context: Major Developments in Global Politics and Climate Science

Where the Copenhagen summit had complicated relations between players in global climate politics, the overall context in which actors operated throughout the years 2010 to 2012 had not significantly altered as compared to the period 2007 to 2009, with the exception of an observable decrease in the sense of urgency regarding climate change. In terms of *external events*, ever since 2008, global, EU and national leaders had been attempting to attenuate the shocks of the economic and financial crises, which would hit the Eurozone particularly hard, diverting attention from climate change and impacting on the finance discussions in global climate talks (Reuters 2011a). What is more, following a devastating earthquake and tsunami on 11 March 2011, Japan witnessed a major accident at its Fukushima nuclear power plant, resulting in nuclear meltdowns and releases of radioactive material contaminating the surroundings of the site (Goldenberg/Elder 2011). Among the political repercussions of this disaster was a reconsideration of nuclear power in Japan, which would influence the country's position in global climate talks (Meltzer 2011: 6). Other countries equally re-assessed their energy mixes. While Germany enacted the accelerated phasing out of its nuclear power plants, for instance, France decided, after a reflection period, that it fully trusted the technology. Finally, compared to the proliferation of fora witnessed in the mid-2000s, this period was characterized by a higher degree of institutional stability as well as a comparatively lower degree of involvement of the higher political echelons. The *issue of climate change* itself was hotly debated following the reactions to the "Climategate" scandal, evoked in the introduction to this book. Sceptics used the aftermath of the Copenhagen summit to attack scientists whose research had been cited in the IPCC's FAR, apparently containing mistakes. Though it played no major role in the negotiations, high degrees of media coverage of this scandal in early 2010 led to negative publicity for climate change. Altogether the post-COP 15 cooldown of relations between key players, the related public perceptions that the Copenhagen summit had failed, the economic crisis management as well as the discreditation of climate science contributed to a shift in the attention of policy-makers, the media and the general public away from the issue of climate change for most of the period discussed here. And this despite the fact that the World Energy Outlook 2011 of the International Energy Agency noted that the possibilities to decisively act against a temperature increase of more than 2°C were becoming slimmer in the absence of swift and comprehensive actions (IEA 2011).

Key Actors in the Global Climate Regime and their Positions

Just like the overall context, the main positions of key actors in the UN climate regime remained rather stable from 2010 to 2012.

In the **Umbrella Group**, previous positions and trends were reinforced, notably in the *United States*. In July 2010, the Obama administration's ambition of passing climate legislation through Congress was disappointed when the "Clean Energy and Security Act" introduced in 2009 was not pursued by the Democratic majority in the Senate (Daly 2010). The Republican landslide win in the mid-term elections, in which they gained 63 extra seats in the House and six in the Senate, would then imply a very narrow marge for manoeuvre on internal and external climate policies for the US government prior to (and also after) the presidential elections of 2012. As substantive changes of its position on the issues of targets and responsibilities thus became de facto impossible, the US would adjust its strategy in the course of 2010 and 2011, privileging the search for global solutions outside the realm of the UN, more particularly via opening up a new strand of talks under the Copenhagen Accord in 2010 (Phillips 2010; Bowering 2011; Interview US representative 2). When this was resisted through the Cancun Agreements, the US indicated, for the further procedure, that it was only willing to agree to a comprehensive package (ICTSD 2010), and would resist "cherry pick[ing] some issues and not find[ing] the balance" the talks needed (Watts 2010; de la Vina/Ang 2010). The situation in *Japan* was initially equally similar to the pre-Copenhagen period. As of 2010, the government was "committed to continue its ambitious emission reduction efforts beyond 2012", but unwilling to subscribe to a second commitment period of the Kyoto Protocol, which it considered as "neither fair nor effective" (MOFA 2010). In 2011, following the nuclear disaster and subsequent debate about nuclear power, the commitment came under pressure, since "without an expansion of nuclear power (...) Japan's Copenhagen Accord pledge to reduce its emission by 25 percent below 1990 levels by 2020 [would be] difficult to achieve" (Meltzer 2011: 6). Accordingly, the strategy after 2011 was to keep a low profile short of backtracking, while continuing to oppose a second Kyoto Protocol commitment period.[1] In similar vein, *Canada, Australia and Russia* reinforced previous positions, voicing concerns against a second commitment period of the Kyoto Protocol, and stalling in domestic mitigation efforts. While Canada openly pondered the withdrawal from the Protocol in 2010 and would – in late 2011 – de facto leave the treaty, the Australian government had to temporarily shelve the idea of starting an EU-style cap-and-trade system in 2010 (Reuters 2011b; Young 2010). 2011 became then a year of newly intensified efforts at adopting climate measures in Canberra when a carbon tax transitioning into an emissions trading system was put into place (Johnston/Hudson 2011).[2]

When it comes to the **G-77/China**, the overall positions and cleavages on the key issues discussed here had remained stable, with a few notable

---

[1] The backtracking followed in 2013 when Japan announced to reduce its emissions by 3.8% compared to 2005 by 2020, which would actually result in an increase from 1990 levels.

[2] At the time of writing, this measure was however expected to be repealed, as of 2014, by the subsequent government.

exceptions, however, concerning strategies rather than positions. On the one hand, the big *emerging countries*, China and India, recognized a need to collaborate more intensely around the issue of climate and energy politics, bilaterally and in concert with Brazil and South Africa as part of the BASIC group (Hallding *et al.* 2011). For China, which had taken much of the blame for the failure of the Copenhagen summit (Zhang 2010), but also for India, this would soon be regarded as a vital strategic interest (Michaelowa/ Michaelowa 2011: 22). This loose coalition converged around a certain number of "broad principles, but [could] not settle the finer details needed to translate these into concrete contributions for the international negotiating process" (Hallding *et al.* 2011: 92). BASIC was thus not a "tight negotiating bloc": South Africa (as designated host of COP 17/MOP 7) and Brazil (as host of the Rio+20 summit in 2012), for instance, frequently showed a less aggressive negotiation stance *vis-à-vis* industrialized parties than China or India (Hallding *et al.* 2011: 92–95). On the other hand, *AOSIS* and the *LDCs* seemed to increasingly observe instances of desolidarization of the big emerging countries, which prompted them to look for other ways to find solutions to the climate crisis, like teaming up with the more progressive Annex I parties in a newly created forum, the Cartagena Group/Dialogue, which brought together countries from the G-77 (notably AOSIS and LDCs), the Umbrella Group (Australia, New Zealand) and the EU as well as the European Commission in the post-Copenhagen summit era. What continued to unite the entire G-77/China, however, was its joint support for the UN system. Initially, this meant efforts to support the idea that the Copenhagen Accord should become part of the UN negotiation framework (Phillips 2010). After attaining this minimum common objective at Cancun, the G-77/China would find less common ground, given that – for the BASIC countries – fighting off own legally binding measures was essential, while the LDCs and AOSIS desired fast and profound mitigation advances. This cleavage would show especially at the Durban summit and thereafter.

In between these two negotiating blocs, the **European Union** found itself – alongside a few other players like Norway or Mexico, the host of COP 16/MOP 6 – in a delicate position. In 2010, it took some time to digest what many of its climate negotiators had initially perceived as their marginalization at COP 15, paired to a failure of the global negotiations. Gradually, a set of changes were implemented, affecting its actor capacity, negotiation positions and strategy.

EU actor capacity primarily depends on the existence of legal competences for climate change activities, of decision-making and coordination procedures as well as of external representation arrangements. Major changes after the Copenhagen summit resulted primarily from the modified legal-institutional framework for European external action (on climate change) and were directly related to the entry into force of the

Lisbon Treaty on 1 December 2009. Treaty modifications regarding the issue of climate change represented mainly an institutionalization of previous practices, with Article 191 TFEU stating that "Union policy on the environment shall contribute to (...) the following objectives: (...) promoting measures at international level to deal with regional or worldwide environmental problems, and in particular combating climate change", which remained a shared competence (Lieb/Maurer 2009: 89). Major Lisbon Treaty provisions regarding foreign policy were related to the creation of the post of a High Representative of the Union for Foreign Affairs and Security Policy[3] (HR) (Art. 18 TEU) and of the European External Action Service (EEAS) (Art. 27 TEU). These institutions and the granting of legal personality to the EU (Art. 47 TEU) were accompanied by an introduction of new rules on who was to do what in the conduct of EU foreign policy, with repercussions for the functioning of its external representation. In general terms, while the "Commission shall ensure the Union's external representation" outside the CFSP, the HR was primarily responsible for CFSP (Articles 17.1 and 18.2 TEU). Nonetheless, in a programmatic speech at the launch event of the EEAS, the first HR – the UK's Catherine Ashton – expressed her desire that climate change would also become one of the priorities for the EU diplomats' future work (Rettmann 2010). This ambiguity about role definitions in EU foreign climate policy could also be found back in the treaty provisions on the conduct of international negotiations leading to agreements: Article 218.3 TFEU stipulated that the "Commission, or the [HR] where the agreement envisaged relates exclusively or principally to the [CFSP], shall submit recommendations to the Council, which shall adopt a decision (...) and, depending on the subject of the agreement envisaged, nominate the Union negotiator or the head of the Union's negotiating team". The provision sparked a conflict between the Commission, demanding the right to negotiate on behalf of the EU, and the Council about questions of representation in global environmental fora. It was initially settled to the advantage of the latter.

De facto, the EU's coordination and representation arrangements therefore underwent only minor modifications. Although efforts were made, notably by the Foreign Ministers of the 27 and in Coreper 1 meetings, to enable a more coherent EU appearance in global climate talks, this was not a major concern for the Environment Ministers. Despite the fact that strategic debates were regularly on the agenda of the WPIEI-CC in 2010, and in spite of exchanges between the EEAS and the Commission about a "renewed and strengthened climate diplomacy" (EEAS/Commission

---

[3] The function came with a double-hat: being a Vice-President of the Commission, the Representative also presides over the newly created Foreign Affairs Council and has to ensure the coordination of EU foreign policy (Art. 18, para. 4, 3 TEU).

2011) – leading to minimalistic "Council conclusions on EU Climate Diplomacy" in July 2011[4] –, a diffuse desire to act more coherently was not translated into any significantly altered outreach. Rather, only a minimum degree of integrated representation was agreed to for the Cancun, Durban and Doha summits, where the Commission played a prominent role alongside the Council Presidencies (2010 Belgium, 2011 Poland, 2012 Cyprus). The Commission's standing had been boosted by its decision to create, in 2010, the post of a Commissioner for "Climate Action", supported by a small DG.[5] For the rest, the status quo – a system relying on lead negotiators and issue leaders – largely prevailed for all negotiations under the UN umbrella. Outside the UN, the Lisbon Treaty had led to several ambiguities about the EU's representation, necessitating pragmatic interinstitutional deals. At G-20 summits, for instance, Commission President Barroso would speak for the EU on climate change, whereas the President of the European Council would be responsible for other topics (Pop 2010).

When it comes to the key contours of the Union's negotiation position, no major changes were observed for the period 2010 to 2012 with regard to its positions on the issues of targets and differentiation. Despite a May 2010 Commission communication that argued for stepping up the Union's reduction pledge from 20 to 30% (European Commission 2010) and repeated attempts by, e.g., the Environment Ministers of France, Germany and the UK to convince their colleagues of the necessity to move in that direction (Willis 2010a), no sufficient momentum could be generated in the midst of strong resistance notably from the new member states. The conclusions of the June 2010 Environment Council therefore also welcomed Commission efforts, but intended to buy time by "stress[ing] that the abovementioned communication covers a wide range of issues which need to be discussed in-depth in order to prepare the EU for the medium- and longer-term climate change challenges (…) in the international climate negotiations" (Council 2010a: 9). A few days later, the European Council stated that it would "revert to climate change in the autumn, in advance of the Cancun conference" (European Council 2010). This and other key issues remained unsettled in the negotiation directives adopted by the Environment Council and endorsed by the European Council in October 2010 for the Cancun summit. Following similar inconclusive summits in 2011 (Council 2011a), this situation persisted well into 2012 (Agence Europe 2012a).

Where the EU did not manage to advance its positions on these items, also because the substance on the key issues was not expected to

---

[4] The Council debate was held on the request of Germany and the UK and resulted in the commitment "to address climate change at all political levels", employing also the services of the EEAS (German Foreign Ministry 2011).

[5] In early 2010, the European Parliament would approve the former Danish President of COP 15, Connie Hedegaard, as first Climate Commissioner.

be central to global negotiations during these years, it did continue to consider its internal climate policies as well as its positions on regime development per se, i.e. the questions where and how to proceed with global talks. Internally, a March 2011 Commission communication entitled "A Roadmap for moving to a competitive low-carbon economy in 2050" set the agenda for discussions about how to meet the long-term target of reducing domestic emissions by 80 to 95% by 2050 (European Commission 2011), but decisions were blocked essentially by Poland in 2011 and 2012 (Council 2011b). On the shape of future global negotiations, the Environment Ministers clarified, before COP 16, "the need to anchor all countries' pledges in Cancun, whether made pursuant to the Copenhagen Accord or otherwise, in the context of the (…) UNFCCC" (Council 2010b: 4). This position represented a re-affirmation of EU commitment to the UN process. When the integration of the Copenhagen Accord into the UN climate regime had been achieved, 2011 saw greater EU efforts to concentrate again on the means of advancing the global negotiation process so as to determine the future of the Kyoto Protocol and close the gap of collective commitments. To address those issues, the Environment Council's October 2011 negotiation mandate for the Durban summit suggested the adoption of "a roadmap, including a timeline with a final date and process taking into account the 2013–2015 review" of the IPCC for the talks under the Convention (Council 2011a: 2–3). Only if this could be agreed, it confirmed "its openness to a second commitment period under the Kyoto Protocol as part of a transition to a wider legally-binding framework" (Council 2011a: 2–3).

The Negotiation Process and the EU's Influence Attempts

Following the Copenhagen summit, the **year 2010** "was a time of recovery and soul-searching" for global climate politics (La Vina/Ang 2010: 1). As a first act of good will, parties that had made informal pledges at Copenhagen notified, in line with the Accord, the UNFCCC secretariat of their voluntary reduction targets and/or actions. By 31 January 2010, ten Annex I and 20 non-Annex I parties had formally communicated their commitment/action pledges to the secretariat (UNFCCC 2010a). Despite the apparent commitment to honour the Accord, many parties had, however – due to the uncertain legal status of the document and continued distrust – attached conditions to their pledges. The EU had pledged a 20% reduction by 2020 (from 1990 levels) and reiterated its conditional offer of moving to 30%. The US recalled its pre-Copenhagen pledge and made it contingent on the domestic legislative process. Japan had offered 25% reductions by 2020 "premised on the establishment of a fair and effective international framework in which all major economies participate" (UNFCCC 2010a: Japan). China had pledged voluntary and autonomous domestic mitigation actions: it "will endeavor to lower its carbon dioxide emissions per unit of GDP by 40–45% by 2020 compared to the 2005 level, increase the share of non-fossil fuels in primary energy consumption to around 15% by 2020" (UNFCCC 2010a: China). India "will endeavour to reduce the emissions intensity of its GDP by 20–25% by 2020 in comparison to the 2005 level" (UNFCCC 2010a: India). Existing pledges fell way short of the 25–40% range for industrialized countries to meet the 2°C target stipulated in the Accord (Ecofys 2010; Curtin 2010: 6).

The COP 15/MOP 5 decisions to continue negotiations under both the AWG-LCA and AWG-KP tracks meant that parties reconvened in the **UN regime** to try and find common ground. The first post-Copenhagen meeting of the AWG-LCA 9/AWG-KP 11 between 9 and 11 April 2010 in Bonn took place in a charged atmosphere, and resulted mainly in a common understanding about continuing the process in the UN, while leaving the aim of the negotiations for 2010 undefined. Where some parties, including the EU, expressed hope for Cancun to achieve what Copenhagen had not, namely a legally binding outcome with stronger emission reduction efforts, others were looking for far more modest results (ENB 2010a: 12). In the *AWG-LCA*, the Chair was given the mandate to prepare a new negotiation text for the June session, while the *AWG-KP* continued its debates on QELROs (ENB 2010a). At this session, the EU, while arguing for using the Copenhagen Accord outcome as guidance for the UN process, kept a rather low profile aimed at trust-building (ENB 2010a: 12). Already at the next session of the two bodies (AWG-LCA 10/AWG-KP 12, 31 May–11 June, Bonn), text-based negotiations were pursued. The *AWG-LCA* Chair's negotiation text was gradually updated, but remained controversial, with the G-77/China criticizing it as unbalanced in favour of Annex I countries' interests for seeing stronger mitigation commitment by developing countries (ENB 2010b: 22). To overcome this imbalance, parties like AOSIS, but also the EU, attempted to join discussions under the two AWGs, but without success (ENB 2010b: 23). In the *AWG-KP*, no major advances on QELROs were reported (ENB 2010b). Two more meetings would follow prior to the Cancun summit. AWG-LCA 11/AWG-KP 13 (2–6 August 2010, Bonn) would witness debates on a novel *AWG-LCA* draft text on how to fill the gap between the end of the first commitment period of the Kyoto Protocol and a new agreement in the AWG-KP (ENB 2010c). Given the slow pace of the progress in these talks, key UN and EU negotiators began to lower expectations for the Cancun COP. The EU's lead negotiators, for instance, referred to the talks as a "fragile process" with an unclear finality, re-iterating however the Union's hope for a balanced set of decisions and a legally binding outcome later on (EU 2010). Finally, at AWG-LCA 12/AWG-KP 14 in Tianjin (4–9 October 2010), talks on the draft negotiation texts continued in both AWGs, but options on crunch issues could not be narrowed down, leaving the bulk of the work to the Cancun summit. Negotiators expected this summit to achieve little, but hopefully *"enough* to send a signal" that the multilateral process was still alive (ENB 2010d: 15).

Outside the UN regime, discussions among key players were held on bilateral and multilateral bases throughout 2010. During the first months after the Copenhagen COP, these were mostly dedicated to keeping the channels of communication around climate change open. Minilateral meetings with potential interest for the topics under investigation here were those of the G-20 and the Major Economies Forum, the latter of which convened four times in 2010, without tangible outcomes (MEF 2012). New initiatives were started, such as the Petersberg Climate Dialogue, which met for the first time in May 2010 and brought together ministers and climate negotiators from 45 countries in Bonn (BMU 2010). Yet, little momentum was generated by these high-level talks. This was arguably different for the newly created Cartagena Group/Dialogue for Progressive Action, which involved around 30 parties from all negotiation coalitions interested in making advances toward a legally binding global climate agreement (Bowering 2011).[6] Founded in the wake of the Copenhagen COP, it held discussions on key issues negotiated in the UN in an effort to identify solutions that could suit both Annex I and non-Annex I parties. Key EU players invested heavily into this Group, which operated out of the limelight during most of 2010, but would be instrumental for the Cancun top and thereafter.

Where none of these meetings had brought any decisive leaps forward, the expectations for **COP 16/MOP 6** in Cancun, Mexico (29 November–11 December 2010), were understandably low (La Vina/ Ang 2010). Nonetheless, the Mexican COP presidency was determined to make the summit a success, adopting a more open communication style than its Danish predecessors, and striving for convergence on minimum outcomes (ENB 2010e). During the first ten days of the talks, this approach did not yield any major advances. In the *COP and AWG-LCA*, discussions focused on issues such as mitigation and MRV, with the familiar controversies between developed and developing countries, but also on topics related to the operationalization of the institutions created in the Copenhagen Accord, notably regarding finance (e.g. Green Climate

---

[6] The Cartagena Dialogue participants in 2010 were: Antigua and Barbuda, Australia, Bangladesh, Belgium, Chile, Colombia, Costa Rica, Denmark, Dominican Republic, Ethiopia, the EU Council Presidency, the European Commission, France, Germany, Ghana, Guatemala, Indonesia, Malawi, Maldives, Marshall Islands, Mexico, Netherlands, New Zealand, Norway, Panama, Peru, Rwanda, Samoa, Spain, Tanzania, Thailand, Timor-Leste, United Kingdom and Uruguay (Casey-Lefkovitz 2010).

Fund) (ENB 2011e). In the *MOP and AWG-KP*, parties continued to search for solutions to adopt emission reduction targets under a second Kyoto Protocol commitment period. From the start of the summit, Japan and Russia, followed by Canada, made it clear that their governments opposed such a second period (Sterk *et al.* 2011a: 6). It was thus up to the EU to build bridges toward the developing countries, both by signalling openness toward a second commitment period and via coalition-building with AOSIS, facilitated through previous exchanges in the Cartagena Group (The Guardian 2010). Against the backdrop of rather controversial discussions, observers on the morning of the final day were still unsure whether modest advances would be made or if the summit would end "in a procrastinating failure" (The Guardian 2010). The dynamics of the meeting would become more positive when the Mexican presidency tabled proposals for COP and MOP decisions summarizing the progress and identifying ways forward for both the AWG-LCA and the AWG-KP in the evening of 10 December (Sterk *et al.* 2011a). Without going into the details of the procedures (for a summary of the discussions, see ENB 2010e), the final plenary demonstrated that these texts could be accepted by all parties, Bolivia excepted. Despite Bolivian resistance, the Mexican presidency gavelled the "Cancun Agreements" through, arguing that one party could not prevent the consensus expressed by all others (ENB 2010e: 28–29). In the end, delegates delivered thus an outcome that essentially anchored previous agreements under the Copenhagen Accord in the UN regime.

In **2011**, a comparatively limited number of three preparatory sessions for COP 17/MOP 7 was held **within the UN climate regime**. The AWG-LCA 14/AWG-KP 16 meeting between 3 and 8 April 2011 in Bangkok focused mainly on discussing a work programme for 2011, notably in the *AWG-LCA* (ENB 2011a). Debates implicated primarily the G-77/China, whose members had diverging views on whether the implementation of the Cancun Agreements represented a sufficient agenda for future debates – a position supported by China and India, but also the US – or if new paths had to be taken to ensure progress toward the "agreed outcome" stipulated by the Bali Roadmap, which was desired by AOSIS (ENB 2011a: 16). In the end, a work plan could be agreed to. It covered the implementation of Cancun decisions as well as issues related to the Bali Roadmap (above all shared vision including temperature limits, legal nature of the outcome). In the *AWG-KP*, inevitable debates opposed developing countries, who desired to obtain clear pledges by Annex I parties for the QELROs to be adopted under a second commitment period, to developed countries, who were primarily interested in debating the modalities of reduction targets (ENB 2011b: 16). The EU clearly emphasized its preparedness to engage in such a second period "if the conditions are right", i.e. if partners from the developing and developed world would engage in activities satisfying the Union's long-standing negotiation position (Ten Kate/Airlie 2011). Canada, Russia and Japan (and the US, as non-party to the KP), in the meantime, reiterated that they did not intend to participate in a novel commitment period, while no longer actively opposing discussions on it (ENB 2011b: 16). The subsequent meeting, AWG-LCA 14.2/AWG-KP 16.2 in Bonn (6–17 June 2011), continued the discussions held at Bangkok, especially in the *AWG-KP* (ENB 2011b). Work in the *AWG-LCA* was dominated by technical issues regarding implementing elements of the Cancun Agreements, most notably the Technology Mechanism, Adaptation Committee and Green Climate Fund, with limited advances. Little progress was also booked in the parallel debates on advancing talks about developed and developing country mitigation and the legal outcome (ENB 2011b: 24). Finally, AWG-LCA 14.3/AWG-KP 16.3, held in Panama City between 1 and 7 October 2011, while advancing on some technical issues related to the implementation of the Cancun Agreements, exacerbated cleavages between key players (Friedman 2011). In the *AWG-LCA*, debates progressed on the operationalization of new institutions such as the Technology Mechanism, and parties reached an understanding of how to match efforts to limit emissions by developing countries with support from developed countries through the creation of a central "registry" (ENB 2011c: 13). At the same time, discussions about the Green Climate Fund and other components for an overall deal based on the Cancun Agreements as well as debates on a new mandate for negotiations toward a legally binding deal remained controversial (ENB 2011c: 13). Where many industrialized countries, headed by the US, but also China and India, opposed the idea of adopting such a mandate, the EU and AOSIS argued in favour of starting a new negotiation process (ENB 2011c: 14). To ensure good relations with the developing countries, the EU also re-iterated its positions on fast start and long-term finance (ENB 2011c: 13; Friedman 2011). In the *AWG-KP*, its offer to accept a second commitment period was re-stated, and closely linked to the condition that parties would agree to a mandate for talks on a legally-binding instrument under the Convention in Durban (ENB 2011c: 14). Prior to COP 17, no real advances on major points related to emission reduction targets and actions had thus been booked, but a loose EU-developing country coalition had begun to form.

In **extra-UN global climate politics**, the year 2011 would witness renewed efforts of bodies like the Major Economies Forum, which brought together leaders in April and September (MEF 2012), or the Petersberg Climate Dialogue attended by Ministers and high-level officials from 35 countries (Berlin, 2–4 July 2011, BMU 2011). Both fora had little impact on the UN talks, however. This was arguably different for the Cartagena Group, which provided fertile grounds for debates among progressive parties also in 2011 (Bowering 2011).

The yearly showdown, **COP 17/MOP 7** in Durban, South Africa (28 November–9 December 2011), had to deal with mainly two sets of issues. The first one concerned the operationalization of decisions taken in Cancun regarding the Green Climate Fund, the Technology Mechanism plus several other, technical items (e.g. CDM). On these points, which are of less interest for the analysis here, the COP took several decisions discussed elsewhere (see ENB 2011d; Sterk *et al.* 2011b). Of greater relevance for this study were debates under both the COP and the MOP on the future of the climate regime and its Kyoto Protocol (for a very comprehensive summary, see ENB 2011d). In a move toward finally linking the debates in the two negotiation tracks, the EU formed a "green coalition" with AOSIS and the LDCs, which had emerged from the Cartagena Group. This coalition demanded the clear "roadmap, including a timeline with a final date and process taking into account the 2013–2015 review" for future negotiations on a "legally binding agreement" under the Convention that had been identified in the EU Environment Council's October negotiating mandate (Council 2011a: 2–3). Only if such a roadmap was adopted would the Union accept a second commitment period for the Kyoto Protocol (Sterk *et al.* 2011b). As it pushed hard for this outcome during the entire duration of the talks, countries that had resisted debates on a legally binding outcome – China, India, but also the US, which had apparently thought that the EU would give in and accept a second commitment period without any concessions from other countries (Sterk *et al.* 2011b: 31) – had to move into the green coalition's direction, if they did not want to be seen as deal-breakers. This led to a memorable show-down in the final COP plenary when Commissioner Hedegaard and the Indian and Chinese negotiators quarrelled over language for the decision that the COP was to adopt. While Hedegaard assertively insisted on phrasing that would underscore the legally binding nature of the outcome, the Indian delegate wanted to soften the wording of the decision (ENB 2011d: 30; for a reconstruction of this exchange, see Sterk *et al.* 2011b: 8). The solution was a compromise formula suggested by the US, namely to speak of a future "agreed outcome with legal force" (Verolme 2012: 4). To gain the support of the emerging economies, another concession was made regarding the year of a possible entry into force of such an "agreed outcome" (2020), whereas finance, particularly the operationalization of the Green Climate Fund, was used as an additional sweetener to win over other G-77 members. The outcome of the debates, the "Durban Package" – with decisions to continue the AWG-LCA and the AWG-KP until 2012, adopt a second Kyoto Protocol commitment period and engage in a new regime reform process under the Ad Hoc Working Group on the Durban Platform for Enhanced Action (ADP) – sent the UN climate regime thus "on the road again", while leaving the magnitude of targets, the sharing of responsibilities and the precise legal form of the outcome undecided (Sterk *et al.* 2011b).

The **year 2012** was marked by two preparatory sessions for COP 18/ MOP 8 **within the UN climate regime**. Initially, it was unclear to many negotiators what the talks under the ADP were going to lead to and how they would relate to the ongoing negotiations under the AWG-LCA, whose mandate the COP had extended for one final year to allow for "identifying a global goal for substantially reducing global emissions by 2050" (UNFCCC 2011a: 1, UNFCCC 2011c: 1). As a result, first *ADP* negotiations held between 14 and 25 May 2012 in Bonn focussed essentially on clarifying procedures and work plans – issues that provoked disputes and could only be settled on the very last day of the session (ENB 2012a; Herold *et al.* 2012). The agreement that was reached foresaw two workstreams: one on the negotiations for a post-2020 agreement in 2015 and one on a work plan aimed at enhancing the level of ambition beyond 2015 (ENB 2012a). The parallel *AWG-LCA 15* debates proved to be equally difficult, as developing country parties stated that they wanted to finalize discussions in this group *only* when all issues on the Group's agenda had been conclusively addressed, which was far from being the case (ENB 2012a). *AWG-KP 17*, by contrast, dealt with very concrete topics regarding the adoption of a second commitment period of the Kyoto Protocol. It did not however succeed in settling key questions (e.g. on the duration of this period) (Herold *et al.* 2012: 16–17). During a second, informal negotiation session organised in Bangkok (30 Aug. – 5 Sept.), discussions were pursued in all three groups. In the *ADP*, parties reflected on the legal form of their future agreement and on the "ambition gap" (ENB 2012b). Moreover, roundtables were organised on the two workstreams. Cleavages emerged between, on the one hand, the EU and some progressive, mostly developing country players interested in employing the ADP as leverage to increase mitigation actions prior to and beyond 2020 and, on the other hand, parties like China who wanted to focus on discussions in the *AWG-LCA* (ENB 2012b). In the latter group, the emerging and many developing countries then again argued against premature closure of debates as long as issues like developed country reduction targets or finance had not been sufficiently addressed (ENB 2012b). In the *AWG-KP 17(bis)*, finally, major pending issues regarding the second Kyoto Protocol commitment period could still not be solved, prompting parties to give the Chair the authority to identify various options so as to facilitate talks in Doha (ENB 2012b).

Outside the UN regime, several major global summits with potential importance for the climate talks were organized in 2012. While the first half of the year was dominated by the preparation and organization of the UN Conference on Sustainable Development (UNCSD or Rio+20 summit, 13–22 June 2012, Rio de Janeiro), the actual outcome of this summit was judged as rather limited, with close to no implications for the climate negotiations (Beisheim *et al.* 2012). Other meetings held in 2012 brought together the Major Economies Forum (Rome, 17 April 2012, New York, 27 September 2012), the Third Petersberg Climate Dialogue (Berlin, 16–17 July 2012, BMU 2012) and the Cartagena Group (Herold *et al.* 2012: 73). Finally, on 21 to 23 October 2012, a pre-COP involving 43 parties was organised in Seoul (Herold *et al.* 2012: 15). While all these meetings did not go much beyond exchanges of positions in which the importance of the UN process was emphasized, interesting interactions outside the multilateral regime occurred in the International Civil Aviation Organization (ICAO). They prominently involved **the EU**. In 2011 the Union had adopted legislation that would impose, as of 2012, a cap on GHG emissions from flights operating to and from EU airports. This implied the inclusion of some 4,000 EU and non-EU-based aircraft operators into its Emissions Trading System. This move initially provoked hostile reactions from 29 member countries of the ICAO, most notably the US and China (Egenhofer/Alessi 2013). In the exchanges that followed, the Union pursued its position with rather unusual firmness.[7] In November 2012, the debate took a new turn when the ICAO Council initiated a discussion on the possibility of globally concerted action on aviation and climate change. In reaction to this debate, the European Commission suggested "stopping the clock" in the application of the legislation so as, in its own words, to "demonstrate goodwill towards the successful conclusion" of the ICAO talks (European Commission 2013). Where the aviation debate opened a second important arena for EU foreign climate policy, the Council conclusions containing the Union's mandate for COP 18/MOP 8 re-called unequivocally that the UN remained the key global climate forum and that "balanced progress on all elements of the package agreed upon in Durban" was its precondition for agreeing to a second commitment period on the Kyoto Protocol at Doha (Council 2012: point 14). The commonly agreed mandate could not mask certain dissonances within the EU, where Poland and a few other Central and Eastern European countries had not only blocked the attempt to step up the Union's mitigation ambitions to 30% by 2020, but had also prevented an agreement on the (non-)use of "hot air" under the second period of the Kyoto Protocol.

The negotiations of the year 2012 culminated in **COP 18/MOP 8**, held in Doha, Qatar (26 November – 7 December 2012). Characterized as a "transitional" or "intermediate" COP (ENB 2012c; Marcu 2012),

---

[7] At the same time, the EU actively attempted to intensify its bilateral climate cooperation with major players, especially China (Belis/Schunz 2013).

the aim of this meeting had from the start been two-fold: closing the negotiation tracks opened in Bali and paving the way for the negotiations toward an agreement in 2015. Under the *ADP*, discussions were pursued under the two workstreams. Workstream 1 focussed on the role of the Convention principles in the further work of the group. Where emerging economies argued that the ADP should be explicitly guided by the CBDR principle, the US was strictly opposed to this (C2ES 2012: 3–4). The EU and others argued for interpreting the Convention principles in an overall evolving global context (ENB 2012c: 16–17). In the end, a soft reference to "the principles of the Convention" was made in the final COP decision (UNFCCC 2013b). In workstream 2, debates concentrated on key issues such as the meaning of some of the concepts under the ADP and on adopting a work plan, which marked also the main result of the talks. In the *AWG-LCA*, lengthy debates opposed emerging and developed countries (ENB 2012c). Where the former wanted to fully finalize some of the discussions started in Bali (especially on issues such as finance and developed country mitigation), the latter saw advantages in having the ADP as the only future negotiation forum under the UN. By way of compromise, and to enable a closure of this strand in line with what had been agreed at Durban, negotiations on topics such as market mechanisms were effectively shifted into the Subsidiary Bodies. Most importantly, the COP/MOP also witnessed the closure of the *AWG-KP*, which had to essentially still decide on three issues that had been pending since 2007: the length of the commitment period (eight years as favoured by the EU and the Umbrella Group vs. five years as requested by many developing countries); the magnitude of targets (with the developed countries pledging individual targets and the developing countries demanding greater overall ambition); and the modalities of the use of AAUs. The first two items were quite easily settled in favour of the developed country group, whereas the third issue required a last-minute deal (ENB 2012c). The EU member states (including the initially reluctant Poland) and Australia politically declared not to buy hot air, while Russia, Ukraine and Belarus protested until the end against restrictions on the use of AAUs. Their opposition was judged by the COP Presidency as insufficient to block consensus (ENB 2012c: 27). This paved the way for adopting the "Doha Gateway".

The Outcomes: the Cancun Agreements, the Durban Package and the Doha Gateway

Between 2010 and 2012, the global climate negotiations produced three outcomes with repercussions for the development of the UN regime and thus, more indirectly than directly, also for the issues of emission reduction targets and differentiation.

The 2010 **Cancun Agreements** represented a set of decisions taken by the COP and the MOP, which – in essence – formalized the outcomes embodied in the 2009 Copenhagen Accord (Sterk *et al.* 2011a). On the issues of greatest interest here, emission reduction targets and differentiation, the Agreements under the COP recognized the need for "deep cuts (...) so as to hold the increase in global average temperature below 2°C above pre-industrial levels" and pledged to re-consider this target "on the basis of the best available scientific knowledge, including in relation to a global average temperature rise of 1.5°C" (UNFCCC 2010c: 2). Industrialized countries engaged themselves to attain "quantified economy-wide emission reduction targets" in line with their January 2010 pledges, and the Agreement "urges them" to consider higher targets (UNFCCC 2010c: 7). For developing countries, NAMAs "aimed at achieving a deviation in emissions relative to business-as-usual by 2020" would be included in a registry so as to recognize all NAMAs and match finance, technology and capacity-building support to NAMAs seeking international support (UNFCCC 2010c: 20, 9). These decisions were coupled to a pledge by developed countries of a total of 30 billion USD in fast-start finance to support climate action in developing countries up to 2012 and the intention to raise 100 billion USD by 2020 (UNFCCC 2010c: 15). Besides these key outcomes, the COP decided that a Green Climate Fund, a Technology Mechanism and a novel Adaptation Framework should be put into place (UNFCCC 2010c). The mandate of the AWG-LCA was prolonged by one year (UNFCCC 2010c: 24). Agreements under the MOP were not so substantial, the consensus on a base year (1990) excepted. The final decision simply noted that "further work is needed to convert emission reduction targets into QELROs", and prolonged the mandate of the AWG-KP indefinitely, stipulating delivery of an outcome "as soon as possible" (UNFCCC 2010b: 2). Although the outcome of the climate negotiations over the year 2010 represented thus an incorporation of decisions taken twelve months earlier in Copenhagen into the UN framework, it was widely celebrated as a success (see, e.g., ENB 2010; Beament 2010; Rajamani 2010b). This prompted some observers to speak of the "Cancun paradox": processes and outcomes of global climate politics that had been considered insufficient and illegitimate in Copenhagen had suddenly become acceptable, even promising only one year later in Mexico (Audet/Bonin 2010). From a long-term incremental institutionalist perspective, and assuming that this meeting's purpose was above all to restore faith in the multilateral process, this may indeed be a fair assessment (Bodansky 2011). The state of the climate, however, was not (sufficiently) advanced with these Agreements, which displayed the same shortcomings already discussed for the Copenhagen Accord, notably the fact that the gap between reduction pledges and prescriptions by the IPCC continued to persist, and that the non-binding "pledge and review" character of the agreement had remained unaltered.

The 2011 **Durban Package** of decisions that had resulted from the showdown in South Africa did not actually go much beyond the Cancun Agreements regarding the key issues under analysis here. It did, however, kick-start a new negotiation process. Besides a few substantial decisions on the implementation of the Green Climate Fund, the Adaptation and Technology Mechanisms, the main *outcome under the COP* was the establishment of the "Durban Platform for Enhanced Cooperation" (UNFCCC 2011a). It essentially decided "to launch a process to develop a protocol, another legal instrument or an agreed outcome with legal force under the Convention applicable to all Parties, through a subsidiary body (...) to be known as the" ADP (UNFCCC 2011a: 1). This body was to "complete its work as early as possible but no later than 2015" so that its output could "come into effect and be implemented from 2020" (UNFCCC 2011a: 1). In its deliberations, it was to take account of the Fifth Assessment Report of the IPCC, to be fully released in 2014 (UNFCCC 2011a: 2). While neither the approach nor the substance of the decision were truly novel, its new features concerned the dates for the procedure as well as the agreement, especially by China and India, to engage in talks on "an agreed outcome with legal force applicable to all parties", including developing countries (Verolme 2012: 4). It is noteworthy, in this context, that the Durban documents do not re-iterate the distinction between Annex I and non-Annex I parties, and do not include specific reference to the CBDR principle (Bodansky 2012: 3). Initially, however, the Platform was – similar to what the Bali Roadmap was for the process until 2011/2012 – "little more than an agreement to discuss" some form of legal outcome (Hultman 2011), through which parties have "decided to decide" on this crucial issue (Dupont 2012). As Hultman (2011) notes, "with this declaration, it is possible (but not required) for both developed and emerging economies to take on some kind of emissions reduction target" – whether and how this is going to happen remained unclear even after 2012. The *decision by the MOP*, based on the work of the AWG-KP, was slightly more concrete: it established "that the second commitment period (...) shall begin on 1 January 2013 and end" either in 2017 or in 2020 (UNFCCC 2011b: 1). Negotiations on this and the concrete QELROs to be adopted were to continue throughout 2012 in the AWG-KP, whose mandate was prolonged until MOP 8 (UNFCCC 2011b: 2). Reactions to the overall outcome of the Durban summit ranged from moderately positive ("milestone" toward final "agreed outcome") (for a small sample, see Reuters 2011c; Sterk *et al.* 2011b) to very critical ("nothing new") (e.g. Carrington 2011; Bruyninckx 2011; Verolme 2012: 3).

The set of outcomes known as the 2012 **Doha Gateway** represented a combination of COP and MOP decisions that concluded a number of processes started with the 2007 Bali Roadmap, while paving the way for further regime discussions. Main accomplishments included: (i) the *closure of the AWG-LCA*, five years after the Bali COP. Some of the questions that this Working Group had been unable to settle (e.g. assessment of progress toward reaching QELROs) were transferred to the two Subsidiary Bodies for further debates (UNFCCC 2013a; see also C2ES 2012); (ii) the *conclusion of the work of the AWG-KP* with the adoption of an amendment to the Kyoto Protocol on a second commitment period (2013–2020). Besides the EU, Australia, Belarus, Iceland, Kazakhstan, Liechtenstein, Luxembourg, Monaco, Norway, Switzerland and Ukraine adopted targets. Canada, Japan, New Zealand and Russia, by contrast, did not. The individual QELROs of the participating parties added up to about 15% of global GHG emissions (C2ES 2012: 2), and the use of AAUs was restricted. Furthermore, parties decided to "revisit" their mitigation pledges in 2014 (UNFCCC 2012: point 7); (iii) *the adoption of a work plan, under the ADP*, towards the 2015 COP. The plan foresaw a more stringent operating modus for 2013 and 2014 "with a view to making available a negotiating text before May 2015" (UNFCCC 2013b: point 9). All in all, these "modest outcomes" had been expected all along the year 2012 and qualify therefore at best as a "consolidation" stage in the process toward a legally binding outcome (C2ES 2012: 1; Agence Europe 2012c; Marcu 2012). Despite this widespread assessment, certain key negotiators, including EU Climate Commissioner Hedegaard, viewed the Doha Gateway slightly more positively. According to her, although the outcome of COP 18 was "not fantastic", it built "the bridge from the old climate regime to the new system. We are now on our way to the 2015 global deal. It was not an easy and comfortable ride (…). But we have managed to cross the bridge" (Hedegaard 2012; Agence Europe 2012c). One of the clearly positive outcomes in her view was the reduction of the number of fora in which negotiations would be held: from three (AWG-LCA, AWG-KP, ADP) to one (ADP) (Hedegaard 2012). While this was certainly a potential improvement in light of the difficulties that notably the European Union experienced with the multiplicity of parallel negotiation tracks throughout the years 2007 to 2012 (see Chapter 5), the Doha decisions also added to the existing complexity already in place. In parallel to the targets adopted under the second commitment period of the Kyoto Protocol, the emission reduction pledges made by industrialized and developing countries through the Cancun Agreements will also run until 2020. Even if they are all fulfilled, "the Copenhagen-Cancun pledges, combined with Kyoto's second commitment period have left a significant ambition gap" when compared to the 2°C objective enshrined in the Copenhagen Accord (European Commission 2013: 6, drawing on calculations by UNEP 2012). While their fulfilment will need to be monitored closely, new targets will thus have to be negotiated. These targets will normally include developed *and* developing countries under a single legal agreement, implying a major modification to the long standing division of responsibilities within the climate regime. Although only time will tell if the re-ignition of the global negotiation process in Durban and Doha delivers what the previous process ending in Copenhagen, with prolongations in Cancun, could not,[8] it is undisputable that, in terms of environmental effectiveness, no substantial advances in the global combat against climate change could be observed by 2012.

Determining and Explaining the EU's Influence during the Period 2010 to 2012

The narrative of the years 2010, 2011 and 2012 demonstrates that the global climate talks were gradually brought "back on track" following the problematic Copenhagen summit and its rather painful aftermath. While they had initially been stalled in much of 2010, the Cancun summit restored confidence and Durban and Doha developed the institutional underpinning for a new round of structured global negotiations.

Throughout this period, and following a rather brief moment of paralysis of its foreign climate policy, the EU largely continued behavioural patterns of the past. An analysis of its influence attempts reveals a short period of hesitation in the first few months of 2010. This hesitation had to do with the way the Union's climate negotiators perceived their own performance and the outcome of the Copenhagen summit, but also with the changes implemented through the Lisbon Treaty. It showed in the Union's deliberations on both its position and its foreign policy strategy for the global climate negotiations. Starting from the late spring of 2010, however, the EU gradually re-engaged, albeit in a more pragmatic manner, based on lowered expectations and a lower-key profile than in 2009. In 2011, it soon resumed its pre-Copenhagen strategy, and from then on figured among the most active parties in the global climate talks again, notably regarding issues related to long-term regime development. Its major influence attempts in this context consisted in proposing the discussed "road map, including a timeline with a final date and process taking into account the 2013–2015 review" of the IPCC and in offering a QELRO for the second commitment period for the Kyoto Protocol.

---

[8] In 2013, the negotiations in the global climate regime were indeed slowly advancing toward the 2015 deadline. Among the key outcomes of COP 19 in Warsaw were parties' pledges to submit national "contributions" towards the envisaged 2020 agreement well in advance of COP 21 (by the first quarter of 2015). The EU, which had proactively argued for a more ambitious "stepwise approach", expressed its content with this intermediate outcome.

These diplomatic attempts based on arguing and bargaining offers insert themselves perfectly into behavioural patterns the EU had displayed prior to 2010 (Geden 2011), and which were discussed at length in Chapter 5. The major novelty was that they were now coupled to partially reinforced efforts at coalition-building and a more assertive diplomatic communication around the need for re-igniting the negotiation process. This slightly adapted foreign policy implementation, as will be demonstrated, also earned it some degree of influence. A significant outlier in its behaviour was arguably its rather tough stance on the inclusion of aviation emissions in the ETS, partially revoked in 2012.

To assess the Union's influence regarding regime development during this three-year period, it is first necessary to identify possible turning points with respect to the key analytical units of emission reduction targets and responsibilities. The three major turning points in global climate talks between 2010 and 2012 arguably occurred during the final stages of the COPs in Cancun, Durban and Doha. A first turning point at COP 16/ MOP 6 consisted in the decision to incorporate the Copenhagen Accord into the UN legal process, which implied that the negotiations on global climate policies would continue under the UN umbrella. Turning point 2 during the last hours of COP 17/MOP 7 in South Africa then brought an agreement on a renewed UN climate negotiation process, initiating a sort of "post-2012 negotiations bis". Turning point 3 at Doha concerned the adoption of a second commitment period of the Kyoto Protocol, through which the EU formally pledged its long-standing 2020 reduction targets and thus contributed to fully ensuring the collective transition to a new negotiation process. The first two turning points thus essentially touched on the issues of *where* (in which fora) and *how* (by what rules) the climate regime is developed, determining the modalities of a debate about emission reduction targets and responsibilities. The final turning point concerned also these modalities, but had repercussions for the magnitude of emissions reduction targets.

To establish EU influence at these turning points, the constitutive dimensions of influence can once again be tested. This delivers a straightforward picture for the first turning point: the EU had desired the incorporation of the Copenhagen Accord into the UN climate regime talks and had argued – after a short period of reflection and not as vocally as before – for such a step to be taken at COP 16 (*purposive behaviour, interaction*). This objective was ultimately attained in Cancun, but not necessarily due to the EU and its actions: it was neither the first, nor the only actor to argue in favour of it. Already at or immediately after COP 15, countries from the BASIC group, AOSIS and the LDCs had expressed their desire in embedding the Accord into UNFCCC structures, and had fiercely defended the UN as the key negotiation arena for global climate policies, which is why EU leverage over this decision cannot be established (test on *goal attainment, temporal sequence*: not fulfilled). In counterfactual analysis, it would indeed be difficult to argue that in the absence of EU advocacy for this objective, the same result would not have been accomplished, given the strong pressure from the G-77/China (test on *absence of auto-causation*: negative). For these reasons, no significant EU influence can be discerned over the major outcome of COP 16/MOP 6. This was also the widely held impression by commentators of the 2010 talks, who highlighted the skilled chairmanship of the Mexican presidency and factors such as the passivity of the US, but, while acknowledging the constructive role played by the EU, did not single out the Union as responsible for the outcome (ENB 2010; Audet/Bonin 2010; Rajamani 2010b; Willis 2010b).

The assessment is quite different for the second turning point. On the decision incorporated in the "Durban Package", namely to start a new negotiation process to reach a global agreement by 2015, the EU arguably did exert influence. First, the Union designed, as of the second half of 2011 and in the hope of setting the agenda on this issue, a clear position on the need for a renewed "roadmap", and reached out to partners on this basis (*purposive behaviour, temporal sequence, interaction*). While some parties (e.g. AOSIS) did not have to be convinced, others – like the BASIC countries and the US – needed to be, which was attained via coalition-building between the EU, AOSIS and the LDCs at COP 17 (*goal attainment*). Even if the final wording of the deal brokered at that summit may not fully correspond to what the EU had desired, notably as concerns the rather loose phrase "an agreed outcome with legal force" and the year of entry into force of a future agreement (2020), EU influence can be affirmed also from a counterfactual perspective. It is rather probable that had the Union not existed or not adopted this strong stance on the necessity of a new process, no other actor would have done so, or, if it had, would have been able to push this through without the EU's strong support. As observed in the narrative, other industrialized players were rather passive and disengaged at COP 17/MOP 7, while China and India had to be convinced – and they were (test on *absence of auto-causation*: positive).[9] It can thus be argued that the EU did exert at least partial influence, together with its coalition partners, over this outcome. Reputation analysis drawing on commentators' views and the self-perception of key EU negotiators largely confirms this observation. Commissioner Hedegaard and her staff praised the EU's role at Durban, claiming that "Europe has

---

[9] The question could be posed how big of a sacrifice this was for players like the US, China or India. Nonetheless, it has to be noted that they subscribed to a procedure suggested by the EU.

brought about a new phase in global climate policy" (Hedegaard 2011; Delbeke 2011). This was echoed by other observers, even if they rightly insisted on placing the Durban outcome and the Union's performance into context: "the role played by the EU can only be considered a success if compared to the abject disappointment of the 2009 COP-15" (Dupont 2012; Vidal/Harvey 2011; ECFR 2012; Verolme 2012; ENB 2011d). If the EU did exert influence over the Durban outcome and thus "re-established itself as a key player in the climate negotiations" (Sterk *et al.* 2011b: 31; Bäckstrand/Elgström 2013), this influence concerned the future negotiation process rather than the substance of emissions reductions and responsibilities, even if the Chinese and Indian willingness to consider an "agreed outcome with legal force" as well as the absence of a reference to differentiated responsibilities may have repercussions for the distribution of responsibilities in the climate regime. If one moves away from considering solely the very limited EU leverage over the substance of the climate regime during this period, the analysis once again suggests a non-negligible Union influence on keeping regime development negotiations within the UN context. This impact on the institutional framework of global climate talks can even be characterized as high and rather enduring, as the Durban Package obliges parties to at least continue to discuss in the UN regime, and ideally to find solutions by 2015.

It is also in this vein that one has to interpret the EU's influence at COP 18/MOP 8. Since 2010, the EU had clearly displayed its willingness to contribute to a second commitment period under the Kyoto Protocol, even if *not* all major industrialized countries would join into such a novel commitment (*purposive behaviour, temporal sequence*). This position was aimed at demonstrating goodwill to both the emerging and developing countries, with the objective of obtaining their green light for a reinvigorated negotiation process in which these players would consider binding commitments for themselves (*interaction*). While the EU's stance at Doha was thus not new, and while it failed to rally some of the major industrialized players like Japan, Russia and Canada behind its position, it managed at COP 18/MOP 8 to forge the ultimate decision on the new commitment period under the Kyoto Protocol. This allowed the EU to *attain its goals* regarding not only some of the key modalities and especially the overall duration of this period (coinciding with the internal timing of its climate policies for 2020), but also in respect of the interlinked transition to a new negotiation process. Without the EU as a strong advocate for this adoption, a second commitment period would not have been very probable, as not only EU negotiators themselves (Hedegaard 2012; Delbeke 2012), but also other commentators have highlighted (Marcu 2012; ENB 2012c) (*absence of auto-causation*). If the EU did exert influence over this outcome, this has above all significance for the continuity of the UN regime, as it enabled a transition to a negotiation process in which the responsibilities between parties might be more equally distributed. Almost as a by-product, the EU also gained some leverage over emission reduction targets of the other countries that made commitments, which is not insignificant given the fact that these latter can be regarded as being of a more binding legal nature than the pledges made in the Cancun Agreements. The EU's influence should not be overstated, however. Its limits can be observed in the fact that (i) major developed countries did not follow its example, (ii) the EU did not really influence the magnitude of other parties' targets, as its own offer had also remained unaltered since 2007, and (iii) even if the EU obtained the start of new talks towards a legally binding outcome and with targets for all parties, it remained unclear what the final result of the ADP would precisely be, as noted above. In this context, the Union's incapacity to continue to form fruitful coalitions with developing countries at Doha could have negative repercussions in the medium term (Marcu 2012: 1).

To account for the influence the Union exerted during the period 2010 to 2012, both enabling and constraining factors are briefly considered. When it comes to *constraining factors*, it is evident that the external context that had not played to the Union's advantage already before and at Copenhagen would not do so during the three years that followed either. However, internal factors further complicated matters during this period: in the first half of 2010, the EU was in a state of turmoil over the process and outcome of COP 15, with an unclear position on how to proceed. Ever since, internal divisions, and notably the fierce opposition by Poland (the Council Presidency during the second half of 2011), have rendered a reinforced internal climate regime and stronger positioning in global climate politics difficult (Verolme 2012: 10).[10] What is more, inter-institutional quarrels over the implementation of the Lisbon Treaty initially did not facilitate deliberations on a stronger foreign (climate) policy implementation and strategy either. As a result, the EU remained rather silent for much of 2010. Only after the Cancun summit did this situation change to some extent: for 2011 and 2012, the EU was at least able to forge a minimum common position, which represented in essence the revival of its long-standing leadership-by-example approach based on reinvigorated old positions and diplomatic means drawing on continued internal policy advances (Geden 2011). When doing so, and this brings the *enabling factors* into the picture, it benefitted from a general window of opportunity for exerting influence, which had opened up after COP 16. Given the

---

[10] For 2012, some observers even remarked that "the climate diplomacy of the EU is faltering due to Poland's obstruction, which is preventing the EU from increasing its level of ambition" (Agence Europe 2012c).

restored faith in the UN process, but limited investments of other industrialized players and the major emerging economies during this period, discussions on the substance of the regime had become less important than the re-ignition of the process toward negotiating a new global agreement. This provided the EU with the opportunity to step in with its "road map" proposal and resort to a diplomatic strategy that had already proven its worth in the mid-1990s. Greater efforts were invested into coalition-building, notably with the LDCs and AOSIS, and particularly through the Cartagena Group in 2010 and 2011 (Verolme 2012; Bowering 2011), but to a lesser extent already in 2012 when it "failed to build new alliances with the poorest countries and those most vulnerable to climate change, by turning down almost all of the[ir] requests", according to Green MEP Sandrine Bélier (Agence Europe 2012c; see also Marcu 2012). In 2011 particularly, and based on the force of numbers (EU-27 plus its allies), a more assertive stance in defending its arguments for the road map approach could be adopted, which resulted in the successful exercise of influence at Durban. Its leverage over the Doha outcome can then be explained by its willingness to make a commitment on an emissions reduction target it would, in times of economic crisis, certainly achieve by 2020.

In sum, as the UN regime moved toward the expiry of the Kyoto Protocol's first commitment period on 31 December 2012 and recovered from the Copenhagen summit, its proceedings were marked by a high degree of continuity. Despite the crisis of the multilateral process right after COP 15, illustrated by debates about the possibilities of moving forward in smaller coalitions of the willing, the UN climate regime proved a high degree of perseverance. Negotiations were gradually re-centred around the Convention, and, in 2012, moved completely "back on track". In many ways, over 20 years of climate regime negotiations have thus demonstrated an astonishing resistance to change, which can best be understood with reference to the very nature of multilateral diplomacy in the UN system. A similar evolution was detected for the EU. Where observers had speculated about the "Copenhagen disaster" as a critical juncture for its climate policies in early 2010, a major overhaul of either the Union's internal or its external climate policy has not occurred. On the contrary, recent developments insert themselves rather well into the overall logic discovered in this longitudinal study: a slow and incremental development of EU actor capacity paired to a gradual expansion of its foreign climate policy based primarily on a leadership-by-example approach.

CHAPTER 7

Explaining EU Influence on the Global Climate Regime

Chapters 2 to 6 of this study analysed EU influence by tracing actors' interactions in the global climate regime through different time periods. In so doing, they generated numerous findings on the Union's foreign policy activities and their effects and provided explanations of its influence for each analysed period. This chapter strives to generalize within the longitudinal case. It does so by, first, synthesizing findings so as to identify patterns of EU influence and of all pre-specified and newly emerged potential explanatory factors over time and to explore associations between them; second, by identifying outliers in these patterns; and third, by engaging in explanation-building on the scope conditions that account for EU influence via arguing or bargaining.

Patterns of EU Influence across Time

Building on the analytical framework developed in Chapter 1, Table 5 synthesizes the evidence for all studied time periods and key variables. In so doing, it allows for a visualization of its main results by incorporating various levels of analysis, different theoretical perspectives considered as complementary (institutional, interest-, power-, value-based) as well as dynamics over time. It thus fully accounts for the complexity of the instances of social reality studied in this work. At the same time, it accomplishes – wherever applicable and not previously achieved – a conversion of empirical data into more abstract categories to facilitate the formulation of explanations. Concepts coded in this manner are treated, if possible, as continual (e.g. regarding the interest constellation: very homogenous-homogeneous-heterogeneous-very heterogeneous) (Miles/Huberman 1994: 57–58).

There are numerous ways in which the table can be read. Three types of interpretation seem crucial to move from descriptive inference to explanation (Miles/Huberman 1994: 91, 119–122): (i) a consideration of each identified concept (cluster) in its temporal sequence so as to identify patterns of change across time; (ii) a focus on outliers in these patterns; and – against that backdrop – (iii) an exploration of associations between independent variables that seem to be most significant in accounting for EU influence, the dependent variable.

Table 5: EU Influence on the Global Climate Regime over Time

		1991–1995 Building the climate regime	1995–1997 Developing the climate regime	1998–2007 Consolidating the climate regime	2007–2009/ 2010–2012 Reforming the climate regime
DEPENDENT VARIABLE: EU INFLUENCE	On norm	medium	high	medium	low
	On principle	none	none	none	none
	On agenda-setting	medium	medium	medium	medium
	On keeping debates in UN	n/a	n/a	high	medium to low
	Overall	medium	medium	medium	(very) low (2009), medium
CAUSAL MECHANISMS		bargaining	bargaining	arguing, bargaining	arguing
EXTERNAL CONTEXT	Fora	very simple (UN)	simple (UN+G-7)	complex (UN, G-8+5, MEM)	very complex (UN, G-8+5, MEF, G-20)
	Trends and events	none	none	rise in number of actors, rise of emerg. Countries	rise in number of actors, rise of emerg. countries
CLIMATE SCIENCE	Scientific knowledge	very uncertain; not compelling	relatively uncertain; fairly compelling	relatively uncertain; fairly compelling	relatively certain; compelling
DOMESTIC LEVEL OF ANALYSIS (NON-EU ACTORS)	Institutional set-up	complex	complex	complex	complex
	Interest constellation	heterogeneous	heterogeneous	heterogeneous	heterogeneous
	Beliefs constellation	heterogeneous	heterogeneous	heterogeneous	heterogeneous

		simple (INC)	simple (AGBM)	complex (AWG-KP, Dialogue)	very complex (AWG-KP, AWG-LCA, ADP)
	Institutional set-up of negotiations	simple (INC)	simple (AGBM)	complex (AWG-KP, Dialogue)	very complex (AWG-KP, AWG-LCA, ADP)
	Interest constellation (key cleavages)	heterogeneous (North-South, US vs. other indust. countries)	heterogeneous (North-South, EU vs. JUSSCANNZ)	heterogeneous (North-South, EU vs. Umbrella Group/US)	very heterogeneous (BASIC vs. US, EU vs. BASIC/US, LDCs vs. BASIC)
	Beliefs constellation	heterogeneous	heterogeneous	heterogeneous	more homogeneous
INTERNAT LEVEL OF ANALYSIS – REGIME DYNAMICS	Power constellation	hegemonic, bipolar (US vs. other OECD [incl. EU] and G-77, North vs. South)	tripolar (US – Japan – EU; JUSSCA-NNZ vs. G-77/EU)	multipolar (disengagement US; EU vs. Umbrella Group; Umbrella Group vs. G-77)	centrally bipolar; multipolar (US vs. CH(+IND); US/BASIC vs. EU/LDCs)
	Positions on regime development: scope	wide (US – rest)	limited (pro treaty, legally binding targets)	wide (US – EU – emerg. countries – AOSIS)	wide (US/BASIC – other ind. countries – G-77)
	Positions on emission target: scope	wide (US vs. OECD vs. G-77)	wide (US – EU – G-77)	wide (US – EU – G-77 – AOSIS)	wide (US – EU/J – G-77 – AOSIS)
	Positions on differentiation: scope	limited	wide (JUSSCA-NNZ-G-77+EU)	wide (US/Umbrella Group vs. G-77)	wide (US/Umb. Gr.-EU-AOSIS-G-77)
	Logic of action	arguing, then bargaining	arguing, then bargaining	arguing, then bargaining	arguing, then bargaining
	Competence	low	high	high	high

		1991–1995 Building the climate regime	1995–1997 Developing the climate regime	1998–2007 Consolidating the climate regime	2007–2009/ 2010–2012 Reforming the climate regime
EU ACTOR CAPACITY	Foreign policy making and int. coordination	complex	complex	complex	complex
	• Interests	heterogeneous	heterogeneous	heterogeneous	heterogeneous
	• Beliefs	heterogeneous	heterogeneous	heterogeneous	heterogeneous
	Representation	fragmented	centralized (Troika)	centralized ("Climate Troika")	centralized, delegated (Climate Troika, lead negotiators), later stage: fragmented (high level)
	Objectives	relatively clear, focused (a treaty, target)	relatively clear, focused, ambitious on target	relatively clear, diffuse (maintenance and consolidation of regime)	clear, specific (on regime development), relatively ambitious (on target)
EU FOREIGN POLICY IMPLEMENTATION	Overarching strategy	limited, ad hoc	limited, leading by example	leading by example, reinforced outreach	leading by example, broad coalition-building (2010/1)
	Instruments (primarily) used: scope	limited (diplomatic)	limited (diplomatic)	extended (diplomatic, economic)	wide (diplomatic, economic, [coercive])
	Actors (primarily) targeted: scope	very limited (US)	limited (US, Japan, G-77 as bloc)	extended (US (from 2005), Russia (2002–2004), emerging countries, LDCs (from 2005))	wide (US, emerging countries (BASIC), other ind. countries, LDCs)

When tracing patterns across time, several continuities and some significant discontinuities can be detected. Emphasis is placed on each broader concept cluster.[1] As all other variables will need to be interpreted in relation to the dependent variable, it is useful to begin with the latter: **EU influence**. Medium EU *overall influence* could be detected for the entire climate regime history up to the period 2007–2012, with the exception of the year 2009, which constituted a major break with previous trends because the Union's influence suddenly became much lower. This overall influence had been attributed on the basis of an analysis of its influence on two key pillars of the climate regime. EU impact on the key norm of the regime (*emissions reduction targets*) was found to be fairly consistently moderate until 2009, and particularly high during 1995–1997. By contrast, its influence on the *principle of responsibilities* was permanently non-existent. EU *influence over the agenda* of the UN climate negotiations was consistently moderate, even if it did not systematically lead to leverage over the end product of talks. Moreover, a notion of EU *influence over "where to discuss"* climate policy (i.e. in which fora to conduct negotiations) emerged from the analysis (see Chapters 4–6). Repeated attempts by other players, notably the US, to dislocate climate talks from the multilateral arena into other fora (APP, MEF) were countered quite successfully by the EU (and its allies) during the period after the US withdrew from the Kyoto Protocol ratification process in 2001. Confirming the identified pattern, slightly less impact in this regard was detected for the period 2007–2012. Finally, the findings indicated a clear break for this latter period with regard to the *causal mechanisms* underlying successful EU influence attempts: where bargaining had played a major role when the EU was central to last-minute compromises (as on the UNFCCC, the Kyoto Protocol, the Marrakech Accords) or on Russia's ratification of the Protocol, arguing was decisive in the few major instances of EU influence during the post-2012 talks (especially on the inclusion of the 2°C reference in the Copenhagen Accord).[2] Altogether, the period between 2009 and 2012 constitutes thus a break with a long-term pattern of continuously moderately positive EU influence-wielding on the climate regime. The identification of such an outlier in a long-term pattern allows for engaging in a broader exercise of explanation-building.

Continuing the discussion of Table 5 from top to bottom, the cluster **external context** had originally been designed to trace *external events* both in general and regarding the issue of climate change. No major single event

---

[1] Concept clusters (e.g. external context) consist of two or more sub-categories (in the cited case: fora, trends).

[2] A partial exception may have been the adoption of the second commitment period under the Kyoto Protocol, the agreement to which the EU did also employ as a bargaining chip.

with durable effects on the climate talks was however detected. By contrast, in the course of the research, the cluster had to be modified to also include general *long-term global trends*. For one, a clear tendency of increasing complexity of climate politics was observed (*fora*). This was paired to a trend of rising importance of emerging economies in this arena and in global politics more generally. For the last two analysed periods, these changes were perceived as significant factors impacting EU influence, which merit further exploration below. The **issue of climate change** itself was considered in conjunction with the external context: after initial scientific uncertainty in the first IPCC report (1990), the second report of 1995 led to greater awareness of climate change, but also nourished the belief in its manageability through, e.g., flexible mechanisms. While the third report of 2001 did not receive such wide attention, the findings of its 2007 successor were more compelling and shaped many politicians' and public perceptions of climate change as a topic deserving urgent attention. Given the EU's approach to climate change based on the precautionary principle, science has been interpreted in Chapter 5 as a major enabling condition of its influence through arguing in the post-2012 talks.

The evolving science did not spark major durable change in the regime dynamics, which marks another key cluster. As these dynamics are partially the expression of **domestic contexts** clashing at the global level, the latter first require consideration. During the entire regime evolution, the major non-EU players (US, China, India, Japan, Russia) faced intricate *internal institutional contexts* (e.g. in the US through high institutional hurdles for international treaty-making, in other countries through deficient and/or competing institutions involved in climate policy-making). Moreover, the domestic *interest* and *beliefs constellations* on climate change remained consistently heterogeneous. The observed complexity represents a rather stable long-term trend. As these circumstances usually complicated decision-making on climate change, favouring conservatism, they certainly did not facilitate the EU's attempts at decisively tackling the problem globally. Yet, as this cluster of factors remained largely unaltered over time, it can also be ruled out as key explanatory factor for cross-time variation in EU influence.

At the aggregate level of **regime dynamics**, these complicated domestic contexts found their expression in a complex *interest constellation* among key actors in the climate regime since the early 1990s. During each time period, the regime was characterized by very divergent preferences not only on major topics such as targets or responsibilities, but also on related issues like the modalities of emission reductions. Fundamental cleavages opposed the industrialized and the developing world, but also regularly the US (often with allies) and the EU (sometimes with other industrialized countries), and doubtlessly qualify as significant continuities of the climate regime. While such divides may restrain the Union's ability to change others' behaviour and preferences in general, the fact that they remained largely unchanged throughout the evolution of the regime implies that they cannot account for variation in its influence. The same can be said for the *logic of action* under which the regime operated due to these contradicting interests: arguing in early stages of negotiation processes was systematically substituted by bargaining for balancing out players' positions at later stages. Although this might help to account for the difference between EU influence on agenda-setting as opposed to over final outcomes, as further explored below, it cannot aid in explaining its variation across time. A slightly different picture emerges when considering the *beliefs constellation*: more compelling climate science after the fourth IPCC report led to a moderately more homogeneous constellation, when all players became slightly more inclined to engage in reforms of the regime. They did not, however, fundamentally change their key positions, which is why it is also improbable that this factor can account for variation of EU influence over time. A major discontinuity with regard to regime dynamics concerned then the *power constellation* among actors in the climate regime.[3] Clearly, this constellation evolved most remarkably since the early 1990s. While the US was the largest emitter and hegemon of sorts in the early 1990s, the US-Japan-EU triangle was dominant in the regime during the mid-1990s. In the 2000s, gradual socioeconomic transformations, notably in the emerging countries, rendered the world – and the climate regime – politically more multipolar. In 2009 then, tendencies of a new bipolarity could be observed with the central importance of the G-2 regrouping the two major, economically closely tied emitters US and China (sometimes via the BASIC group).[4] For the EU, the evolving power constellation meant a discontinuity with regard to its own relative power, which may have strong explanatory value for its influence on the regime: fairly powerful in the 1990s, the Union seized the opportunity when the US disengagement during the George W. Bush era opened a window of opportunity for greater exercise of influence. This has come to a gradual halt since 2007, and most notably in 2009, when it partially lost its clout to the emerging economies. These power-related cleavages had repercussions on many discussions in the regime, and also found their

---

[3] Power is understood as relying on the material and immaterial resources of an actor, and it was found to be closely linked to its emission profile: countries with the highest emissions and/or mitigation possibilities dispose of enormous potential to exert influence in the climate regime.

[4] At the same time, the continuously unaltered opposition of the two major blocs – the G-77/China and the developed countries (OECD, later Umbrella Group and EU, whereby the EU would at times side with the developing countries) – was also marked by a power struggle.

expression in an ever-increasing complexity of the *institutional set-up* of the negotiations (with the two tracks under the UNFCCC/Kyoto Protocol plus many fora outside the UN until 2012). This observation about how the Union fit into the overall external environment with regard to power, self-evidently a central variable for explaining its influence, also necessitates a consideration of how it was situated *vis-à-vis* other players regarding its *positions*. In the course of the evolution of the climate regime, it was not only increasingly less powerful in relative terms, but also often defended outlier positions. This was particularly the case regarding the overall issue of regime development during the period after 1997, when the EU was calling for a legally binding approach. By contrast, the Union's influence was highest when key actors' positions on the overarching aim of the negotiations (i.e. adopting legally binding QELROs) were closest (in 1997). Defending outlier positions renders the exercise of influence difficult.

If one focuses next on the EU-related variables summarized in Table 5, a distinction needs to be made between the Union's actor capacity and foreign policy activities. Regarding **actor capacity**, several continuities stand out: the EU's *competence* in primary law to act on climate change internally and externally was continuously high since the entry into force of the Maastricht Treaty in 1993. At the same time, the *interest and beliefs constellations* among member states were consistently heterogeneous, rendering *foreign policy-making* especially complex. This did not prevent the Union from acting comparatively ambitiously on climate change and defending increasingly articulate *objectives* regarding the future shape of the UN climate regime, but only on the basis of delicate internal compromises. Such compromises, in turn, hampered flexibility and required greater coordination efforts to the detriment of foreign policy implementation. A major discontinuity in the whole picture is the *institutional set-up* of internal decision-making and representation, which evolved through the Commission's participation in the Troika, and the ever-increasing specialisation of the WPIEI-CC combined with the emergence of a system of lead negotiators and issue leaders. Although intended to improve the Union's unity as a global player, these institutional changes – and with them, an assumed overall improvement in its actor capacity – did not coincide with greater EU influence in the global climate talks. The improved actor capacity did, however, come with a change in its **foreign policy activities**. The EU's outreach activities grew in quantity and changed in quality (broader geographical scope, use of wider range of instruments) over time. While the *overarching strategy* since at least the Kyoto COP can best be described as "leading-by-example" and problem-solving, diplomatic *instruments* were supplemented by economic tools in the period after Kyoto. Moreover, legal instruments were used in both an incentivizing and (threatened) coercive form, evoking border adjustment mechanisms and aviation levies. Regarding *actors*, the EU's focus on the US during the entire regime history represented a major continuity. Gradually, however, other actors became also significant targets of Union influence attempts, including Japan since 1995, the other industrialized countries since 1998, and the major developing countries, but also LDCs since the early to mid-2000s. These changed patterns arguably promised potential for improved EU influencing. It was thus unexpected to find that – while the overarching strategy bore some fruit between 1997 and 2008 and in 2011 – the Union's regenerated foreign policy implementation did not contribute to influence at crucial talks such as COP 15 in 2009.

Generally, from this exercise of extracting cross-time patterns, several outliers and/or significant, unexpected findings need to be highlighted. Regarding the variation across time of EU influence, the latest analysed periods (2007–2009/2010–2012) – and here especially the months before COP 15 and the summit itself – mark the key outlier. At the same time, this phase constituted also an exception with regard to (i) relative scientific certainty about the *issue of climate change*, (ii) *global trends* regarding the *power constellation*, which led to the described proliferation of fora beyond the UN, the complex institutional set-up of the regime debates and a drop in relative power of the EU also within the UN climate regime, (iii) more developed *EU actor capacity* and (iv) *wider EU foreign policy activities*. Although patterns do not "prove" that these factors actually determined EU influence, they provide strong cues that the conditions for variations in EU influence on the climate regime across time need to be in the first placed searched for in these analytical categories and their inter-relations. Rather than at the domestic level, which was complex (possibly in different ways, but with the same effect of complicating EU influence-wielding) for all major players during the entire history of the climate regime, the results of the study suggest that the crux for understanding and accounting for EU influence lies in fact in the *interplay between the international and the EU levels of analysis*. Especially the detected pattern indicating that the EU's evolving actor capacity (EU level) did not yield the desired effects in the changed global regime context (international level) during the late 2000s calls for further exploration.

Comparing EU Influence Attempts to its Actual Influence: the "Goodness of Fit" Puzzle

Central to exploring the crucial interplay between the evolving external context and the EU's foreign policy capacity is focussed attention on the notion of EU foreign policy implementation. A further exploration and development of the concept of influence attempts constitutes therefore a key means of moving toward explanations of its influence across time. Such concept development becomes possible by assessing its influence attempts against their success, i.e. whether an instrument actually yielded influence or not, as far as such an evaluation is possible.[5] This ultimately allows for detecting further patterns of EU activity in relation to its environment, which can facilitate explanation.

Table 6 compiles, rather than compares across time, all major (types of) EU influence attempts. Building on Table 2 developed in Chapter 1, it distinguishes between the main two types of influence attempts and key (categories of) actors in the climate regime: the US, other industrialized countries and the G-77/China, the latter split here – following the latest developments – into the BASIC countries and the LDCs/AOSIS. As a result of tabulating the EU's influence attempts in this way, further patterns emerge on the crucial relationship between its acts and the external context in which it operates. Clearly, EU foreign policy implementation considerably expanded over time to cover almost all countries (with the exception of some LDCs) and instruments (coercive tools excepted). Yet, imbalances existed: *vis-à-vis* the US and major developed countries, influence attempts were mainly focused on persuasion-based tools during the entire history of the climate regime. With regard to the developing world by and large, the major tools were concrete economic policy instruments to engage those players in climate mitigation policies, paired to argumentation-based influence attempts. Despite these differences, the table illustrates first and foremost that the EU really did try out quite a range of tools *vis-à-vis* different actors. Yet, a large number of its attempts remained without tangible results.

This empirical finding calls even more for an investigation into the causes of the apparent ineffectiveness of much of EU foreign climate policy. Obviously, numerous reasons related to the external environment could account for this (e.g. the relative power of other actors). A closer inspection of the Union's successful influence attempts suggests, however, that EU influence-wielding also has to do with the precise way in which it implements its foreign policy decisions and, more specifically, the extent to which it takes into account the external context. It is thus more often than not about the choice of the right instrument at the right time. The Union's – comparatively – greatest successes, besides a somewhat more diffuse influence over agenda-setting, concerned the bargaining proposal of 15% emissions reductions in 1997 (a politically set, fairly arbitrary diplomatic proposal aimed at major industrialized countries), bargaining

---

[5] Not all of the Union's manifold influence attempts have been assessed in detail for their success, but overall trends on how certain types of attempts contributed to EU influence can be derived from the study.

Table 6: EU Influence Attempts and their Success in the Global Climate Regime (1991–2012) – a Compilation of Instruments[1]

ACTORS	USA	UMBRELLA GROUP	BASIC	LDCs/ AOSIS
CATEGORY OF INFLUENCE ATTEMPTS: ARGUING				
Issue démarches, declarations, statements Visit Make proposals Initiate political dialogue Send envoys, experts Sponsor conferences Support action	Issue démarches before COPs (/), declare positions, demand more ambition (/) Political Dialogue (/) Visits, send experts, organise conferences (/)	Issue démarches before COPs (/), declare positions, demand more ambition (/) Political Dialogue (/) Visits, send experts, organise conferences (/)	Issue démarches before COPs (/), declare positions, demand more ambition (/) Political Dialogue (/) Visits, send experts, organise conferences (/)	Declare positions (/) Visits, send experts, organize/ support conferences (/)
	Proposals: argumentation for ambitious climate mitigation based on climate science, economic modelling and own legislation (ETS-linkage) – search for mutual gains (/)	Proposals: argumentation for ambitious climate mitigation based on climate science, economic modelling and own legislation (ETS-linkage) – search mutual gains (+: Australia, Japan)	Proposals: argumentation for climate mitigation necessity on basis of science (+ on 2°C) and economic modelling to aid them in assuming bigger role (−)	Proposals: argumentation for climate policies in general (/)
CATEGORY OF INFLUENCE ATTEMPTS: BARGAINING				
Offer diplomatic recognition Offer membership		Offer help on WTO membership (Russia, +)		
Offer trade, cooperation or association agreement Reduce tariffs Increase quota	15% by 2010 target proposal (1995–7, +) Offer to go to 30% reductions (2007–12,−)	15% by 2010 target proposal (1995–7, +) Offer to go to 30% reductions (2007–12,−)	Offer to go to 30% reductions (2007–12,−)	

ACTORS	USA	UMBRELLA GROUP	BASIC	LDCs/ AOSIS
Grant inclusion in the general system of preferential treatment Provide aid Extend loans Threaten with embargo or boycott	Offer energy/ technological partnership (−)	Offer energy/ technological partnerships (/) Offer other partnerships (Russia, +)	Offer technological and energy partnerships (-)	Offer technological and energy partnerships, aid (+)
Grant diplomatic recognition Grant membership Conclude trade, cooperation or association agreement Reduce tariffs Increase quota Grant inclusion in the GSP Provide aid Extend loans	Conclude partnership (e.g. creation of Transatlantic Energy Council) (−)	Conclude partnerships (+/−) Conclude other partnerships (Russia, +)	Conclude energy/ technological partnerships (/)	Provide aid in terms of capacity-building, technology transfer, financing of energy transitions and adaptation etc. (e.g. initiative launched during COP 14, +)
Threaten diplomatic sanction Threaten to refuse recognition Threaten with embargo (ban on exports) or boycott (ban on imports) Threaten to… increase tariffs, decrease quota withdraw GSP reduce/ suspend aid delay conclusion of agreements suspend agreements	Threaten not to attend Major Economies Meeting (at COP 13) (/) Inclusion into ETS (aviation) (/)	Inclusion into ETS (aviation) (/)	Threaten border adjustment taxes, exclusion from CDM (/) Inclusion into ETS (aviation) (/)	Inclusion into ETS (aviation) (/)

[1] (+) indicates that an influence attempt contributed to EU influence; (−) indicates that an influence attempt did not result in influence; (+/−) signals a mixed outcome. (/) indicates that the effects of an influence attempt on EU influence were not clear.

proposals made to Russia in 2004 (a combined diplomatic and economic tool targeted at one actor), arguing for the 2°C target since 1996 (a science-based, long-term diplomatic proposal aimed at everyone) and, to some extent, for a mitigation approach built around emissions trading since the mid-2000s (a proposal based on economic rationale and targeted at industrialized countries). Seen from the perspective of the influence-wielder, the influence attempts these successes were based on were quite different in nature. No clear pattern suggests itself, which implies that no one-size-fits-all foreign policy approach that will consistently yield influence in global climate negotiation processes appears to be at the EU's disposal. If considered from the perspective of the outcome (i.e. exerted influence), these different attempts must have somehow *matched* the external parameters set by the global climate regime at the given points in time to trigger actual EU influence, however. In other words, at times the Union seems capable of designing foreign policies that fit well enough with the external conditions to result in influence, while at other moments there appears to be a mismatch. To better be able to articulate this finding, it can be beneficial to give it a name. The term "goodness of fit", borrowed from the literature on Europeanization, may adequately capture the notion of match between EU activities and the external context (Börzel/Risse 2000). If this goodness of fit is high, the Union's chances for exerting influence are apparently increased. Whenever the degree of "goodness of fit" seems low, EU influence appears to be less probable. As a – for this research context – new concept that emerged from the findings, the notion of goodness of fit requires further attention.

Determinants of EU Influence over Time: Propositions on Causal Mechanisms and their Scope Conditions

Having re-considered each potential explanatory factor of EU influence as well as the main interactions between these factors from a longitudinal perspective, it is now time to move from descriptive inference to explanation and provide clearer answers, based on cross-time consideration, to research question 3: *why did the European Union actually exert influence on the negotiations pertaining to the development of the global climate regime?* Answering this question requires a combination of causal mechanism and conditional causal analysis, which acknowledges that "causal effects depend on the interaction of specific mechanisms with aspects of the context within which these mechanisms operate", and that causes are conjunctural (Falleti/Lynch 2009: 1144). The two causal mechanisms arguing and bargaining indicate why an (EU) influence attempt led to an outcome: through agency involving either reason-giving or strategic behaviour. Yet, whether the one or other can operate in a given social context depends on certain scope conditions. For that reason, besides specifying this context – or, in the terms of conditional causal analysis, the "causal field" – to which the suggested causal statements apply, the multiple conditions under which EU acts of arguing or bargaining result in influence need to be exposed through a constant comparison of instances of EU influence (success) and non-influence (failure) across time by drawing on the empirical evidence gathered in the study (Mackie 1974: 34). Scope conditions will be assembled in the form of plausible propositions, formulated in the language of conditionality. Just like for the narrative process trace this study relied on, "the degree of belief in a causal hypothesis depends on the strength of the evidence available to support it" (Marini/Singer 1988: 348). The propositions account for the EU's influence on the climate regime over time – and represent thus the key theoretical conclusions of this study – but could, slightly altered, also serve as hypotheses for future research on the EU's activities in this regime. To situate the propositions into the broader context provided by the scientific community and detect possible inconsistencies and/or rival accounts, they are related to the literature on EU foreign policy and international regimes.

The Causal Field and the Causal Mechanisms: EU Influence in Different Phases of Climate Regime Negotiations

For any conditional causal analysis, the specification of the "causal field" to which statements apply is of central importance (Mackie 1974: 34). It is this field that provides the context in which conditions need to come together to allow for the generation of a particular result (e.g. EU influence through bargaining). In the present study, this causal field was specified as the global climate negotiations covering the UN climate regime and a limited number of clearly circumscribed external analytical units (context, science). In this imaginary field, many variables were identified that may serve as conditions enabling or constraining EU influence. These conditions do, however, in the reasoning adopted here, not directly account for EU influence, but only through the causal mechanisms arguing or bargaining. Causal mechanisms are patterns or ultimate causes that can be transferred from one causal field to the next. In function of the causal mechanism operating in a given field, the logic of agency and interaction in that field is altered for the given instance of influencing: arguing presupposes interaction based on reason-giving, while bargaining implies that players interact strategically.

Against this backdrop, it is interesting to observe that UN climate negotiations apparently function according to different logics at different stages. Although, as noted earlier, communicative acts based on arguing and bargaining can co-exist in negotiation contexts (Kleine/Risse 2005), the findings suggest a straightforward cross-time pattern: climate negotiations that operate with a predetermined deadline (e.g. COP 3 and COP 15 were designated as end dates of regime reform processes respectively with the Berlin Mandate and the Bali Roadmap) typically function predominantly according to an arguing mode during earlier stages, when key actors do not yet possess stable preferences on many issues or deliberately hold back positions, while shifting into a bargaining mode towards the end. This was strongly indicated by the technique of "backloading" frequently employed by the major actors, by proposals unambiguously aimed at horse-trading and by the regular last-minute compromise outcome embodied in the idea "that nothing is agreed until everything is". The findings further point to a certain congruence between these phases and the causal mechanisms through which the EU exerts influence: the Union was consistently influential through arguing by proposing problem-solving solutions to the climate challenge when it came to agenda-setting during early stages in the negotiation processes, while it had influence on the final outcomes of negotiation episodes only if it was engaged in bargaining. This leads to a not so trivial observation: employing arguing-based tools in a context in which other actors look for give-and-take does not work, as the EU's performance in the endgame of COP 15 demonstrated. Making bargaining proposals when others are still building their positions is equally dysfunctional.

This observation provides for a clear distinction between stages in – and thus types of – UN climate negotiations through two logics of action, which already indicates that EU influence through the one or other mechanism will depend on divergent conditions. It finds itself in congruence with the theoretical proposals of scholars applying communicative action theory in International Relations, e.g. through regime analysis (Hasenclever *et al.* 1996: 205–206; Müller 1994), or to multilateral (including climate) negotiations (Ulbert *et al.* 2004; Steffek 2005). In this literature, the finding that the importance of arguing per se decreases when negotiations have passed the agenda-setting, problem-definition and early positioning phases is largely confirmed (Ulbert *et al.* 2004: 7–8; Kleine/Risse 2005: 18). The window of opportunity for any actor to exert influence through arguing in a regime thus closes as talks move toward a concluding COP. In this context, Ulbert *et al.* (2005: 15) and Checkel (2005: 813) also stress, in the words of the former, that "the less certain actors are about the nature of the problem and about their own interests and preferences, the more they are likely to be open to persuasion and arguing". In early stages of negotiation processes, this "uncertainty" provides an indispensable "prerequisite" of successful arguing (Kleine/Risse 2005: 19), which the EU – through a proactive problem-solving strategy – can (and has) exploit(ed) to its advantage.

Necessary Conditions: Actor Capacity and Foreign Policy Implementation as Explanatory Factors of EU Influence

Before exploring the differences regarding the scope conditions between a negotiation context governed by the rules of arguing and a bargaining situation, general observations can be made about any type of climate negotiation situation when it comes to the *internal* conditions for EU influence. They concern both actor capacity and foreign policy implementation. At the same time, the notion of goodness of fit between EU policies and external contexts is further specified.

Actor capacity, i.e. the legal and institutional framework for and practice of the EU's internal decision-making, coordination and representation, constitutes a prerequisite for its influence because it provides the Union with the necessary abilities to act independently on the world stage. Without possessing a minimum degree of actor capacity, the EU could not define a position, let alone a strategy of foreign policy implementation. Yet, by itself, actor capacity does not account for variation in EU influence. The findings for the post-2012 period demonstrate that it was unable to convert power into influence despite a fairly developed actor capacity. Conversely, a low degree of actor capacity does not preclude the EU from exerting influence: during most of the UN climate regime history, the Union was relatively influential despite frequent internal incoherence, ad hoc coordination and cacophonic representation. Hence, no indicators were found to suggest a quantitative correlation between the Union's degree of actor capacity and its chances of exerting influence. By contrast, greater actor capacity does seem to enable the EU to pay more attention to foreign policy implementation, thus indirectly heightening its chances of exerting influence. In the final analysis, a minimum degree of actor capacity appears therefore to be a threshold condition representing a necessary, but not sufficient prerequisite for the EU's exercise of influence. This yields

PROPOSITION 1: *In global climate regime negotiations of any type, the EU can only exert influence if it possesses a minimum degree of actor capacity, which represents thus a necessary, but by no means sufficient condition for successful influencing.*

Indicators for the degree of EU actor capacity are the various components of this concept so that it would be necessary to test, inter alia, whether the Union disposes of sufficient legal bases and instruments to act or if and how well its internal coordination and representation arrangements function. As a minimum, one would expect that it disposes of legal competence and institutions capable of adopting and adapting positions and defining who is to represent these. As the concept of actor capacity captures many of the "usual suspects" identified as explanatory variables (internal decision-making, coordination) of the Union's performance by actor-centric accounts of EU foreign policy analysis (Groen/Niemann 2012; Groenleer/van Schaik 2007), proposition 1 suggests that none of these seem to (independently) play such an important role in explaining the Union's effectiveness in the climate regime after all. This is not to mean that the components of actor capacity are obsolete. Analysts of foreign policy implementation confirm that these elements still matter very much as internal preconditions when it comes to accounting for why the EU was (un)able to define a policy and/or adapt it and/or defend it (Brighi/Hill 2008: 125). Yet, just *being* a foreign policy player, i.e. having the capacity to act, is insufficient for exerting influence. An actor also needs to possess a strategy on how to use its resources, and carry it out to exploit its potential.

These observations inevitably raise the question to what extent *foreign policy implementation* – as a link between actor capacity and the external context – must be considered a condition for EU influence through arguing or bargaining. The concept of foreign policy implementation was originally conceived in a fairly narrow fashion as "the execution of influence attempts" based on certain tools at the moment in time when "actors confront their environment and (…) the environment confronts them" (Brighi/Hill 2008: 118). The importance of an *overarching strategy* – in terms of both political practice and an analytical unit in studies of foreign policy and influence – later emerged as a crucial sub-category of this concept to supplement the notions of "foreign policy instruments used" *vis-à-vis* specific (groups of) actors. Regarding this latter aspect, Table 6 provided an overview of the EU's influence attempts by bringing instruments and actor groups together, exposing patterns of the Union's actual outreach strategy. It was found that in some instances the EU's overarching strategy that informed the choice for particular foreign policy instruments was adequate. In the Kyoto Protocol talks, the choice for formal, diplomatic instruments out of the available range of tools, the frequency (often) and timing (early) of influence attempts, their substance (highly political on the targets and PMs, rigid) and coherence (putting very much all eggs in one basket, i.e. on the target) had an important impact on the finally adopted target proposal because the external conditions were favourable. This front-running strategy also played out positively – via arguing – for the EU's influence on many elements of agenda-setting and the 2°C target adopted with the Copenhagen Accord. By contrast, the same approach was interpreted as not appropriate for the exercise of influence on many other issues and at other crucial points in time when the EU could not benefit from positive external conditions. The most telling example of this was probably provided by the final stages of the post-2012 negotiations in 2009, during which key non-European players were more concerned with geopolitical power games and the protection of their economic self-interest than with solving the problems posed by climate change. This indicates that a strategy that may be fitting for one point in time in the climate regime might not be fitting for another period because important external parameters may have changed. From these observations, one can conclude that the EU's chances for exerting influence in any type of climate regime situation depend on how well its foreign policy approach fits the negotiation context:

> **PROPOSITION 2:** *In global climate regime negotiations of any type, the EU can only exert influence if its foreign policy implementation effectively fits the evolving regime context.*

The degree of goodness of fit between the EU's foreign policy implementation and the regime context therefore qualifies as another necessary condition for its exercise of influence. To test this goodness of fit, strategy, instruments used and actors targeted can be employed as indicators of foreign policy implementation and assessed against the international environment dissected into various components: the logic of action (does the EU act in line with other players' predominant mode of acting?), actors' interests, beliefs and positions in individual and aggregate (are those understood and addressed by the EU?, is the EU in line with them, at least to some extent?), the power constellation (is the Union aware of its own capacities, their potential and limits?) and the institutional set-up of negotiations (is the EU institutionally prepared for and present in all fora?). Comparing proposition 2 to the literature on (EU) foreign policy, research on foreign policy implementation – even though its aim is to understand why an actor acted the way it did rather than with what effects – largely supports the findings of this study. First of all, this body of literature underscores the importance of the restraining effects of the external context, but acknowledges the importance of agency and "strategy" (Brighi/Hill 2008; Webber/Smith 2002). Yet, "neither strategy nor context taken in isolation can explain the success or failure of a certain foreign policy to deliver an intended outcome" (Brighi/Hill 2008: 119). Rather, in "order to be successful in achieving their objectives, actors need to pursue a foreign policy that is compatible with the context" (Brighi/Hill 2008: 125). Linked to this, the same analysts also clarify that implementation has to be considered as quite distinct from foreign policy decision-making: positions adopted by an actor "are not self-executing" (Brighi/Hill 2008: 127, 134). These observations have important theoretical and political-practical implications. On the one hand, there is a clear difference between "the capacity to act and the capacity to get results", which implies that just focusing on actor capacity in academic research on EU foreign policy performance is indeed insufficient for explaining its influence, as already pointed out in the discussion of proposition 1 (Webber/Smith 2002: 80).

On the other hand, just possessing a position is equally insufficient for exerting influence in practical terms. Yet, this is precisely what the EU oftentimes appeared to think when it employed its leading-by-example approach in the global climate negotiations: that possessing a proactive position would necessitate no further action besides its explanation. This lack of strategic thinking may account for many instances in which the EU did not exert influence despite an a priori conducive environment. As a matter of fact, analysts of foreign policy implementation report that foreign policies often fail not due to bad design, but precisely because they were not at all or insufficiently implemented, i.e. essentially not targeted to the context (Brighi/Hill 2008: 123; Webber/Smith 2002: 80).

To take the notion of "compatibility" a bit further, the concept of goodness of fit can be explored some more: scholars of Europeanization employ it as a top down concept to assess the impact the EU has internally, i.e. on its member states, in terms of compatibility between EU and national laws, institutions and policies (Börzel/Risse 2000). In their reasoning, a "misfit" leads to greater pressure on the member states to adapt to the EU: "the lower the compatibility between European and domestic processes, policies and institutions, the higher the adaptational pressure" (Börzel/Risse 2000: 5). By consequence, the goodness of fit is considered as high when a national policy "satisfies the expectations or requirements of European policy and law" (Caporaso/Jupille 2001: 23). Analogically, and turning this reasoning completely around, goodness of fit as a bottom up concept for assessing EU external impact here would be an indicator of how well the Union matches the conditions set by its evolving external environment in institutional and policy terms. A higher degree of fit improves the effectiveness of its foreign policy, yielding more favourable outcomes.

Conditions Triggering the Causal Mechanisms

Specific conditions are necessary for the EU to exert influence through bargaining or arguing.

First, from the rich data gathered and analysed so far, insights can be derived on the conditions under which EU influence on the climate regime becomes possible through *bargaining*. During all negotiation episodes before 2009, the EU was a fairly successful bargainer on one key pillar of the climate regime (targets) during the endgames of talks, but not on other issues, including the CBDR principle. Reconsidering these patterns, one can therefore ask which conditions were present for the EU to exert influence over the emission reduction targets that were not present for other issues. From the analysis of successful instances of EU bargaining (INC, COP 3, COP 7), several insights may be gleaned: the Union was more effective when it was (i) on an equal footing with other major players, (ii) when it did not defend an outlier position with regard to the overall outcome of a negotiation process, i.e. when the interest scope on this was more limited (e.g. COP 3: all players wanted legally binding QELROs; COP 7: all players desired the operationalization of the Kyoto Protocol) as opposed to wide (COP 15: EU argued long for a legally binding outcome, other major actors were in favour of a political agreement), and (iii) when it was lined up for bargaining and behaved coherently on the basis of its position. Counterfactual analysis shows that these conditions were not present when the EU overtly failed to exert influence: on the topic of differentiation, the Union was often unsure about its position (1995–1997, early 2000s) and was effectively sidelined by (relatively) more powerful actors (G-77/China, US) in a context of strongly polarized interests. Further, EU influence never occurred when it was not sufficiently prepared for bargaining and/or in line with what other players aimed for, as with its ineffective 30% conditional offer for emission reductions during the post-2012 negotiations. This latter example also underscores the second major internal condition: the EU's unity as a strategic foreign policy actor. If the EU acts largely as one bloc, its chances for exerting influence are higher, as in Kyoto or in most of the post-Kyoto period until 2005. During the endgame in Copenhagen, by contrast, the EU was represented by many voices sending various messages (the Troika, the big member states), which further diminished its already (through the external context) low chances for successful bargaining. What these observations boil down to is then best captured through the following proposition:

PROPOSITION 3: *In global climate regime negotiations functioning according to a bargaining logic – assuming minimum EU actor capacity – the EU can only exert influence through bargaining if*
– *its relative power is at a comparable level to that of other major players AND IF*
– *its position on the overarching aim of the negotiations and/or on specific items under discussion is close to that of those players AND IF*
– *it is well-prepared for strategic interaction in the specific negotiation context AND IF*
– *it behaves coherently as a foreign policy actor.*

As the findings do not allow for qualifying any of the conditions as necessary, but suggest that all of them together are sufficient for EU influence in the climate regime, they can best be characterized as "INUS", i.e. "**I**nsufficient but **N**on-redundant parts of an **U**nnecessary but **S**ufficient condition", conditions (Mackie 1974: 62). Indicators for the relative power are, as previously remarked, the relative share of global GHG emissions making a party central to mitigation, but also a player's overall relevance to global politics, which can be based on material (e.g. size of the economy) or immaterial resources (diplomatic skills etc.). Its preparedness for bargaining in the specific regime negotiation context and the coherence as an actor can also be dissected into indicators. The former requires not only that the EU sufficiently understands other actors' stances and preferences and possesses an adequate position – which signals elements for give and take and a certain degree of flexibility implying fall-back positions (actor capacity) – but also an appropriate defence of this position. This demands the use of suitable instruments which address other actors' positions, preferences and underlying interests at the right time (foreign policy implementation). Coherence, in turn, implies a stringent external defence of its positions.

Proposition 3 engages literature at both levels of analysis, since it combines external with actor-related determinants of EU influence. Two strands of regime theory with regard to the external conditions need to be considered: the neo-liberal institutionalist interest-based and the power-based strand (Hasenclever *et al.* 1996). Firstly, neo-institutionalist regime theorists point to the fact that the prospects for agreement in a regime context are highest if mutual interests among actors exist (Rowlands 2001: 54–60; Young 1989b: 366). As the EU was consistently among the actors that most urgently desired the climate regime negotiations to succeed, one can assume that its own influence would, to a certain extent, co-depend on the prospects for successful regime negotiations. In that case, regime theory can help to account for EU impact: bargaining for a compromise in which all key actors gain something necessitates positions that are close, with some overlapping interests, at least on the overarching aim of the negotiations (e.g. wanting to reach an agreement of a certain type). During the UNFCCC and Kyoto Protocol negotiations, all major players were interested in a meaningful agreement so that such a convergence of positions on the overarching aim of the negotiations existed. In counterfactual perspective, regime negotiations largely failed (and with them the EU) when key players had no overarching mutual interests, and certain actors were particularly unwilling to play by the rules of bargaining, following inflexible, incommensurable take-it-or-leave approaches on key issues in the talks (e.g. the G-77/China on the issue of differentiation throughout the entire regime history, the US in 1992, China, India and the US in 2009). As a result, from an institutionalist regime perspective – taking the EU's desire to get to a multilateral agreement for granted – the second component of proposition 3 (positional proximity) is largely covered by the relevant theories. Secondly, as strategic interaction presupposes that all actors are willing to "play by the rules" of bargaining, the issue of relative power, engaging a different strand of regime theory, comes into play. Only those actors who are powerful enough can afford not to play by these rules. In this respect, the first part of the proposition that emerged from the empirical evidence is in line with hypotheses of the power-based strands of regime theory, which assume that greater symmetry in the distribution of power will heighten the prospects of regime formation or reform (Hasenclever et al. 1996; Krasner 1991). This, in turn, implies that the EU's chances for exerting influence will be relatively higher if it is among the most powerful players. Conversely, under conditions of unanimity, if an asymmetry of power exists in regime negotiations and the EU is not among the powerful players, its chances for exerting influence (together with the chances that a regime negotiation process will be successful) will be considerably diminished. The evidence from the climate regime evolution suggests that the EU was regularly not powerful enough *vis-à-vis* other industrialized countries as a bloc and/or the US and/or the G-77/China, especially on the issue of differentiation. By contrast, when it was among the most powerful players, as during the UNFCCC and Kyoto Protocol negotiations, its influence was relatively pronounced. Moreover, the EU was also comparatively more influential on the broad lines of talks when the relatively more powerful US had effectively disengaged from the regime (between 2001 and late 2007). Neither on these occasions nor during the period after 2001 could it have exerted influence without the presence of the other INUS conditions, however. In the face of this multiple conditionality, the blunt realist hypothesis that "power is the central feature of regime formation and survival" and, by extension, of the influence of individual players in regime negotiations is not confirmed by the findings (Little 2008: 299). Altogether, and despite this discord between the findings expressed through proposition 3 and power-based regime theory, the congruencies between the two should not be understated. Both suggest that relative power is one central determinant of its influence. This is per se not new or surprising: obviously power is a crucial capacity an actor draws on to exert influence. Yet, and this is what the findings unambiguously demonstrate, there is no simple quantitative correlation between power and influence ("the bigger an actor's power, the greater its influence"): an actor's power as a resource is to be conceived in relative (*vis-à-vis* others in a given context) and never in absolute terms. And the use of this limited, relative power is then also further restricted or enabled through conditions related to the external context as well as to the actor wanting to employ it. This implies that even a less powerful actor can be influential, while a very powerful one can be an ineffective foreign policy player. In the final analysis, EU power per se cannot explain its influence, but its relative power in a given context represents one significant explanatory factor in a broader picture.

With regard to the actor-related conditions for EU influence through bargaining, their importance is partially confirmed when considering the literature on actors' bargaining capacity in multilateral negotiations more widely and on the EU's capacity more specifically. In general terms, this capacity is supposed to be highest when an actor not only has high relative power and acts coherently, but also operates proactively ("making offers, rather than responding to them") and directly addresses others' preferences (Muthoo 2000: 165; Fisher et al. 1991). As a matter of fact, EU influence was at its highest when it proactively made its 15% emissions reduction bargaining proposal in 1997 and stuck with it in an otherwise favourable context. Yet, as seen from this research, the exercise of influence through proactive behaviour is only possible under certain enabling conditions and can therefore not be taken for granted. The fact that a lack of coherence and of preparedness for bargaining restrains influence is confirmed by the literature on EU bargaining power, which points to the recurrent problems of EU inflexibility and incapacity to define fallback positions, rendering strategic behaviour impossible (unless used itself strategically, which does not usually seem to work) (Meunier 2000: 105–106; Rhinard/Kaeding 2006). Considering the fourth component of the proposition, the necessity to behave coherently as a foreign policy actor in bargaining contexts is more generally confirmed by a small subdomain of the EU foreign policy analysis literature (Nutall 2005). This literature parts from the discussion of provisions in the Treaty on European Union on "coherence and consistency" in EU foreign policy (above all, Article 3 TEU). Although the two terms are regularly used interchangeably (Nuttall 2005: 92), a useful analytical distinction can actually be made between consistency referring to an "absence of contradiction" and coherence comprising notions of synergy and added value (de Jong/Schunz 2012). Once coherence has been determined this way, two different types can be identified: horizontal coherence between the EU's (foreign) policies and vertical coherence between EU and member state activities (de Jong/Schunz 2012; Nuttall 2005: 92, 97). The underlying assumption is that both types of coherence ensure EU unity and heighten its chances for impact. Proposition 3 clearly uses coherence in the sense of vertical coherence by stating that a synergetic relationship between the EU and its member states' foreign climate policies across time is an enabling condition of its influence.

In contrast to the conditions identified for EU influence through bargaining, whenever the Union defends, as during much of the global climate regime evolution, an outlier position, somewhere in between a resourceful group of non-European industrialized countries and the G-77/China, and when it possesses a low degree of relative power *vis-à-vis* other major players, *arguing* seems to be the only promising avenue through which to exert influence. As influence through persuasion implies a change of preferences or beliefs in the influenced, the analyst needs to enquire under which external and actor-related conditions the

EU could actually shape those. Besides the uncertainty during early stages of negotiations identified above, a successful change of other actors' preferences or beliefs apparently necessitates an elaborate position on key issues, based on consistent arguments, which are coherently advanced. These arguments seem to be most convincing if based on climate science and related economic models and/or on internal structures grounded in norms that can be accepted by other players and thus directly address their beliefs. Oftentimes, the Union invoked the IPCC and its science as indispensable sources of authority and knowledge. Especially on the 2°C target proposal, which had been a long-standing EU position, this strategy was successful: its proposal was accepted by others at a point (in 2009) when climate science had become more compelling, thus providing for a favourable external context. On other instances of EU influence through arguing, e.g. over the mitigation targets and the emissions trading approach in some countries (Australia), internal structures (climate and energy legislation, the ETS) created with reference to some generally accepted principles (e.g. cost-effectiveness) provided the ground for successful persuasion. The role of timing was essential during all periods, with the EU regularly acting far in advance of other major actors. In counterfactual analysis, when the EU was unclear about its arguing strategy (as during the immediate post-Kyoto period), came too late (ditto), had no compelling science or internal structures based on widely accepted norms that it could invoke (as during most of the regime until the mid-2000s), influence through arguing was impossible. This yields a proposition on the conditions triggering the causal mechanism "arguing":

PROPOSITION 4: *In global climate regime negotiations functioning according to an arguing logic – assuming EU actor capacity – the EU can only exert influence through arguing if*
- *other actors are uncertain about their preferences, AND IF*
- *the EU directly addresses other actors' preferences or beliefs, AND IF*
- *the EU can invoke external sources of knowledge/authority and/or internal structures based on widely accepted norms, AND IF*
- *it behaves proactively, coherently and consistently as a foreign policy actor.*

Once again, the conditions qualify as INUS. Indicators of uncertainty were provided above. To assess the EU's proactivity and coherence, process-tracing seems to be in order so as to search for timing and contradictions in the Union's position and its implementation or between these two. When comparing these findings to the relevant literature, certain overlaps can be found. For the first component of the proposition (uncertainty), it was already discussed above how studies on communicative action in IR corroborate the finding (Kleine/Risse 2005). In a similar vein, the notion of coherence and consistency (component 4) – as used in EU foreign policy analysis (Nuttall 2005) – was discussed in the context of bargaining, for which they were identified as equally central prerequisites. As seen, coherent foreign policy behaviour in bargaining contexts implies vertical coherence between the EU and its member states, which need to send out the same signals to all third parties and make concessions as a group. In arguing contexts, it implies also such vertical coherence, in the sense of a synergetic, repeated defence by all EU actors, including the member states, of a set of logically stringent arguments. Yet, the notion of consistency further underscores the absence of contradictions in this arguing process. In distinguishing between the two, proposition 4 is in line with the relevant EU foreign policy literature (de Jong/Schunz 2012). Regarding the remaining components of the proposition, scholars of communicative action in IR suggest that early and persistent reason-giving, appealing to beliefs, is central for successful arguing (when others are still uncertain about their positions), and that knowledge is key (here the IPCC science) to heightening the influence-wielder's credibility (Ulbert *et al.* 2004: 17; Kleine/Risse 2005: 13). They also point to the fact that an actor who is considered as a legitimate and knowledgeable moral authority has higher chances of successfully persuading others (Ulbert *et al.* 2004: 16; Checkel 2005: 813). The EU's influence through arguing regularly coincided with enhanced efforts at being perceived as more legitimate through internal preparation and/or legislation as well as the use of IPCC science or some other external source of legitimacy (such as widely respected norms like cost-effectiveness embodied in the flexible mechanisms). Yet, certain examples of incongruence between some of the findings and the theoretical propositions from this body of literature can also be detected. Checkel argues that an entity that wants to exert influence through persuasion does not "lecture or demand, but, instead, acts out principles of serious deliberative argument" (2005: 813). Clearly, in the EU case, this normative condition was rather unfulfilled and apparently unnecessary in many instances. Partially due to its active lobbying for the 2°C target, the EU was joined by others in its support for this aim in 2009, to name but the most evident example. Altogether, however, the fourth proposition is largely in concordance with broader theoretical considerations made in the literature on arguing in international relations.

Conclusion

This study examined the activities and impact of the European Union as a foreign policy actor in one central domain of global politics. In so doing, it touched upon a range of crucial academic and political debates about the opportunities for a single actor to make a difference on the world stage in an age of globalization. This concluding chapter synthesizes the key insights of the study by explicitly answering the three research questions that guided it, before setting the results into a broader context. To do so, a brief sketch of the contribution this study makes to existing research is linked to the identification of research desiderata. The work closes with a reflection on the normative implications of the study, projecting itself into the future of EU foreign climate policy.

Major Findings of the Study and their Significance

Understanding the EU as a Foreign Policy Actor in the Global Climate Regime

When it comes to answering the first research question of *how the European Union attempts to exert influence on the global climate regime*, the findings of this study were fairly clear-cut across time. As a foreign policy actor, the EU makes use predominantly of formal, diplomatic tools aimed at solving the problem of climate change through arguing *within* the multilateral arena. Its main influence attempts are geared toward the UN negotiation process, and regularly come in the form of written or oral submissions to the UNFCCC, based on internal political agreements or legislation. In advancing its positions, timing is frequently crucial: the EU regularly tries to be the first major player to make far-reaching substantial proposals. Across time, a clear tendency toward complementing this arguing approach with incentivizing economic tools, for example through bilateral technological partnerships with emerging countries and clean energy aid programmes for LDCs, could be discerned. The Union has also increasingly engaged in broader outreach activities in recent years, mostly to explain its position to partners outside the UN arena. At the same time, its foreign climate policy positions were regularly found to be rather inflexible, resulting from decision-making processes that focus on internal policy preferences with limited reflection about the external political context. Frequently, extensive internal coordination is needed to adjust the external behaviour, which implies less time for foreign policy implementation.

In the final analysis, a systematic mapping of its activities suggests that the EU can best be qualified as a *multilateral, diplomacy-focused and policy-oriented foreign policy actor* in the area of climate change. By contrast, strategic thinking in terms of politics, characteristic of many other players' approaches to global climate talks, clearly represents the more neglected dimension of the Union's foreign policy in this domain.

Specifying the EU's Influence on Global Climate Politics

The second research question guiding this work was designed to allow for *the assessment of EU influence* so as to come to statements about whether the Union actually makes a difference through its actions in world politics. In answer to this question, the EU was found to have exerted medium overall influence during the entire evolution of the global climate regime and very low influence during the December 2009 Copenhagen summit and its immediate aftermath. This overall influence was attributed on the basis of an analysis of its leverage over the two key pillars of the climate regime: (i) the core norm of the regime (emissions reduction target), which was consistently moderate until 2012 but low at COP 15, and on (ii) the principle of common but differentiated responsibilities, which was permanently limited. Moreover, the EU was found to have influenced the agenda of the climate talks at various points in time as well as the decision on the fora in which to discuss global climate policies, notably through the early to mid-2000s and in 2011. While the EU is thus capable of shaping world affairs if it meets with favourable conditions, the most striking finding of the study is certainly the decline in its influence over key pillars of the climate regime across time, which underscores the need for yet better understanding and, eventually, explaining its external impact.

Explaining EU Influence on the Global Climate Regime

While understanding the Union's influence attempts and their effects was a major objective of this study, it did not stop with a description and analytical reflection of its foreign policy acts. Rather, the third research question stipulated a further investigation into the *determinants of the effects of EU actions* in the context of the studied case. The responses to this question are necessarily complex, as the Union's influence co-depends on multiple exogenous factors at the international level of analysis as well as on actor-specific variables. A central result of the study is that a minimum degree of actor capacity represents only a necessary, but by no means a sufficient condition for the EU to exert influence. The study demonstrated that while the EU's actor capacity and activity level generally increased over time, this did not clearly correlate with heightened degrees of impact. Whether or not it is able to shape world affairs thus does not depend so much on what the Union *is* and on the full development of its formal preconditions for acting (i.e. possessing competence, speaking with a single voice), but rather on what it *does* and how it does it, *given the external context*. By consequence, a concept was introduced into this research context to grasp a finding strongly suggested by the evidence: the *goodness of fit*, i.e. a certain degree of compatibility between the Union's action and its external environment, which is regarded as a second necessary condition for EU influence. Finally, more specific propositions were formulated on the basis of the two causal mechanisms and in the form of INUS conditions. The EU is capable of influencing global climate politics through *bargaining* when it possesses sufficient actor capacity and a well-prepared negotiation strategy, which it coherently defends in a global context of relative power homogeneity and positional proximity between major players. It is able to wield influence through *arguing* when global climate negotiations are at an early stage and other actors still uncertain about their preferences, when the EU possesses at least minimum actor capacity and addresses other actors' beliefs, acts proactively, coherently and consistently as a foreign policy actor with a strategy invoking external sources of knowledge and/or norms.

Research and Normative Implications of the Study

Following the synthesis of the study's main insights, this final section of the work invites the reader to consider the broader implications of the findings for the future, in both academic and political-practical terms. Regarding academic research, the analysis generated a few findings requiring further clarification as well as a range of interesting new questions. Concrete suggestions for future research are embedded into a discussion of what the insights of this work signify for the different bodies of literature it aimed to contribute to. For political practice, the publication of this book does not coincide with the end of global climate negotiations or of the EU's participation therein, far from it. To remain within the problem-driven logic of the study, a brief outlook into the future of EU foreign climate policy is therefore in order.

How the Study Relates to Existing Bodies of Literature: Suggestions for Future Research

An analysis of EU foreign policy activities and influence in global affairs necessarily had to draw on different bodies of literature at both levels of analysis. By consequence, the findings of this study hold a range of implications for several fields of research: besides the contribution it makes to IR research pertaining to the object of the study (global climate politics), where it helps to better understand the dynamics of actors' interaction in – and thus the very essence of the politics of – the global climate regime, the work contributes above all to research on the main subject of the study, the EU. It inserts itself primarily into debates held (i) on the Union's foreign climate policy and (ii) in EU foreign policy analysis as a sub-discipline of integration studies more generally.

First and foremost, the work concretely contributes to **research on EU external climate policy**. For this body of literature, it not only holds a significant amount of new empirical material, gathered from a longitudinal perspective with a special focus on the most recent periods of negotiations in the UN climate regime (2007–2009 and its aftermath), but also makes a critical conceptual-theoretical contribution to the debate. Applying the logic and concepts of foreign policy and influence analysis to the EU's performance in the climate regime over time helps to nuance the notion of EU climate leadership as an empirical reality. While the findings confirm its leadership for isolated issues and specific points in time, especially the period after the US withdrawal from the Kyoto Protocol in 2001, they also show that it would be much more accurate to speak of *attempted* EU leadership for most of the history of the global climate regime: the EU tried hard, but seldom succeeded in mobilising any followers. For academic research on this topic, this implies that leadership as an analytical concept, and more specifically the way this concept has been applied in most studies, indeed displays limitations. While this work made an empirical and conceptual contribution toward rectifying this situation, similar investigations may be needed to corroborate the picture that emerged from it on the type of foreign policy actor the Union is in this domain. Turning to broader research desiderata for scholars working on EU climate policy, three elements of the Union's foreign climate policies require specific attention: its *strategic behaviour* as a foreign policy actor in general terms, its approaches to and *relations with key actors* in global climate politics (the US, China, India, Brazil, South Africa, the African Group, AOSIS, etc.) and its *influence on topics* that were not yet explicitly touched upon in this study (e.g. the EU's capacity to shape the financial architecture of global climate politics). Secondly, as the Union's foreign policy activities and their effects in the domain of global climate politics can be regarded as a critical case for its activities and influence more generally, the study also contributes – empirically, conceptually, theoretically and methodologically – to the **research on EU foreign policy and the EU's role in UN bodies** as such. Empirically, the study provides insights into the analyzed case as an example of its participation in UN treaty-based regimes more widely. Conceptually, the study helps to clarify how the Union acts as a foreign policy player by introducing the concept of EU influence attempts and systematically linking it to specific foreign policy instruments this actor has at its disposal. The result of this exercise is a broader conceptualization of the Union's foreign policy implementation, which provides the foundations for eliminating the blind spot that exists around this issue in the discipline of EU foreign policy analysis. In terms of theory, the broad conceptualization and subsequent explanation of the Union's impact provide further ground for hypothesizing on the capabilities underlying EU influence-wielding. In this regard, the findings of this analysis point to the explanatory power of factors at various levels of analysis and their interplay, which could inspire studies on similar cases of EU participation in global (environmental) politics. Questions that persist after this study concern firstly the concept of EU foreign policy implementation: while the tools that the Union employs were identified, the question – again from the classical foreign policy analysis perspective – *why* it chooses the one above the other at given points in time remains unsettled. Further, the link between a specific tool and the capabilities it relies on needs to be worked out more precisely (Brighi/Hill 2008: 127). Finally, in terms of methodology, this work has developed a method capable of combining a broad mapping of a foreign policy actor's influence attempts with the determination of this influence and its explanation. In so doing, it provides an example of how a significant methodological research gap for EU foreign policy studies can be filled (Smith 2007, 2010: 335). Following its application, lessons can be drawn on its operability and usefulness. On the one hand, the study does demonstrate that an influence analysis method can be designed and successfully applied to determine a foreign policy actor's influence. On the other hand, further development can help to still improve the operability of the method. First, a yet stronger pre-framing of influence analyses – reducing the temporal, actor-specific or thematic scope of such studies – could yield even more precise results. Second, the reputational analysis of EU influence could be strengthened by exploiting synergies with another important research domain investigating "how others see the EU" (Lucarelli/Fioramati 2010), which could help to determine the role of existing – and detect further – external explanatory factors of the Union's impact.

The EU's Future Foreign Climate Policy: Policy-relevant Findings of the Study

The discussion of its key findings highlighted that one value added aspect of the study lies in its longitudinal character, which allows for identifying the (dis)continuities in the EU's foreign climate policy over time. An extrapolation of the discovered trends into the future makes it possible to discuss the major policy-relevant insights that emerged from this audit of the Union's external effectiveness in global climate politics.

Following the – for European climate diplomats – quite disappointing experience of the 2009 Copenhagen summit, reactions to the EU's under-performance seemed to go into two directions, which could well be described as the poles of a continuum. On the one hand, concerns were voiced about the limits of the Union's achievements in light of its considerable investments in the climate domain. This led some policy-makers and commentators to prescribe the EU a more modest and pragmatic approach to global climate politics, which was also reflected in its foreign policy actions right after 2009, and especially in the run-up to the Cancun COP (see Chapter 6). Another group of policy-makers and observers drew, on the other hand, the opposite conclusion: for them, the EU had not quite done enough yet to "show the way" and convince other parties of the well-foundedness of its positions. Continued efforts and patience were needed, and it was then only a matter of time until the Union would eventually book successes in global climate politics (see, for example, Wurzel/Connelly 2010). While the policy discussions have been – and continue to be – held between the former proponents of greater pragmatism and scaled-down ambitions and the latter – arguably more successful – advocates of a more-of-the-same approach, a promising future development of EU foreign climate policy might actually lie in a combination of the two: while maintaining a high level of aspiration as a final *objective* of its foreign policy action, the EU could use a more pragmatic *approach* to attempt to reach this aim. The crux obviously lies in delimiting aims from means, and striking the right balance between the two. Several suggestions, based on the empirical findings of this study, may provide starting points for reflecting about necessary changes. Key concepts used in this study – EU actor capacity, EU foreign policy implementation and the notion of goodness of fit – are employed to structure these thoughts. The insights are based on the premise that the EU is bound to remain active in global climate politics. It has invested too much, and has harmonized internal climate policies to such an extent that a simple disengagement from the global negotiation process is no option. Of course, a major precondition for continued engagement is a solid internal climate regime. In this respect, recent problems encountered by the ETS will need to be settled, both by reinforcing this flagship policy and by complementing it further through policies and measures that will ensure that the Union attains its emissions reduction objectives for 2020 and beyond, or else the credibility of its climate policy might suffer a severe blow (Verdonk et al. 2013). Parting from the assumption that the EU is capable of solving these domestic problems, the study's findings concretely suggest that its activism will need to rely on greater strategic capacities for it to become more effective. Importantly, it must empower itself to be a more flexible foreign policy player (for the more detailed argument, see Schunz 2011, see also Torney 2013). This flexibility depends on an improved actor capacity and a more adequate foreign policy implementation that ensures a procedural and a policy fit between the EU's activities and the context it operates in.

An improvement of its actor capacity would require a number of adaptations that could be made at a fairly low cost, but require political willingness. To begin with, the EU could use the provisions of the Lisbon Treaty as a basis for reforming its system of internal coordination and external representation to ensure a more coherent position-building and outreach strategy. In recent years, EU external climate policy has often run on two parallel tracks: the multilateral negotiations under the UN, which were the responsibility of the lead negotiators and issue leaders and, at the higher level, of Environment Ministers; and the climate-relevant negotiations at bilateral summits or in fora such as the G-20, which were a matter for Foreign Ministers and/or Heads of State, the HR and the Commission President and their staff, including the EEAS. This study suggests that this repeatedly caused high transaction costs. Not only for that reason, the two tracks should be more systematically integrated, providing ownership of EU foreign climate policy to both constituencies. This would require a more stringent intra-EU task-sharing and coordination that would assign roles to all actors.[1] While the WPIEI-CC and the system of lead negotiators and issue leaders, involving member states' environmental experts and the Commission's DG Climate Action, could continue to prepare the EU's positions and play a role in technical negotiations in all fora, the strategic outreach could be coordinated by the HR and the EEAS. Regular joint sessions of the Environment and Foreign Affairs Councils could define the EU's mandate for climate (and other environmental) negotiations (see also Van Schaik/Egenhofer 2003). EU positions could then be represented by the EU HR in all fora outside the UN climate regime and through EU delegations in key countries, while a Troika of the Commissioner for Climate Action, the Environment Minister of the Council Presidency and the HR could conduct the UN climate negotiations. Key conditions for this task-sharing to succeed would be a symmetrical access to information of all actors involved, ensured via close cooperation between the WPIEI-CC and the EEAS's department for global and multilateral issues. Such collaboration between experts on the subject matter and diplomats would exploit synergies and ensure coherence between positions. Moreover, it would allow for information about third parties' preferences to be systematically fed back into the EU's decision-making machinery to adapt

---

[1] While discussions have been held, ever since 2010, on reforms that could point into the direction of this scenario (see EEAS/Commission 2011; Council 2011c, 2013), no significant advances have been made with regard to integrating environmental expertise and diplomatic skills. At the time of writing, only two diplomats in the EEAS are in charge of dossiers directly related to climate change.

positions and coordinate between outreach activities. This would enable strategic behaviour placing the Union on an equal footing with other parties in all climate fora. The resultant approach would be characterized by a greater feeling of solidarity among member states, based on the insight that the EU can only be an effective foreign climate policy player if it acts collectively and coherently – and risks repeated instances of failure and a damaged reputation, if it does not.

With its actor capacity enhanced in this way, the Union's foreign policy implementation could also be improved in several respects. The findings of this study suggest that the EU fares better when it can provide for a procedural and policy fit, i.e. possesses a clear vision of (i) which actors to approach (who?) (ii) via what type of channels (where?) and with the help of (iii) what type of instruments (how?), and if it manages to implement this approach coherently (Schunz 2011). In terms of *channels*, the EU has always had a default strategy of acting through the multilateral system. This may not in each case be the most suitable choice. Other fora, such as the Major Economies Forum or the Cartagena Group can be used to advance global climate talks, and the EU needs to think about ways of smartly integrating those into the negotiations under the UN. First steps into this direction have been taken after COP 15, especially via the Cartagena Group. The prospect that an EU member state will host the COP in 2015 (France) could be a strategic advantage in this regard. When it comes to the *actors* the Union reaches out to, the main lesson from the Copenhagen experience is that it not only needs to diversify its outreach by benefiting from its impressive diplomatic network, but should also listen more attentively to other players and gear its positions more adequately toward their preferences and underlying interests. To that end, relationships with a wide variety of actors need to be reinforced and transcend pure exchanges of positions. Judged again on the basis of the example of the Cartagena Group, the Union seems to have learned part of the lesson, but it can certainly still do better, integrating country-/group-specific approaches into one overarching strategy. To do so, it will have to think more and more multilaterally while acting bilaterally, i.e. adopt an approach marked by "effective multiple bilateralism" (Keukeleire/Bruyninckx 2011). Although it seems indispensable to resort to bilateral relations to build trust and ensure continuous exchanges, it does not make sense to address AOSIS or the United States without thinking about the impact certain positions and decisions would have on China or India, and vice-versa. With regard to foreign policy *instruments*, diversification equally seems to be the key. To date, the EU's focus on diplomatic instruments has been fairly technical and policy-oriented. The findings from this study underscore the fact that the Union cannot afford to stop at this level of purely argumentation-based influence attempts, but needs to more systematically "think the other" and reason more in terms of politics than policy. Although it has tried to use economic tools like bilateral partnerships in recent negotiation rounds, it has not often effectively employed issue-linkage and conditionality to reinforce its argumentation strategy. For the future, such intelligently conceived linkage seems to hold greater potential: if the Union manages to mainstream climate change into its development aid programmes and international trade negotiations, for instance, leverage may be gained over key partners' positions on global climate politics (Curtin 2010; Purvis/Stevenson 2010). Other than the nexus between diplomatic and economic tools, the EU could also more systematically consider the use of coercive foreign policy instruments. The analysis showed that it was quite reluctant to actively employ mechanisms like border adjustment taxes, even as threats, during the post-2012 negotiations until 2009. By contrast, in 2011/2012, it adopted and initially enforced legislation that imposed a cap on GHG emissions from flights operating to and from EU airports. The hostile reactions from especially the US and China did not make the Union falter. Only when the International Civil Aviation Organization finally evoked that it could decide on globally concerted action on aviation and climate change at its September 2013 Assembly, the European Commission proposed "stopping the clock" in the application of its legislation – a proposal that was later endorsed within the Council and the European Parliament (Agence Europe 2013). Although the EU's behaviour did not deliver the desired advances in the short term,[2] its position on aviation could be a sign that it is willing to adopt a more assertive strategic stance *vis-à-vis* key parties in the global climate talks, aligning ambitions and strategic capacities.

In the final analysis, although it may practically be far from straightforward to implement these changes, following the policy-relevant insights of the study could help the Union to improve its effectiveness as a foreign policy player in the climate change regime. While the global context for climate politics is without doubt intricate and cannot be altered by the EU, it does have the possibility to address the main weaknesses that have characterized its activities to date: its inflexibility and its reliance on a policy- rather than politics-based approach. Internal consolidation to provide for better preconditions for foreign policy implementation should be paired to a more strategic, flexible approach and a diversification of its outreach to ensure that the EU lives up to its full potential in global climate politics.

---

2   In September 2013, the ICAO committed to starting negotiations on a global market-based mechanism on aviation emissions that would only take effect as of 2020. At the same time, it refuted the EU's plans to impose ETS rules outside its own airspace. In reaction to this development, intra-EU debates on whether to maintain the original provision, apply it only to EU airspace or abandon it altogether are ongoing at the time of writing.

References

Primary sources

UN sources

AGBM. 1995a. *Report of the AGBM on the work of its first session* (doc. FCCC/AGBM/1995/2). 28 September.

AGBM. 1995b. *Implementation of the Berlin Mandate* (doc. FCCC/AGBM/1995/MISC.1/Add.1). 29 September.

AGBM. 1995c. *AGBM. Second session* (doc. FCCC/AGBM/1995/4). 6 Oct.

AGBM. 1995d. *Implementation of the Berlin Mandate* (doc. FCCC/AGBM/1995/Misc.1/Add.3). 2 November.

AGBM. 1995e. *Report of the AGBM on the work of its second session* (doc. FCCC/AGBM/1995/7). 21 November.

AGBM. 1996a. *Implementation of the Berlin Mandate. Comments from Parties* (doc. FCCC/AGBM/1996/MISC.1). 15 February.

AGBM. 1996b. *Report of the AGBM on the work of its third session* (doc. FCCC/AGBM/1996/5). 23 April.

AGBM. 1996c. *Implementation of the Berlin Mandate. Comments from Parties. Addendum* (doc. FCCC/AGBM/1996/MISC.1, Add.3). 27 June.

AGBM. 1996d. *Implementation of the Berlin Mandate. Proposals from Parties. Addendum* (doc. FCCC/AGBM/1996/Misc.2/Add.2). 15 November.

AGBM. 1996e. *Synthesis of proposals by Parties* (doc. FCCC/AGBM/1996/10). 19 November.

AGBM. 1996f. *Implementation of the Berlin Mandate. Proposals from Parties. Addendum* (doc. FCCC/AGBM/1996/Misc.2/Add.4). 10 December.

AGBM. 1996g. *Report of the AGBM on the work of its fifth session* (doc. FCCC/AGBM/1996/11). 17 February 1997.

AGBM. 1997a. *Framework compilation of proposals from Parties for the elements of a protocol or another legal instrument* (doc. FCCC/AGBM/1997/2). 31 January.

AGBM. 1997b. *Implementation of the Berlin Mandate. Proposals from Parties* (doc. FCCC/AGBM/1997/Misc.1). 19 February.

AGBM. 1997c. *Framework compilation of proposals from Parties for the elements of a protocol or another legal instrument. Addendum* (doc. FCCC/AGBM/1997/2. Add. 1). 27 February.

AGBM. 1997d. *Report of the AGBM on the Work of Its Sixth Session. Negotiating text by the Chairman* (doc. FCCC/AGBM/1997/3/Add.1). 22 April.

AGBM. 1997e. *Implementation of the Berlin Mandate. Additional proposals from Parties. Addendum* (doc. FCCC/AGBM/1997/Misc.1/Add.3). 30 May.

AGBM. 1997f. *AGBM: Seventh session – Implementation of the Berlin Mandate. Additional proposals from Parties. Addendum* (doc. FCCC/AGBM/1997/MISC.1/Add.2). 30 May.

AGBM. 1997g. *AGBM: Seventh session – Implementation of the Berlin Mandate. Comments from Parties. Addendum* (doc. FCCC/AGBM/1997/MISC.2/Add.1). 27 June.

AGBM. 1997h. *Reports by the chairmen of the informal consultations conducted at the seventh session of the AGBM* (doc. FCCC/AGBM/1997/INF.1). 22 Sept.

AGBM. 1997i. *Completion of a protocol or another legal instrument. Consolidated negotiating text* (doc. FCCC/AGBM/1997/7). 13 October.

Audio. 1996. *Recording of the Proceedings of the Ministerial Segment of the Second Conference of the Parties to the UNFCCC. Intervention by U.S. Under-Secretary of State Timothy Wirth.* 17 July.

Audio. 1997a. *Recording of the 6th meeting of the Working Group "QELRO I" at the eighth session of the AGBM.* 29 October.

Audio. 1997b. *Recording of the final meeting of the Committee of the Whole at the Third Conference of the Parties to the UN Framework Convention on Climate Change.* 10–11 December.

Ban Ki-moon. 2009. *Letter by the UN Secretary-General Ban Ki-Moon.* 20 March (available at http://media.ft.com/cms/1f749b5e-194c-11de-9d34-0000779fd2ac.pdf, last accessed 25 March 2009).

INC. 1991a. *Report of the INC for a Framework Convention on Climate Change on the work of its first session* (doc. A/AC.237/6). 8 March.

INC. 1991b. *Report of the INC for a Framework Convention on Climate Change on the work of its third session* (doc. A/AC.237/12). 25 October.

INC. 1992. *Report of the INC for a Framework Convention on Climate Change on the work of its sixth session* (doc. A/AC.237/24). 6 January.

INC. 1993a. *Report of the INC for a Framework Convention on Climate Change on the work of its seventh session* (doc. A/AC.237/31). 27 April.

INC. 1993b. *Report of the INC for a Framework Convention on Climate Change on the work of its eighth session* (doc. A/AC.237/41). 20 October.

INC. 1994a. *Matters relating to commitments* (doc. A/AC.237/L.23). 27 Sept.

INC. 1994b. *Matters relating to commitments* (doc. A/AC.237/L.23/Add.1). 27 Sept.

INC. 1995. *Report of the INC for a Framework Convention on Climate Change on the work of its eleventh session. Part one: Proceedings* (doc. A/AC.237/91). 10 January.

UN Framework Convention on Climate Change (UNFCCC) of 9 May 1992.

UN. 2009. *UN High-Level Meeting on Climate Change – Programme.* New York, 22–23 September.

UNEP. 2012. *The emissions gap report 2012*. Nairobi: UNEP.

UNFCCC (Kyoto) Protocol: *Protocol to the United Nations Framework Convention on Climate Change of 11 December 1997*.

UNFCCC. 1994. *Matters relating to commitments. Comments from Parties or other member states* (doc. A/AC.237/MISC.43). 7 December.

UNFCCC. 1995a. *First Conference of the Parties to the UNFCCC. Directory of Participants* (doc. FCCC/1995/Inf.5/Rev.2). April.

UNFCCC. 1995b. *Matters relating to commitments. Comments from Parties and other member states* (doc. FCCC/CP/1995/Misc.1). 9 March.

UNFCCC. 1995c. *Review of the adequacy of Article 4, paragraph 2 (a) and (b). Proposed elements of a mandate for consultations on commitments in Article 4, paragraph 2 (a) and (b)* (doc. FCCC/CP/1995/CRP.1). 2 April.

UNFCCC. 1995d. *Report of the COP on its first session* (doc. FCCC/CP/1995/7). 24 May.

UNFCCC. 1995e. *Report of the COP on its first session. Addendum* (doc. FCCC/CP/1995/7/Add.1). 6 June.

UNFCCC. 1996. *Report of the COP on its second session – The Geneva Ministerial Declaration* (doc. FCCC/CP/1996/15/Add.1). 29 Oct.

UNFCCC. 1997a. *Adoption of a protocol or another legal instrument: Fulfilment of the Berlin Mandate. Revised text under negotiation* (doc. FCCC/CP/1997/2). 12 November.

UNFCCC. 1997b. *Non-paper by the Chairman of the Committee of the Whole* (doc. FCCC/1997/CRP.2). 7 December.

UNFCCC. 1997c. *Draft text by the Chairman of the Committee of the Whole* (doc. FCCC/CP/1997/CRP.4). 9 December.

UNFCCC. 1997d. *Kyoto Protocol to the United Nations Framework Convention on Climate Change. Final draft by the Chairman of the Committee of the Whole* (doc. FCCC/CP/1997/CRP.6). 10 December.

UNFCCC. 1997e. *Report of the COP on its third session. Addendum* (doc. FCCC/CP/1997/7/Add.1). 25 March 1998.

UNFCCC. 1999a. *Report of the COP on its fourth session. Addendum Part Two* (doc. FCCC/CP/1998/16/Add.1). 25 January.

UNFCCC. 1999b. *Mechanisms Pursuant to Articles 6, 12 and 17 of the Kyoto Protocol. Synthesis of proposals by Parties on principles, modalities, rules and guidelines* (doc. FCCC/SB/1999/8). 28 September.

UNFCCC. 1999c. *Mechanisms Pursuant to Articles 6, 12 and 17 of the Kyoto Protocol. Synthesis of proposals by Parties on principles, modalities, rules and guidelines. Addendum* (doc. FCCC/SB/1999/8 Add.1). 24 October.

UNFCCC. 1999d. *Draft report of the Conference of the Parties on its fifth session* (doc. FCCC/CP/1999/L.1). 3 November.

UNFCCC. 2000. *Note by the President of COP 6*. 23 November.

UNFCCC. 2001a. *Implementation of the Buenos Aires Plan of Action* (doc. FCCC/CP/2001/L.7). 24 July.

UNFCCC. 2001b. *The Marrakech Accords and the Marrakech Declaration* (available at http://unfccc.int/cop7/documents/accords_draft.pdf, last accessed 24 July 2009).

UNFCCC. 2002a. *The Delhi Ministerial Declaration on Climate Change and Sustainable Development. Informal Proposal by the President.* 28 October.

UNFCCC. 2002b. *The Delhi Ministerial Declaration on Climate Change and Sustainable Development* (doc. FCCC/CP/2002/L.6/Rev.1). 1 November.

UNFCCC. 2003. *Report of the Conference of the Parties on its ninth session. Part one: Proceedings* (doc. FCCC/CP/2003/6). 30 March 2004.

UNFCCC. 2005a. *Dialogue on long-term cooperative action to address climate change by enhancing implementation of the Convention* (doc. FCCC/CP/2005/L.4/Rev.1). 9 December.

UNFCCC. 2005b. *Consideration of commitments for subsequent periods for Parties included in Annex I to the Convention under Article 3, paragraph 9, of the Kyoto Protocol* (doc. FCCC/KP/CMP/2005/L.8/Rev.1). 9 December.

UNFCCC. 2005c. *Consideration of commitments for subsequent periods for Parties included in Annex I to the Convention under Article 3, paragraph 9, of the Kyoto Protocol* (doc. Decision-/CMP.1). 9 December.

UNFCCC. 2006a. *Five-year programme of work on impacts, vulnerability and adaptation to climate change* (doc. FCCC/SBSTA/2006/L.26). 14 Nov.

UNFCCC. 2006b. *Further commitments for Annex I Parties and Programme of work* (doc. FCCC/KP/AWG/2006/L.4). 14 November.

UNFCCC. 2006c. *Review of the Kyoto Protocol pursuant to its Article 9. Proposal by the President* (doc. FCCC/KP/CMP/2006/L.7). 17 November.

UNFCCC. 2007a. *Thirteenth Conference of the Parties. List of Participants.* 14 December.

UNFCCC. 2007b. *Report of the AWG-KP on its resumed fourth session* (doc. FCCC/KP/AWG/2007/5). 17 February 2008.

UNFCCC. 2007c. *Dialogue on long-term cooperative action to address climate change by enhancing implementation of the Convention: Fourth workshop* (doc. Dialogue WP 6 (2007)). 13 July.

UNFCCC. 2007d. *Dialogue on long-term cooperative action to address climate change by enhancing implementation of the Convention: Fourth workshop – Submission from Portugal on behalf of the European Community and its member states* (doc. Dialogue Working Paper 10 (2007)). 16 August.

UNFCCC. 2007e. *Dialogue on long-term cooperative action to address climate change by enhancing implementation of the Convention: Fourth workshop – Submission from AOSIS* (doc. Dialogue WP 14 (2007)). 24 August.

UNFCCC. 2007f. *Dialogue on long-term cooperative action to address climate change by enhancing implementation of the Convention: Fourth workshop – Submission from Portugal on behalf of the European Community and its member states* (doc. Dialogue WP 16 (2007)). 30 August.

References

UNFCCC. 2007g. *AWG-KP, Fourth Session – Analysis of mitigation potentials and identification of ranges of emission reduction objectives of Annex I Parties* (doc. FCCC/KP/AWG/2007/L.4). 31 August.

UNFCCC. 2007h. *Report on the dialogue on long-term cooperative action to address climate change by enhancing implementation of the Convention* (doc. FCCC/CP/2007/4). 19 October.

UNFCCC. 2007i. *Further views on the development of a timetable to guide the completion of the work of the AWG-KP – Submission from Portugal on behalf of the European Community and its Member States* (doc. FCCC/KP/AWG/2007/MISC.6). 22 November.

UNFCCC. 2007j. *Scope and content of the second review under Article 9 of the Kyoto Protocol and the preparations required for conducting the review. Addendum* (doc. FCCC/KP/CMP/2007/MISC.1/Add.2). 1 December.

UNFCCC. 2007k. *Review of work programme, methods of work and schedule of future sessions. Draft conclusions* (doc. FCCC/KP/AWG/2007/L.6/Rev.1). 13 December.

UNFCCC. 2007l. *Review of the Kyoto Protocol pursuant to its Article 9. Proposal by the President* (doc. FCCC/KP/CMP/2007/L.8). 14 December.

UNFCCC. 2007m. *The Bali Roadmap – Address to closing plenary by His Excellency Mr. Rachmat Witoelar*. Bali, 15 December.

UNFCCC. 2007n. *Bali Action Plan* (doc. Decision-/CP.13). December.

UNFCCC. 2008a. *Views regarding the work programme of the AWG-LCA under the Convention* (docs. FCCC/AWGLCA/2008/MISC.1 and ADD 1–3). Various dates.

UNFCCC. 2008b. *Views and information on the means to achieve mitigation objectives of Annex I Parties. Submissions from Parties* (docs. FCCC/AWG/2008/MISC.1 and ADD 1–3). Various dates.

UNFCCC. 2008c. *Ideas and proposals on the elements contained in paragraph 1 of the Bali Action Plan. Submissions from Parties* (docs. FCCC/AWGLCA/2008/MISC.2 and Add. 1). Various dates.

UNFCCC. 2008d. *Ideas and proposals on paragraph 1 of the Bali Action Plan – Revised note* (doc. FCCC/AWGLCA/2008/16/Rev.1). 15 January 2009.

UNFCCC. 2008e. *AWG-LCA. Scenario note on the first session – Note by the Chair* (doc. FCCC/AWGLCA/2008/2). 13 March.

UNFCCC. 2008f. *Analysis of means to reach emission reduction targets and identification of ways to enhance their effectiveness and contribution to sustainable development. Views and information on the means to achieve mitigation objectives of Annex I Parties. Submissions from Parties* (doc. FCCC/KP/AWG/2008/MISC.1/Add.2). 20 March.

UNFCCC. 2008g. *AWG-LCA First session. Development of a work programme* (doc. FCCC/AWGLCA/2008/L.2). 4 April.

UNFCCC. 2008h. *Analysis of means to reach emission reduction targets and identification of ways to enhance their effectiveness and contribution to sustainable development* (doc. FCCC/KP/AWG/2008/L.2). 4 April.

UNFCCC. 2008i. *Enabling the full, effective and sustained implementation of the Convention through long-term cooperative action now, up to and beyond 2012* (doc. FCCC/AWGLCA/2008/L.5). 5 June.

UNFCCC. 2008j. *Work programme for 2009 – Draft conclusions* (doc. FCCC/AWGLCA/2008/L.4). 11 June.

UNFCCC. 2008k. *Enabling the full, effective and sustained implementation of the Convention through long-term cooperative action now, up to and beyond 2012 – Draft conclusions* (doc. FCCC/AWGLCA/2008/L.7). 27 August.

UNFCCC. 2008l. *Consideration of a work programme for 2009 – Draft conclusions* (doc. FCCC/AWGLCA/2008/L.8). 27 August.

UNFCCC. 2008m. *Ideas and proposals on paragraph 1 of the Bali Action Plan* (doc. FCCC/AWGLCA/2008/16). 20 November.

UNFCCC. 2008n. *Work programme for 2009 – Draft conclusions* (doc. FCCC/AWGLCA/2008/L.10). 10 December.

UNFCCC. 2008o. *Work programme for 2009 – Draft conclusions* (doc. FCCC/KP/AWG/2008/L.19). 10 December.

UNFCCC. 2008p. *Report to the Conference of the Parties at its fourteenth session on progress made – Draft conclusions* (doc. FCCC/AWGLCA/2008/L.11). 10 December.

UNFCCC. 2008q. *Means, methodological issues, mitigation potential and ranges of emission reduction objectives, and consideration of further commitments – Draft conclusions* (doc. FCCC/KP/AWG/2008/L.18). 10 Dec.

UNFCCC. 2009a. *Outcome of the work of the AWG-LCA. Draft conclusions and Addendums* (doc. FCCC/AWGLCA/2009/L.7+ add. 1–9). Various dates.

UNFCCC. 2009b. *Fulfilment of the Bali Action Plan and components of the agreed outcome* (doc. FCCC/AWGLCA/2009/4). 18 March.

UNFCCC. 2009c. *Report of the AWG-LCA on its fifth session* (doc. FCCC/AWGLCA/2009/5). 13 May.

UNFCCC. 2009d. *Report of the AWG-KP on its seventh session* (doc. FCCC/KP/AWG/2009/5). 13 May.

UNFCCC. 2009e. *Draft protocol to the Convention prepared by the Government of Japan for adoption* (doc. FCCC/CP/2009/3). 13 May.

UNFCCC. 2009f. *A proposal for amendments to the Kyoto Protocol pursuant to its Article 3, paragraph 9* (doc. FCCC/KP/AWG/2009/7). 14 May.

UNFCCC. 2009g. *Negotiating text*. (doc. FCCC/AWGLCA/2009/8). 19 May.

UNFCCC. 2009h. *Ideas and proposals on the elements contained in paragraph 1 of the Bali Action Plan – Submissions from Parties. Addendum* (doc. FCCC/AWGLCA/2009/MISC.4/Add.1). 22 May.

UNFCCC. 2009i. *Ideas and proposals on the elements contained in paragraph 1 of the Bali Action Plan – Submissions from Parties. Addendum* (doc. FCCC/AWGLCA/2009/MISC.4/Add.2). 31 May.

UNFCCC. 2009j. *Revised negotiating text* (doc. FCCC/AWGLCA/2009/INF.1). 22 June.

References

UNFCCC. 2009k. *Documentation to facilitate negotiations among Parties. Addendum. Proposed amendments to the Kyoto Protocol pursuant to its Article 3, paragraph 9* (doc. FCCC/KP/AWG/2009/10/Add.1). October.

UNFCCC. 2009l. *Draft report of the AWG-LCA on its seventh session* (doc. FCCC/AWGLCA/2009/L.4). 3 November.

UNFCCC. 2009m. *Adoption of the Copenhagen Agreement. Draft Decision 1/CP.15 of 27 November* (available at http://www.guardian.co.uk/environment/2009/dec/08/copenhagen-climate-change, last accessed 25 June 2010).

UNFCCC. 2009n. *Chair's Proposed Draft Text on the Outcome of the Work of the Ad Hoc Working Group on Long-term Cooperative Action under the Convention.* Version 11 Dec., 8:30 am (on file with author).

UNFCCC. 2009o. *Chair's Proposed Draft Text on the Outcome of the Work of the Ad Hoc Working Group on Further Commitments for Annex I Parties under the Kyoto Protocol.* Version 11 Dec., 9:30 am (on file with author).

UNFCCC. 2009p. *AWG-KP draft texts. Chair's Proposed Draft Text on the Outcome of the Work of the Ad Hoc Working Group on Further Commitments for Annex I Parties under the Kyoto Protocol.* Version 15 Dec., 5:30 pm (on file with author).

UNFCCC. 2009q. *Untitled draft negotiation text of 18 December*, 1:00 pm (on file with author).

UNFCCC. 2009r. *Copenhagen Accord.* Draft of 18 December, 4:30 pm (on file with author).

UNFCCC. 2009s. *Copenhagen Accord.* Draft of 18 December, 7:30 pm (on file with author).

UNFCCC. 2009t. *Copenhagen Accord.* Draft of 18 December, 10 pm (on file with author).

UNFCCC. 2009u. *Copenhagen Accord* (doc. Decision-/CP.15). 19 December.

UNFCCC. 2009v. *Outcome of the work of the AWG-LCA* (doc. Draft decision-/CP.15). 19 December.

UNFCCC. 2009w. *Outcome of the work of the AWG-KP* (doc. Draft decision-/CMP.5). 19 December.

UNFCCC. 2010a. *Information provided by Parties to the Convention relating to the Copenhagen Accord.* 2 Feb. (available at http://unfccc.int/home/items/5262.php, last accessed 10 February 2010).

UNFCCC 2010b. *Outcome of the work of the AWG-KP at its fifteenth session* (doc. -/CMP.6). 10 Dec.

UNFCCC 2010c. *Outcome of the work of the AWG-LCA* (doc. -/CP.16). 10 Dec.

UNFCCC. 2011a. *Establishment of an Ad Hoc Working Group on the Durban Platform for Enhanced Action* (doc. Decision 1/CP.17). 11 Dec.

UNFCCC. 2011b. *Report of the COP serving as the meeting of the Parties to the Kyoto Protocol on its seventh session* (doc. FCCC/KP/CMP/2011/10/Add.1). 12 March 2012.

UNFCCC. 2011c. *Outcome of the work of the AWG-LCA* (doc. 2/CP.17). 11 Dec.

UNFCCC. 2012. *Amendment to the Kyoto Protocol pursuant to its Article 3, paragraph 9* (doc. FCCC/KP/CMP/2012/L.9). 8 Dec.

UNFCCC 2013a. *Agreed outcome pursuant to the Bali Action Plan* (doc. FCCC/CP/2012/8/Add. 1 – 1/CP18). 28 Feb.

UNFCCC. 2013b. *Advancing the Durban Platform* (FCCC/CP/2012/8/Add.1–2/CP18). 28 Feb.

UNFCCC. *Provisional Rules of Procedure of the UNFCCC COP* (available at http://unfccc.int/resource/docs/cop2/02.pdf, last accessed 10 October 2011)

UNGA (United Nations General Assembly). 1988. *Resolution 43/53*. 6 Dec.

UNGA. 1989a. *Protection of Global Climate for Current and Future Generations of Mankind* (doc. Resolution A/RES/43/53). 27 January.

UNGA. 1989b. *Protection of Global Climate for Present and Future Generations of Mankind* (doc. Resolution A/44/862). 19 December.

UNGA. 1990. *Protection of Global Climate for Present and Future Generations of Mankind* (doc. Resolution 45/212). 21 December.

UNGA. 1990. *Resolution 45/212*. 21 December.

UNGA. 2007. *UNGA 61st session: Informal Thematic Debate – Climate Change as a Global Challenge*. 1 August.

UNSD. 2009. *United Nations Statistics Division, Millennium Development Goals Indicators: Carbon dioxide emissions (CO_2), thousand metric tons of CO_2*. New York: United Nations.

Video 12 Dec. 2007. *Intervention by H.E. Mr. Jean-Louis Borloo*. Joint High-level Segment, 12 Dec. (available at rtsp://webcast.un.org/ondemand/conferences/unfccc/2007/hl/unfccc071212pm1-hls-orig.rm?start=03:06:50&end=03:13:49, last accessed 25 June 2010).

Video 28 Sept. 2009. *AWG-LCA Opening Plenary*. Bangkok. (available at http://unfccc2.meta-fusion.com/kongresse/090928_AWG_Bangkok/templ/ply_page.php?id_kongresssession=2018&player_mode=isdn_real, last accessed 25 June 2010).

Video 2 Oct. 2009. *AWG-KP Plenary*. Bangkok. (available at http://unfccc2.meta-fusion.com/kongresse/090928_AWG_Bangkok/templ/ply_page.php?id_kongresssession=1991&player_mode=isdn_real, last accessed 25 June 2010)

Video 15 Dec. 2009, part 1. *COP plenary session*. Bali. (available at rtsp://webcast.un.org/ondemand/conferences/unfccc/2007/cop/unfccc-071215am-cop13-eng.rm, last accessed 25 June 2010).

Video 15 Dec. 2009, part 2. *COP plenary session*. Bali. (available at rtsp://webcast.un.org/ondemand/conferences/unfccc/2007/cop/unfccc-071215am3-cop13-eng.rm, last accessed 25 June 2010).

EU sources

Barroso, J. 2008. *Speech by the President of the European Commission – Europe's Climate Change Opportunity*. Lehman Brothers. London. 21 Jan.

Council of the EU. 1989. *Council Resolution on the greenhouse effect and the Community* (doc. 89/C 183/03). Luxembourg, 21 June.

Council of the EU. 1994a. *Council Decision of 15 December 1993 concerning the conclusion of the UNFCCC* (doc. 94/69/EC).

Council of the EU. 1994b. *1817th Council meeting – Environment*. Brussels, 15–16 December.

Council of the EU. 1996. *1939th Council meeting – Environment*. Brussels, 25–26 June.

Council of the EU. 1997. *Meeting Document* (doc. CONS/ENV/97/1 Rev.1). Brussels, 3 March.

Council of the EU. 2002. *Decision of 25 April 2002 concerning the approval, on behalf of the EC, of the Kyoto Protocol to the UNFCCC and the joint fulfilment of commitments thereunder* (doc. 2002/358/EC).

Council of the EU. 2006. *2773rd Council Meeting – Environment*. Brussels, 18 December.

Council of the EU. 2007a. *EU objectives for the further development of the international climate regime beyond 2012*. Brussels, 21 February.

Council of the EU. 2007b. *2826th Council meeting – Environment*. Luxembourg, 30 October.

Council of the EU. 2008a. *2874th Council Meeting – Environment*. Luxembourg, 5 June.

Council of the EU. 2008b. *2008 EU-US Summit Declaration*. Brdo, Slovenia, 10 June.

Council of the EU. 2008c. *2898th Council Meeting – Environment*. Luxembourg, 20 October.

Council of the EU. 2008d. *Energy and climate change – Elements of the final compromise*. Brussels, 12 December.

Council of the EU. 2009a. *South Africa-EU Strategic Partnership. Joint Communiqué from the Ministerial Troika Meeting*. Kleinmond, 16 January.

Council of the EU. 2009b. *2928th Council meeting – Environment*. Brussels, 2 March.

Council of the EU. 2009c. *The Prague meeting of EU Heads of State and Government with the US President – EU Press Lines*. Prague, 5 April.

Council of the EU. 2009d. *Council adopts climate-energy legislative package*. Press release. Brussels, 6 April.

Council of the EU. 2009e. *12th Africa-EU Ministerial Troika meeting*. Luxembourg, 28 April.

Council of the EU. 2009f. *Submission by the Czech Republic on behalf of the EC and its member states: A negotiation text for consideration at AWG-LCA 6*. Prague, 28 April.

Council of the EU. 2009g. *18th EU-Japan Summit. Joint Press Statement*. Prague, 4. May.

Council of the EU. 2009h. *Joint Statement EU-Rio Group*. Prague, 13 May.

Council of the EU. 2009i. *11ᵗʰ EU-China Summit. Joint Press Communiqué*. Prague, 20 May.

Council of the EU. 2009j. *Republic of Korea-EU Summit. Joint Press Statement*. Seoul, 23 May.

Council of the EU. 2009k. *2948ᵗʰ Economic and Financial Affairs Council meeting – Council conclusions on international financing for climate action*. Luxembourg, 9 June.

Council of the EU. 2009l. *2967ᵗʰ Council meeting – Economic and Financial Affairs*. Luxembourg, 20 October.

Council of the EU. 2009m. *2968ᵗʰ Environment Council meeting – Council Conclusions on EU position for the Copenhagen Climate Conference*. Luxembourg, 21 October.

Council of the EU. 2009n. *Joint Statement of the 12ᵗʰ EU-China Summit*. Nanjing, 30 November.

Council of the EU. 2010a. *3021ˢᵗ Council meeting – Environment*. Luxembourg, 11 June.

Council of the EU. 2010b. *3036ᵗʰ Council meeting – Environment*. Luxembourg, 14 October.

Council of the EU. 2011a. *3118ᵗʰ Council meeting – Environment*. Luxembourg, 10 October.

Council of the EU. 2011b. *Commission Communication on a Roadmap for moving to a competitive low carbon economy in 2050 – Presidency conclusions*. Brussels, 22 June.

Council of the EU. 2011c. *3106ᵗʰ Foreign Affairs – Presidency conclusions*. Brussels, 18 July.

Council of the EU. 2012. *3194ᵗʰ Environment Council Meeting*. Luxembourg, 25 Oct.

Council of the EU. 2013. *Council conclusions on EU Climate Diplomacy. Foreign Affairs Council Meeting*. Luxembourg, 24 June.

EEA. 1997. *Climate Change in the European Union*. Environmental Issues Series, No. 2. Copenhagen: European Environment Agency.

EEA. 2007. *Graph: Emissions per capita by country, 2007* (available at http://dataservice.eea.europa.eu/pivotapp/pivot.aspx?pivotid=475, last accessed 22 April 2010).

EEA. 2008. *Greenhouse gas emission trends and projections in Europe 2008. Country profiles* (available at http://www.eea.europa.eu/publications/eea_report_2008_5, last accessed 22 April 2010).

EEA. 2009. *Change in greenhouse gas emissions in Europe between the base years and 2005, compared to Kyoto targets for 2008–2012* (available at http://www.eea.europa.eu/data-and-maps/figures/change-in-greenhouse-gas-emissions-in-europe-between-the-base-years-and-2005-compared-to-kyoto-targets-for-2008-2012, last accessed 22 April 2010).

EEAS and European Commission. 2011. *Towards a renewed and strengthened EU climate diplomacy*. Joint reflection paper. Brussels, 9 July.

EU. 2010. *Press Conference at Tianjin UN Climate Negotiations*. 6 Aug. 2010.

European Commission. 1997. *Communication: Climate Change – The EU approach to Kyoto* (doc. COM (97) 481 final). Brussels, 1 October.

European Commission. 2003. *Communication from the Commission to the Council and the European Parliament: European Union and United Nations: the choice of multilateralism* (doc. COM (2003) 526). Brussels, 10 Sept.

European Commission. 2005. *Communication from the Commission: Winning the Battle against Global Climate Change* (doc. COM (2005) 35 final). Brussels, 9 February.

European Commission. 2006a. *European Commission, External Relations Directorate General, Panorama du Service extérieur*. Brussels.

European Commission. 2006b. *Multilateral environmental agreements to which the EC is a contracting party or a signatory*. Brussels, 27 October.

European Commission. 2007a. *Working with Developing Countries to tackle Climate Change*. Luxembourg: Office for Official Publications of the EC.

European Commission. 2007b. *Communication: Limiting Global Climate Change to 2° Celsius: The way ahead for 2020 and beyond* (doc. COM (2007) 2). Brussels, 10 January 2007.

European Commission. 2008a. *European Climate Change Programme. Introduction* (available at http://ec.europa.eu/environment/climat/eccp.htm, last accessed 25 July 2009).

European Commission. 2008b. *Visit of President Barroso to China* (24–26 April 2008) (available at http://ec.europa.eu/environment/climat/china.htm, last accessed 23 March 2010).

European Commission. 2008c. *Proposal for a Decision of the European Parliament and of the Council on the effort of Member States to reduce their greenhouse gas emissions to meet the Community's greenhouse gas emission reduction commitments up to 2020*. (doc. COM (2008) 17). 23 January 2008.

European Commission. 2008d. *Proposal for a Directive of the European Parliament and of the Council amending Directive 2003/87/EC so as to improve and extend the greenhouse gas emission allowance trading system of the Community* (doc. COM (2008) 30 final). Brussels, 23 January.

European Commission. 2008e. *EU-India Joint Press Communiqué*. Marseilles, 29 September.

European Commission. 2008f. *Joint Work Programme, EU-India Co-Operation On Energy, Clean Development And Climate Change*. Marseilles, 29 Sept.

European Commission. 2009a. *The EU Green Diplomacy Network* (available at http://ec.europa.eu/external_relations/environment/gdn/docs/gdn_more_en.pdf, last accessed 5 August 2009).

European Commission. 2009b. *Bilateral relations – Canada*. (available at http://ec.europa.eu/environment/international_issues/relations_canada_en.htm, last accessed 5 August 2009).

European Commission. 2009c. *Bilateral relations – USA.* (available at http://ec.europa.eu/environment/international_issues/relations_usa_en.htm, last accessed 5 August 2009).

European Commission. 2009d. *Bilateral relations – Japan.* DG (available at http://ec.europa.eu/environment/international_issues/relations_japan_en.htm, last accessed 5 August 2009).

European Commission. 2009e. *Bilateral and regional cooperation – China* (available http://ec.europa.eu/environment/international_issues/relations _china_ en.htm, last accessed 5 August 2009).

European Commission. 2009f. *Bilateral and regional cooperation – India.* (available at http://ec.europa.eu/environment/international_issues/relations_ india_ en.htm, last accessed 5 August 2009).

European Commission. 2009g. *Regional dialogue and cooperation: ASEM.* (available at http://ec.europa.eu/environment/international_issues/relations_ asem_en.htm, last accessed 5 August 2009).

European Commission. 2009h. *Non-Paper: Green Elements from Member States' Recovery Plans.* Undated (available at http://ec.europa.eu/environment/integration/pdf/recovery_plans.pdf, last accessed 23 October 2009).

European Commission. 2009i. *Communication from the Commission: Towards a comprehensive climate change agreement in Copenhagen* (doc. COM (2009) 39 final). Brussels, 28 January.

European Commission. 2009j. *EU-Russia summit.* Khabarovsk, 21–22 May.

European Commission. 2009k. *Communication from the Commission: Stepping up international climate finance: A European blueprint for the Copenhagen deal* (doc. COM (2009) 475/3). Brussels, 10 September.

European Commission. 2009l. *2009 EU-U.S. Summit Declaration.* 3 November.

European Commission. 2009m. *India-EU Joint Statement.* Brussels, 6 Nov.

European Commission. 2010. *Analysis of options to move beyond 20% greenhouse gas emission reductions and assessing the risk of carbon leakage.* Brussels, 25 May.

European Commission. 2011. *Communication from the Commission: A Roadmap for moving to a competitive low carbon economy in 2050* (doc. COM(2011) 112 final). Brussels, 8 March.

European Commission. 2013a. *The 2015 International Climate Change Agreement: Shaping international climate policy beyond 2020* (doc. COM(2013) 167 final). Brussels, 26 March.

European Commission. 2013b. *Reducing emissions from the aviation sector.* Website (available at http://ec.europa.eu/clima/policies/transport/aviation/index_en.htm, last accessed 23 February 2013).

European Council. 1988. *Presidency conclusions – Annex I: Declaration on the Environment.* Rhodes, 2–3 December.

European Council. 1990. *Conclusions of the Presidency.* Dublin, 25–26 June.

European Council. 2003. *A Secure Europe in a Better World, European Security Strategy*. Brussels, 12 December.

European Council. 2005. *Presidency conclusions*. Brussels, 22–23 March.

European Council. 2007. *Presidency conclusions*. Brussels, 8–9 March.

European Council. 2008a. *Climate Change and International Security. Paper from the High Representative and the European Commission to the European Council*. Brussels, 14 March (Doc. S 113/08).

European Council. 2008b. *Report on the Implementation of the European Security Strategy – Providing Security in a Changing World*. Doc. S407/08. Brussels, 11 December.

European Council. 2009a. *Presidency conclusions*. Brussels, 19–20 March.

European Council. 2009b. *Presidency conclusions*. Brussels, 29–30 October.

European Council. 2009c. *Presidency conclusions*. Brussels, 10–11 December.

European Council. 2010. *Presidency conclusions*. Brussels, 17 June.

European Parliament. 2003. *Summary Note by the Secretariat. Kyoto Protocol Visit to Moscow. 17–19 September 2003*. Brussels. 1 October. (available at http://www.europarl.europa.eu/comparl/envi/pdf/delegations/1999-2004/moscow.pdf, last accessed 23 July 2009).

European Parliament. 2007. *Climate Change: EP temporary committee to be set up*. 19 April.

European Parliament. 2008. *Overview of texts adopted on 17 December 2008*.

European Parliament/Council of the European Union. 2001. *Directive of 27 September 2001 on the promotion of electricity produced from renewable energy sources in the internal electricity market* (doc. 2001/77/EC).

European Parliament/Council of the European Union. 2003. *Directive of 13 October 2003 establishing a scheme for greenhouse gas emission allowance trading within the Community and amending Council Directive 96/61/EC* (doc. 2003/87/EC).

European Parliament/Council of the European Union. 2004. *Directive of 27 October 2004 amending Directive 2003/87/EC establishing a scheme for greenhouse gas emission allowance trading within the Community, in respect of the Kyoto Protocol's project mechanisms* (doc. 2004/101/EC).

European Parliament/Council of the European Union. 2006. *Directive of 5 April 2006 on energy end-use efficiency and energy services* (doc. 2006/32/EC).

European Parliament/Council of the European Union. 2009. *Directive of 23 April 2009 amending Directive 2003/87/EC so as to improve and extend the greenhouse gas emission allowance trading scheme of the Community* (doc. 2009/29/EC).

Dimas, S. 2009. *Climate Change in times of economic crisis – the path to a successful climate conference in Copenhagen*. Speech by the EU Environment Commissioner. Humboldt University, Berlin. 6 July.

German Foreign Ministry. 2011. *Climate change – a foreign policy challenge*. (available at http://www.auswaertiges-amt.de/EN/Aussenpolitik/GlobaleFragen/Klima/Aussenpolitische-Dimension-node.html, last accessed 20 Jan. 2012).

Swedish Presidency. 2009. *Comparability of developed countries mitigation efforts*. Undated paper (available at http://www.se2009.eu/polopoly_fs/1.10668!menu/standard/file/COM%20Note%20on%20comparability%20%20IMEnvM%20%C3%85re.pdf, last accessed 25 June 2010).

Other primary sources

AOSIS. 2009. *Climate Change – The Science*. Website (available at http://www.sidsnet.org/aosis/issues.html, last accessed 20 October 2009).

APEC. 2009. *Leaders' Declaration. The 17th APEC economic leaders' meeting: "Sustaining growth, connecting the region"*. Singapore, 14–15 Nov.

APP (Asia-Pacific Partnership). 2009. *Website* (available at http://www.asiapacificpartnership.org/english/default.aspx, last accessed 24 July 2009).

ASEM. 2008. *Beijing Declaration on Sustainable Development*. Beijing, 24–25 October.

ASEM. 2009. *First ASEM Ministerial Conference on Energy Security*. Brussels, 17–18 June (available at http://www.aseminfoboard.org/Calendar/MinisterialMeetings/?id=256, last accessed 13 March 2010).

Australian Government. 2008. *Carbon Pollution Reduction Scheme: Australia's Low Pollution Future White Paper Executive Summary*. Canberra, 15 December.

Brown, G. 2009. *Transcript of the Prime Minister's press conference at the Copenhagen summit on climate change*. 18 Dec. (available at http://www.number10.gov.uk/Page21850, last accessed 22 Dec. 2009).

CHOGM (Commonwealth Heads of Government Meeting). 2009. *Port of Spain Climate Change Consensus: The Commonwealth Climate Change Declaration*. 28 November.

Clinton, W., and A. Gore. 1993. *The Climate Change Action Plan*. Washington, October (available at http://www.gcrio.org/USCCAP/toc.html, last accessed 15 April 2009).

Clinton, H. 2009. *Remarks at the United Nations Framework Convention on Climate Change Conference*. Copenhagen, 17 December.

De Boer, Y. 2010. *Letter to the UNFCCC secretariat*. December 2009 (on file with author).

Department of State of the United States. 1994. *Climate Action Report. Submission of the United States of America Under the United Nations Framework Convention on Climate Change*. Washington, September.

Environment Canada. 2007. *Regulatory Framework for Air Emissions*. Website (available at http://www.ec.gc.ca/doc/media/m_124/toc_eng.htm, last accessed 22 June 2009).

G-5. 2008. *G5 Statement Issued by Brazil, China, India, Mexico and South Africa on the occasion of the Hokkaido Toyako Summit*. Sapporo, 8 July.

G-8. 1997. *Summit of the Eight: Communiqué*. Denver, 20–22 June.

G-8. 2005. *The Gleneagles Communiqué*. Undated.

G-8. 2007. *Chair's Summary*. Heiligendamm, 8 June.

G-8. 2008. *Chair's summary*. Hokkaido, Toyako, 9 July.

G-8. 2009a. *The Summit*. Website of the Italian G-8 Presidency.

G-8. 2009b. *Responsible Leadership for a Sustainable Future – G-8 Leaders' Declaration*. L'Aquila, 8 July.

G-20. 2008. *Declaration – Summit on financial markets and the world economy*. Washington, DC, 15 November.

G-20. 2009a. *Global plan for recovery and reform: the Communiqué from the London Summit*. London, 2 April.

G-20. 2009b. *Leaders' Statement: The Pittsburgh Summit*. Pittsburgh, 24–25 September.

G-20. 2009c. *Meeting of Finance Ministers and Central Bank Governors – Communiqué*. London, 7 November.

Hague Conference. 1989. *Declaration of The Hague*. 11 March.

International Energy Agency (IEA). 2011. *World Energy Outlook 2011*. Paris: IEA.

IPCC. 1990. *IPCC First Assessment Report*, ed. J.T. Houghton, G.J. Jenkins, and J.J. Ephraums. Cambridge: Cambridge UP.

IPCC. 1995a. *IPCC Second Assessment: Climate Change 1995. A Report of the Intergovernmental Panel on Climate Change*. Cambridge: Cambridge UP.

IPCC. 1995b. WG I: *The Science of Climate Change. Summary for Policy-makers*, ed. J.T. Houghton, L.G. Meira Filho, B.A. Callender, N. Harris, A. Kattenberg, and K. Maskell. Cambridge: Cambridge UP.

IPCC. 1995c. *WG III: Economic and Social Dimensions of Climate Change*, ed. J.P. Bruce, H. Lee, and E.F. Haites. Cambridge: Cambridge UP.

IPCC. 2001a. *Climate Change 2001: Synthesis Report. A Contribution of Working Groups I, II, and III to the Third Assessment Report of the IPCC*. Cambridge: Cambridge UP.

IPCC. 2001b. *Climate Change 2001: The Scientific Basis. Contribution of Working Group I to the Third Assessment Report of the IPCC*. Cambridge: Cambridge UP.

IPCC. 2006. *Principles Governing IPCC Work*. April 2006 (available at http://www.ipcc.ch/pdf/ipcc-principles/ipcc-principles.pdf, last accessed 28 April 2008).

IPCC. 2007a. *Summary for Policymakers of the Synthesis Report of the IPCC Fourth Assessment Report*. Draft of 16 Nov. 2007 (available at http://www.ipcc.ch/pdf/assessment-report/ar4/syr/ar4_syr_spm.pdf, last accessed 22 April 2008).

IPCC. 2007b. *Climate Change 2007: Contribution of Working Groups I, II and III to the Fourth Assessment Report of the IPCC*. Geneva: IPCC.

IPCC. 2007c. Annex I: Glossary. In *Climate Change 2007: The Physical Science Basis. Contribution of Working Group I to the Fourth Assessment Report of the Intergovernmental Panel on Climate Change*, ed. S. Solomon, D. Qin, M. Manning, Z. Chen, M. Marquis, K.B. Averyt, M. Tignor, and H.L. Miller. Cambridge: Cambridge UP.

IPCC. 2008. *About IPCC. Introductory comments on IPCC Homepage* (available at http://www.ipcc.ch/about/index.htm, last accessed 18 June 2008).

Jintao, H. 2009. *Join Hands to Address Climate Change. Statement at the UN Summit on Climate Change.* New York, 22 September.

Joint Declaration. 2005. *EU and China Partnership on Climate Change.* Memo 05/298.

MCE. 2009a. *The Greenland Dialogue – A ministerial gathering on climate change* (available at http://www.kemin.dk/enUS/COP15/Greenland_dia logue/Sider/Forside.aspx, last accessed 29 March 2010).

MCE. 2009b. *The Greenland Ministerial Dialogue on Climate Change. Chair's Summary.* Ilulissat, Greenland, 30 June–3 July.

MEF (Major Economies Forum on Energy and Climate). 2009a. *Chairman's Summary from the First Preparatory Meeting of the Major Economies Forum on Energy and Climate.* Washington, DC, 29 April.

MEF. 2009b. *Second Preparatory Meeting of the Major Economies Forum on Energy and Climate Chairs' Summary.* Paris, 26 May.

MEF. 2009c. *Chairs' Summary: Third Preparatory Meeting of the Major Economies Forum on Energy and Climate.* Jiutepec, Mexico, 23 June.

MEF. 2009d. *Declaration of the Leaders. The Major Economies Forum on Energy and Climate.* L'Aquila, 9 July.

MEF. 2009e. *Chair's Summary: Fourth Meeting of the Leaders' Representatives of the MEF.* Washington, 21 September.

MEF. 2009f. *Fifth Meeting of the Leaders' Representatives of the MEF – Co-Chair's Summary.* London, 19 October.

MEF. 2012. *Overview of Past Meetings* (available at http://www.majoreco nomiesforum.org/past-meetings last accessed 22 February 2012).

MEM. 2008. *Declaration of Leaders Meeting of Major Economies on Energy Security and Climate Change.* Toyako, 9 July.

Merkel, A. 2009. *Pressekonferenz der Bundeskanzlerin auf der UN-Klimakonferenz in Kopenhagen.* 19 Dec.

MOE Japan. 2008a. *Gleneagles-Dialogue on Climate Change, Clean Energy and Sustainable Development – 4th Ministerial Meeting. Outline of the Gleneagles Dialogue.* Chiba, Japan, 14–16 March.

MOE Japan. 2008b. *Gleneagles-Dialogue – 4th Ministerial Meeting. Chairs' Conclusions.* Chiba, Japan, 14–16 March.

Ministry of Foreign Affairs (MOFA). 2010. *Japan's position regarding the Kyoto Protocol.*

National Geographic Society. 1997. *President Clinton speaks on global climate change.* Speech at the National Geographic Society. Washington, 22 Oct.

References

NDRC. 2007. *China's National Climate Change Programme*. June.

Noordwijk Conference. 1989. *The Noordwijk Declaration on Atmospheric Pollution and Climatic Change*. 7 November.

Obama, B. 2009. *Remarks by the President during press availability in Copenhagen*. 18 Dec.

Senate of the United States of America. 1997. *Resolution 98 of the 105th Congress, 1st session*. 25 July.

Stern, T. 2009a. *Opening Plenary Session of the AWG-LCA – Prepared Opening Plenary Intervention*. Bonn, 29 March.

Stern, T. 2009b. *Press Briefing by Mr. Todd Stern, Special Envoy for Climate Change*. Copenhagen, 9 December.

SWCC (Second World Climate Conference). 1990. *Ministerial Declaration*. 7 November.

Toronto Conference. 1988. *The Changing Atmosphere: Implications for Global Security*. Conference Statement. 27–30 June.

US EPA (Environmental Protection Agency). 2008. *Inventory of U.S. Greenhouse Gas Emissions and Sinks: 1990–2006*. Washington, DC: EPA.

US House. 2009. *Chairmen Waxman, Markey Release Discussion Draft of New Clean Energy Legislation*. 31 September.

US Mission (to the EU). 2008. *EU, U.S. Advance Climate Change, Clean Energy and Sustainable Development Dialogue*. Press Release. 12 March.

US. 2009. *U.S. Submission on Copenhagen Agreed Outcome*. 4 May.

White House. 1965. *Restoring the Quality of Our Environment*. Report of the President's Science Advisory Committee. Washington.

White House. 2001a. *Text of a letter from the President to Senators Hagel, Helms, Craig and Roberts*. Press Release. 13 March.

White House. 2001b. *President Bush discusses global climate change*. Press Release. 11 June.

White House. 2009a. *Remarks by President Barack Obama. Suntory Hall, Tokyo*. Press Release. 14 Nov.

White House. 2009b. *U.S.-China Joint Statement*. Beijing, 17 November.

White House. 2009c. *Joint Statement between Prime Minister Dr. Singh and President Obama*. Press Release. 24 November.

White House. 2009d. *President to Attend Copenhagen Climate Talks*. Press Release. 25 November.

Secondary sources

Adam, D., J. Watts, and P. Wintour. 2009. *Copenhagen climate talks: No deal, we're out of time, Obama warns*. The Guardian. 15 Nov. (available at http://www.guardian.co.uk/environment/2009/nov/15/copenhagen-climate-deal-obama, last accessed 4 February 2010).

Afionis, S. 2008. From Montreal to Bali: The 2005–2007 European Union Strategy for Reengaging the United States in UNFCCC Negotiations. *In-Spire: Journal of Law, Politics and Societies* 3 (2): 1–14.

Afionis, S. 2011. The European Union as a negotiator in the international climate change regime. *International Environmental Agreements* 11 (3): 341–360.

AFP (Agence France Presse). 2008a. *Réunion sur le climat à Paris: les Etats-Unis sous le feu des critiques*. 17 April (available at http://afp.google.com/article/ALeqM5j13AKIUzqnEDvnQIv9LUqCLrwUng, last accessed 25 October 2009).

AFP. 2008b. *Obama and McCain both offer new departure for US*. 15 June (available at http://afp.google.com/article/ALeqM5j5n5VTirfbfjWR12VN X3ENcK3Kg, last accessed 25 October 2009).

AFP. 2008c. *Financial crisis must not slow talks on CO_2 emissions: UN*. 14 Oct. (available at http://afp.google.com/article/ALeqM5ipd9Fpx4gIqIH tqT-lN9vcmD8qfw, last accessed 22 March 2010).

AFP. 2009a. *US, China wrap up climate talks*. 11 June (available at http://www.spacedaily.com/2006/090611080408.obhrb33t.html, last accessed 15 March 2010).

AFP. 2009b. *What's going Bonn?* 3 April (available at http://blogs.nature.com/news/thegreatbeyond/2009/04/whats_going_bonn.html, last accessed 29 March 2010).

Agence Europe. 1996. *Council is not able to fix commitments for the reduction of CO_2 emissions*. 11 December.

Agence Europe. 1997. *Breakthrough on climate change*. 5 March.

Agence Europe. 2007a. *Union calls for opening up of Chinese markets, but fails to convince Beijing to participate in joint climate change efforts*. 30 May.

Agence Europe. 2007b. *EU/Japan: Broad consensus on climate and adoption of joint action plan on intellectual property at Berlin summit*. 7 June.

Agence Europe. 2007c. *G8 compromise acknowledges need for substantial emissions reductions for after 2012 without setting binding target, but opens way for international negotiations under aegis of UN*. 9 June.

Agence Europe. 2007d. *EU well on way to achieving its Kyoto objective if Member States implement all measures being examined*. 28 November.

Agence Europe. 2008a. *EU/Latin America summit – climate change and sustainable development high on agenda*. 8 February.

Agence Europe. 2008b. *Member states support structure and ambition of climate/energy package – all want political agreement by end 2008/beginning 2009 at latest*. 4 March.

Agence Europe. 2008c. *EU is determined to adopt climate package by spring 2009 at latest and to protect energy-intensive industries*. 15 March.

Agence Europe. 2008d. *Europeans and Japanese want binding objectives on climate change and food crisis emergency measures*. 24 April.

Agence Europe. 2008e. *EU/Latin America and Caribbean summit expresses deep concern about rise in food prices*. 20 May.

Agence Europe. 2008f. *Full support for architecture of climate/energy package and for it to be concluded by end 2008 – some delegations call for flexibility for necessary adjustments.* 6 June.

Agence Europe. 2008g. *Progress, but much still to be done before Copenhagen, says Mr Barroso after Major Economies Meeting.* 10 July.

Agence Europe. 2008h. *Europe eyeing Africa for energy security.* 9 September.

Agence Europe. 2008i. *European and Asian leaders commit to work together determinedly at UN conference in Poznan.* 28 October.

Agence Europe. 2008j. *EU hopes Poznan conference will provide solid foundations for ambitious agreement in Copenhagen in December 2009.* 2 Dec.

Agence Europe. 2008k. *Four days before European Council, flurry of diplomatic activity and step up in technical talks on climate change/energy package.* 9 December.

Agence Europe. 2008l. *Commission announces €22 million investment in clean and renewable energy projects in Africa and Asia.* 12 December.

Agence Europe. 2008m. *"Historic agreement" on climate change package, retaining targets despite great exemptions on paid allowance.* 13 December.

Agence Europe. 2009a. *Commission and Chinese government willing to work hand in hand in response to global crisis and challenges.* 31 January.

Agence Europe. 2009b. *G8 Agreement on reducing global emissions by half by 2050 is step forward, says EU, even though there is no 2020 target.* 10 July.

Agence Europe. 2009c. *EU speaks of beneficial but insufficient breakthrough by Major Economies Forum in l'Aquila.* 11 July.

Agence Europe. 2009d. *EU Troika in Washington to kick-start climate talk.* 26 August.

Agence Europe. 2009e. *As UN and G20 summits approach, Barroso calls on heads of state to move – Germany and France call for carbon tax at borders.* 22 September.

Agence Europe. 2009f. *EU concerned at feeble progress in Bangkok – Hopes pinned on final talks in Barcelona before Copenhagen summit.* 13 Oct.

Agence Europe. 2009g. *Climate ministers want far-reaching agreement in Copenhagen with immediate effect – ball now in the court of United States and China.* 24 November.

Agence Europe. 2010. *The State of the Union in 2010.* 21 January.

Agence Europe. 2012a. *Poland rejects 2050 roadmap and vetoes progress at Council.* 13 March.

Agence Europe. 2012b. *ETS – EU calm in face of planned retaliatory measures.* 22 February.

Agence Europe. 2012c. *Doha will not save planet, but it has saved face, says EU.* 10 Dec.

Agence Europe. 2013. *ETS "Stop the Clock" – ENVI committee endorses agreement.* 27 March.

Aggarwal, V.K. 1985. *Liberal protectionism. The International Politics of Organized Textile Trade*. Berkeley: University of California Press.

Allen, D. and M. Smith. 1990. Western Europe's Presence in the Contemporary International Arena. *Review of International Studies* 16 (1): 19–38.

Andonova, L.B. 2009. The climate regime and domestic politics: the case of Russia. In *The Politics of Climate Change. Environmental Dynamics in International Affairs*, ed. P.G. Harris, 29–50. London: Routledge.

Andresen, S. 1998. *The Development of the Climate Regime: Positions, Evaluations and Lessons*. FNI Report 3/98. Lysaker: Fridtjof Nansen Institute.

Andresen, S., and S. Agrawala. 2002. Leaders, pushers and laggards in the making of the climate regime. *Global Environmental Change* 12 (1): 41–51.

Andresen, S., and S.H. Butenschøn. 2001. Norwegian Climate Policy: From Pusher to Laggard? *International Environmental Agreements* 4 (3): 337–356.

Arts, B., and P. Verschuren. 1999. Assessing Political Influence in Complex Decision-Making: An Instrument Based on Triangulation. *International Political Science Review* 20 (4): 411–424.

Audet, R., and P. Bonin. 2010. Les Accords de Cancún face aux enjeux des négociations internationales sur le climat. *VertigO-La revue en sciences de l'environnement, Débats et Perspectives* 4: 1–12.

Baker, S. 2006. Environmental Values and Climate Change Policy: Contrasting the European Union and the United States. In *Values and Principles in European Union Foreign Policy*, ed. S. Lucarelli, and I. Manners, 69–87. London: Routledge.

Ball, J., and K. Johnson. 2010. *Push to Oversimplify at Climate Panel*. Wall Street Journal. 26 Feb. (available at http://online.wsj.com/article/SB10001424052748704188104575083681319834978.html, last accessed 26 Oct. 2010).

Barrett, J. 1991. The Negotiation and Drafting of the Climate Change Convention. In *International Law and Global Climate Change*, ed. R. Churchill, and D. Freestone, 183–200. London: Graham and Trotman.

Bausch, C., and M. Mehling. 2006. 'Alive and Kicking': The First Meeting of the Parties to the Kyoto Protocol. *RECIEL* 15 (2): 193–201.

Bäckstrand, K., and O. Elgström. 2013. The EU's role in climate negotiations: from leader to 'leadiator'. *Journal of European Public Policy* (online first).

BBC. 2008. *Australia sets new climate target*. 15 Dec. (available at http://news.bbc.co.uk/2/hi/science/nature/7782919.stm, last accessed 29 March 2010).

BBC. 2009a. *'Urgent action' needed on climate*. 22 Sept. (available at http://news.bbc.co.uk/2/hi/science/nature/8268077.stm, last accessed 22 March 2010).

BBC. 2009b. *APEC leaders drop climate target*. BBC online. 15 Nov. (available at http://news.bbc.co.uk/2/hi/asia-pacific/8360982.stm, last accessed 4 February 2010).

Beament, E. 2010. *Global warming deal hopes revived after Cancun agreement*. The Independent. 11 December (available at http://www.independent.co.uk/environment/climate-change/global-warming-deal-hopes-revived-after-cancun-agreement-2157688.html, last accessed 26 January 2012).

Becker, M., and R. Nelles. 2009. *Gipfel-Minimalisten feilen am Klimakompromiss*. Spiegel Online. 18 Dec. (available at www.spiegel.de/wissenschaft/natur/0,1518,668037,00.html, last accessed 24 February 2010).

Beisheim, M., B. Lode, and N. Simon. 2012. *Rio+20 Realpolitik and Its Implications for "The Future We Want"*. Berlin: German Institute for International and Security Affairs.

Belis, D., and S. Schunz. 2013. China and the European Union: emerging partners in global climate governance? *Environmental Practice* 15 (3): 190–200.

Bennett, A., and A.L. George. 2005. *Case studies and theory development in the social sciences*. Cambridge, MA: MIT Press.

Benwell, R. 2009. Linking as leverage: emissions trading and the politics of climate change. In *The Politics of Climate Change. Environmental Dynamics in International Affairs*, ed. P. Harris, 91–107. London: Routledge.

Bergin, T. 2009. *Climate change funding talks stall at G20*. Reuters. 5 Sept. (available at http://www.reuters.com/article/latestCrisis/idUSL5203919, last accessed 20 May 2010).

Betsill, M. 2005. Global Climate Change Policy: Making Progress or Spinning Wheels? In *The Global Environment. Institutions, Law, and Policy*, ed. R.S. Axelrod, D.L. Downie, and N.J. Vig, 103–124. Washington: CQ Press.

Betsill, M., and E. Corell. eds. 2008. *NGO Diplomacy. The Influence of Nongovernmental Organizations in International Environmental Negotiations*. Boston: MIT Press.

Biermann, F. 2005. Between the USA and the South: strategic choices for European climate policy. *Climate Policy* 5 (3): 273–290.

Black, R. 2009. *Why did Copenhagen fail to deliver a climate deal?* BBC online. 22 Dec. (available at http://news.bbc.co.uk/2/hi/8426835.stm, last accessed 27 March 2010).

BMU. 2010. *Minister Röttgen: New momentum for international climate negotiations* (available at http://www.bmu.de/english/current_press_releases/pm/45968.php, last accessed 10 June 2011).

BMU. 2011. *Petersberg Climate Dialogue II* (available at http://www.bmu.de/english/petersberg_conference/doc/47568.php, last accessed 22 Feb. 2012).

BMU. 2012. *Petersberg Climate Dialogue III* (available at http://www.bmu.de/en/topics/climate-energy/climate/international-climate-policy/petersberg-climate-dialogue/petersberg-climate-dialogue-iii/, last accessed 12 April 2013).

Bodansky, D. 1993. The United Nations Framework Convention on Climate Change: A Commentary. *Yale Journal of International Law* 18 (2): 451–558.

Bodansky, D. 2001. International Law and the Design of a Climate Change Regime. In *International Relations and Global Climate Change*, ed. U. Luterbacher, and D. Sprinz, 201–220. Cambridge, Mass.: MIT Press.

Bodansky, D. 2011. *A Tale of Two Architectures: The Once and Future U.N. Climate Change Regime*. Arizona State University. 7 March.

Bodansky, D. 2012. *The Durban Platform Negotiations: Goals and Options*. Cambridge, Mass.: Harvard University.

Bodeen, C. 2009. *US delegation mixed on China climate change pact*. AP. 29 May (available at http://www.google.com/hostednews/ap/article/AleqM5hLcZ2jQ-4mu4rd7XlB3hetiVn1qbAD98FBCCG0, last accessed 24 February 2010).

Bollen, A., and P. van Humbeeck. 2002. *Klimaatbeleid en Klimaatverandering: een leidraad*. Gent: Academia Press.

Borione, D., and J. Ripert. 1994. Exercising Common but Differentiated Responsibility. In *Negotiating Climate Change: The Inside Story of the Rio Convention*, ed. I.M. Mintzer, and J.A. Leonard, 77–96. Cambridge: Cambridge Studies in Energy and Environment.

Börzel, T., and T. Risse. 2000. When Europe Hits Home: Europeanization and Domestic Change. *European Integration Online Papers* 4 (15).

Botzen, W.J.W., J.M. Gowdy, and J.C.J.M. Van Den Bergh. 2008. Cumulative CO_2 emissions: shifting international responsibilities for climate debt. *Climate policy* 8 (3): 569–576.

Bowering, E. 2011. *After Kyoto: the Cartagena Dialogue and the future of the international climate change regime*. Working Paper. Griffith University (available at http://www.globalvoices.org.au/wp-content/uploads/UNFCCC-Research-Report-Ethan-Bowering.pdf, last accessed 14 December 2011).

Braam, G.P.A. 1975. *Invloed van bedrijven op de overheid. Een empirische studie over de verdeling van maatschappelijke invloed*. Meppel: Boom.

Brambilla, P. 2004. *Europäisches Umweltrecht und das internationale Klimaregime der Vereinten Nationen*. Ph.D. dissertation. Konstanz: University of Konstanz.

Brand, C. 2009. *EU presses US on climate change*. 1 Sept. (available at http://www.google.com/hostednews/ap/article/ALeqM5jQtHqEAgBtWoCdvEW-sykmyTvY7DwD9AEH0J80, last accessed 24 February 2010).

Brenton, T. 1994. *The Greening of Machiavelli: The Evolution of International Environmental Politics*. London: Earthscan.

Brighi, E., and C. Hill. 2008. Implementation and behaviour. In *Foreign Policy: Theories, actors, cases*, ed. S. Smith, A. Hadfield, and T. Dunne, 117–136. Oxford: Oxford UP.

Broder, J. 2009. *Many Goals Remain Unmet in 5 Nations' Climate Deal*. NY Times. 18 Dec. (available at http://www.nytimes.com/2009/12/19/science/earth/19climate.html?_r=3&pagewanted=all, last accessed 22 Dec. 2009).

Broder, J., and E. Rosenthal. 2009. *Obama Has Goal to Wrest a Deal in Climate Talks*. NY Times. 17 Dec. (available at http://www.nytimes.com/2009/12/18/science/earth/18climate.html?_r=4&hp, last accessed 18 March 2010).

Bruyninckx, H. 2005. EU Environmental Policy. In *Handbook of EU Affairs*, ed. A.M. Dobre, M. Martins-Gystelinck, R. Polacek, and R. Hunink, 213–23. Bucharest: Institutul European din Romania.

Bruyninckx, H. 2011. *De zogenaamde doorbraak van Durban*. De Morgen. 12 Dec. (available at http://www.demorgen.be/dm/nl/2461/De-Gedachte/article/detail/1361599/2011/12/12/De-zogenaamde-doorbraak-van-Durban.dhtml, last accessed 24 January 2012).

Bryner, G. 2000. Congress and the Politics of Climate Change. In *Climate Change and American Foreign Policy*, ed. P.G. Harris, 111–130. New York: St. Martin's Press.

Buchner, B., and S. Dall'Olio. 2005. Russia and the Kyoto Protocol: The long road to ratification. *Transition Studies Review* 12(2): 349–382.

Buckley, C. 2009a. *China flexible on rich nations' greenhouse gas cuts.* Reuters. 22 May (available at http://in.reuters.com/article/environmentNews/idINTRE54P4ON20090526?sp=true, last accessed 26 February 2010).

Buckley, C. 2009b. *China steps up climate diplomacy as Copenhagen looms.* Reuters. 29 Oct. (available at http://in.reuters.com/article/worldNews/idINIndia-43515920091029?sp=true, last accessed 26 March 2010).

Bush, G.W. 2007. *President Bush Delivers State of the Union Address.* Washington, D.C., 23 January.

C2ES (Centre for Climate and Energy Solutions). 2012. *Outcomes of the UN climate change conference in Doha, Qatar.* Arlington, VA: C2ES.

Caporaso, J., and J. Jupille. 1998. States, Agency, and Rules: The European Union in Global Environmental Politics. In *The European Union in the World Community*, ed. C. Rhodes, 213–249. Boulder: Lynne Rienner.

Caporaso, J., and J. Jupille. 2001. The Europeanization of Gender Equality and Domestic Structural Change. In *Transforming Europe: Europeanization and domestic change*, ed. M. Green Cowles, J. Caporaso, and T. Risse, 21–43. Ithaca: Cornell UP.

Carlsnaes, W. 2002. Foreign Policy. In *Handbook of International Relations*, ed. W. Carlsnaes, T. Risse, and B.A. Simmons, 331–349. London: Sage.

Carrington, D. 2011. *Climate deal: A guarantee our children will be worse off than us.* The Guardian. 11 Dec. (available at http://www.guardian.co.uk/environment/damian-carrington-blog/2011/dec/11/durban-climate-change-conference-2011-climate-change, last accessed 24 January 2012).

Casey-Lefkovitz, S. 2010. *New international agreement to fight climate change found spirit for consensus in Cartagena Dialogue countries.* 11 Dec. (available at http://switchboard.nrdc.org/blogs/sclefkowitz/new_international_agreement_to.html, last accessed 24 February 2012).

Cass, L.R. 2007. The indispensable awkward partner: the United Kingdom in European climate policies. In *Europe and Global Climate Change. Politics, foreign policy and regional cooperation*, ed. P.G. Harris, 63–86. Cheltenham: Edward Elgar.

Cass, L.R. 2009. A climate of obstinacy: symbolic politics in Australian and Canadian policy. In *The Politics of Climate Change. Environmental Dynamics in International Affairs*, ed. P.G. Harris, 11–28. Routledge: London.

CBS. 2009. *G-20 Declared Top Global Economic Council.* 25 Sept. (available at http://www.cbsnews.com/stories/2009/09/25/business/main5338518.sh tml, last accessed 20 Nov. 2009).

Charlton, A. 2009. *Help for poor countries at Paris climate talks.* AP. 27 May (available at http://www.google.com/hostednews/ap/article/ALeqM5gamKFpfPBUXl2x85j_benvY8_1oQD98E20980, last accessed 15 June 2009).

Charter, D., and P. Webster. 2009. *Gordon Brown's climate change finance package hangs in balance*. Times Online. 26 Oct. (available at http://www.timesonline.co.uk/tol/news/environment/article6889867.ece#, last accessed 22 March 2010).

Chasek, P.S. 2001. *Earth Negotiations. Analyzing Thirty Years of Environmental Diplomacy*. Tokyo: UN UP.

Checkel, J. 2005. International Institutions and Socialization in Europe: Introduction and Framework. *International Organization* 59 (3): 801–826.

Chinability. 2009. *GDP growth in China 1952–2009*. Oct. (available at http://www.chinability.com/GDP.htm, last accessed 15 Oct. 2009).

CIA (Central Intelligence Agency). 2009. *The World Factbook. India. Economy* (available at https://www.cia.gov/library/publications/the-world-factbook/geos/in.html, last accessed 15 Nov. 2009)

Coloma, R. 2009. *Washington eyes compromise at Copenhagen talks*. AFP. 11 Nov. (available at http://www.google.com/hostednews/afp/article/ALeqM5ilnRJe9HSpLbiwAXE1I4 aKWa328w, last accessed 4 February 2010).

Costa, O. 2007. *Who decides on EU foreign climate policy and why it matters. Actors, alliances and institutions*. BISA Annual Conference 2007. University of Cambridge, 17–19 December.

Costa, O. 2009. Is Climate Change Changing the EU? The Second Image Reversed in Climate Politics. In *The Politics of Climate Change. Environmental Dynamics in International Affairs*, ed. P. Harris, 72–89. London: Routledge.

Cox, R., and H.R. Jacobson. 1973. The Framework for Inquiry. In *The Anatomy of Influence. Decision-Making in International Organization*, ed. R. Cox, and H. Jacobson, 1–36. New Haven: Yale UP.

Curtin, J. 2010. *The Copenhagen Conference: how should the EU respond?* Dublin: Institute of International and European Affairs.

Dahl, R.A. 1957. The concept of power. *Behavioural Science* 2 (3): 201–215.

Dahl, R.A., and B. Stinebrickner. 2003. *Modern Political Analysis*. 6th ed. Upper Saddle River, NJ: Prentice Hall.

Daly, M. 2010. *Climate Bill: Senate Democrats Abandon Comprehensive Energy Bill*. Huffington Post. 22 July (available at http: http://www.huffingtonpost.com/2010/07/22/climate-bill-senate-democ_n_656175.html, last accessed 16 September 2010).

Damro, C., and P.L. Mendez. 2003. Emissions Trading at Kyoto: from EU Resistance to Union Innovation. *Environmental Politics* 12 (2): 71–94.

Dasgupta, C. 1994. The Climate Change Negotiations. In *Negotiating Climate Change: The Inside Story of the Rio Convention*, ed. I. M. Mintzer, and J. A. Leonard, 129–148. Cambridge: Cambridge Studies in Energy and Environment.

Davies, R. 2006. 2006: The Year of Climate Change. *The Voice of San Diego*. 18 September.

Dee, M. 2013. *Challenging Expectations: A study of European Union performance in multilateral negotiations*. Ph.D. dissertation. Glasgow: University of Glasgow.

De Jong, S., and S. Schunz 2012. Coherence in European Union External Policy before and after the Lisbon Treaty: The Cases of Energy Security and Climate Change. *European Foreign Affairs Review* 17 (2): 165–186.

Dehousse, F., and N. Bekkhus. 2007. Energy and climate change in the European Union. *Studia Diplomatica* (LX) 2.

Delbeke, J. 2011. *Climate change policy post-Durban – Keynote lecture.* 14 Dec. Brussels: Centre for European Policy Studies.

Delbeke, J. 2012. *The International Climate Regime after Doha.* 20 Dec. Brussels: Centre for European Policy Studies.

Delreux, T. 2008. *The European Union negotiates multilateral environmental agreements: an analysis of the internal decision-making process.* Ph.D. dissertation. Leuven: Katholieke Universiteit Leuven.

Depledge, J. 2005. *The Organization of Global Negotiations. Constructing the Climate Change Regime.* London: Earthscan.

Depledge, J. 2006. A Breakthrough for the Climate Change Regime? *Environmental Policy and Law* 36 (1): 14–19.

Dessai, S. 2001. Why did The Hague climate conference fail? *Environmental Politics* 10 (3): 139–144.

Dessai, S., and L. Schipper. 2003. The Marrakech Accords to the Kyoto Protocol: analysis and future prospects. *Global Environmental Change* 13 (2): 149–153.

Dessai, S., N.S. Lacasta, and K. Vincent. 2003. International political history of the Kyoto Protocol: from The Hague to Marrakech and beyond. *International Review for Environmental Strategies* 4 (2): 183–205.

Dessai, S., L. Schipper, E. Corbera, B. Kjellén, M. Gutiérrez, and A. Haxeltine. 2005. Challenges and Outcomes at the Ninth Session of the Conference of the Parties to the United Nations Framework Convention on Climate Change. *International Environmental Agreements* 5 (1): 105–124.

Douma, W.T. 2006. The European Union, Russia and the Kyoto Protocol. In *EU Climate Change Policy: The Challenge of New Regulatory Initiatives*, ed. M. Peeters, and K. Deketelaere, 51–68. Cheltenham: Edward Elgar.

Downie, D.L. 2005. Global Environmental Policy: Governance Through Regimes. In *The Global Environment. Institutions, Law, and Policy*, ed. R.S. Axelrod, D.L. Downie, and N.J. Vig, 64–82. Washington: CQ Press.

DPA. 2009a. *EU sagt Milliarden für Klimaschutz zu.* 11 Dec. (available at http://de.news.yahoo.com/26/20091211/tpl-eu-sagt-milliarden-fr-klimaschutz-zu-a70ba75.html, last accessed 28 March 2010).

DPA. 2009b. *Neue Hoffnung vor Klima-Schlussrunde.* 18 Dec. (available at http://de.news.yahoo.com/26/20091218/tpl-neue-hoffnung-vor-klimaschlussrunde-a70ba75.html, last accessed 18 March 2010).

Drexhage, J., D. Murphy, and J. Gleeson. eds. 2008. *A Way Forward. Canadian perspectives on post-2012 climate policy.* Winnipeg: IISD.

Dupont, C. 2012. *Leading or muddling through? The EU and Durban.* 16 April (available at http://ecfr.ideasoneurope.eu/2012/04/16/leading-or-muddling-through-the-eu-and-durban, last accessed 22 April 2012).

E3G. 2009. *What the EU Climate Package means for the Global Climate Deal.* E3G Briefing Note. January.

Earthtimes. 2008. *EU's internal laws target global climate deal: analysis.* 23 Jan. (available at http://www.earthtimes.org/articles/show/177361, analysis%C2%A0eus-internal-laws-target-global-climate-deal.html#, last accessed 22 Feb. 2010).

ECFR. 2012. *Scorecard: Multilateral issues – climate change* (available at http://www.ecfr.eu/scorecard/2012/issues/73, last accessed 16 January 2012).

Ecofys. 2010. *Ambition of only two developed countries sufficiently stringent for 2°C.* ClimateActionTracker. 2 February (available at http://www.climateactiontracker.org/news.php, last accessed 2 March 2010).

Eeckhout, P. 2011. *EU external relations law.* 2nd ed. Oxford: Oxford UP.

Egenhofer, C. 2010. A Closer Look at EU Climate Change Leadership. *Intereconomics* 45 (3): 167–170.

Egenhofer, C., and A. Georgiev. 2009. *The Copenhagen Accord. A first stab at deciphering the implications for the EU.* Brussels: Centre for European Policy Studies.

Egenhofer, C., and M. Alessi. 2013. *EU Policy on Climate Change Mitigation since Copenhagen and the Economic Crisis.* Brussels: Centre for European Policy Studies.

Eilperin, J. 2009. *U.S. weighs backing interim international climate agreement. Smaller-scale approach seen as first step toward full pact.* Washington Post. 13 Nov. (available at http://www.washingtonpost.com/wp-dyn/content/article/2009/11/12/AR2009111209127_pf.html, last accessed 4 February 2010).

Eilperin, J., and C. Lynch. 2009. *Nations Appear Headed Toward Independent Climate Goals.* Washington Post. 23 Sept. (available at http://www.washingtonpost.com/wpdyn/content/article/2009/09/22/AR2009092201137.html?Hpid%3Dtopnews, last accessed 24 March 2010).

Eisenhardt, K.M. 1989. Building Theories from Case Study Research. *The Academy of Management Review* 14 (4): 532–550.

Elster, J. 1998. A plea for mechanisms. In *Social Mechanisms: an analytical approach to social theory*, ed. P. Hedström, and R. Swedberg, 45–73. Cambridge: Cambridge UP.

ENB. 1995a. *Curtainraiser – COP 1.* Vol. 12 (12), 28 March. New York: IISD.

ENB. 1995b. *Climate Change Convention COP-1.* Vol. 12 (13). 29 March. New York: International Institute for Sustainable Development (IISD).

ENB. 1995c. *Climate Change Convention COP-1 Highlights: Wednesday, 29 March 1995.* Vol. 12 (14). 30 March. New York: IISD.

ENB. 1995d. *Climate Change Convention COP-1 Highlights: 31 March and 1 April 1995.* Vol. 12 (16). 3 April. New York: IISD.

ENB. 1995e. *Climate Change Convention COP-1 Highlights: 3 April 1995.* Vol. 12 (17). 4 April. New York: IISD.

ENB. 1995f. *Climate Change Convention COP-1 Highlights: 5 April 1995.* Vol. 12 (19). 6 April. New York: IISD.

References

ENB. 1995g. *Report of the AGBM: 21–25 August 1995*. Vol. 12 (22). 28 August. New York: IISD.

ENB. 1995h. *Report of the second session of the AGBM*. Vol. 12 (24). 7 November. New York: IISD.

ENB. 1996a. *Report of the third session of the AGBM*. Vol. 12 (27). 11 March. New York: IISD.

ENB. 1996b. *Summary of the second Conference of the Parties to the UNFCCC*. Vol. 12 (38). 22 July. New York: IISD.

ENB. 1996c. *Report of the Meetings of the Subsidiary Bodies of the UNFCCC*. Vol. 12 (39). 23 December. New York: IISD.

ENB. 1997a. *Highlights from the sixth session of the AGBM, 3 March 1997*. Vol. 12 (41). 4 March. New York: IISD.

ENB. 1997b. *Highlights from the sixth session of the AGBM, 4 March 1997*. Vol. 12 (42). 5 March. New York: IISD.

ENB. 1997c. *Report of the sixth session of the AGBM*. Vol. 12 (45). 10 March. New York: IISD.

ENB. 1997d. *Summary of the Nineteenth UNGA Special Session to Review Implementation of Agenda 21*. Vol. 5 (88). 30 June. New York: IISD.

ENB. 1997e. *Highlights from the FCCC AGBM, 31 July 1997*. Vol. 12 (50). 1 August. New York: IISD.

ENB. 1997f. *Highlights from the Meeting of the FCCC Subsidiary Bodies: 5 August 1997*. Vol. 12 (53). 6 August. New York: IISD.

ENB. 1997g. *Report of the Meeting of the FCCC Subsidiary Bodies to the UNFCCC*. Vol. 12 (55). 11 August. New York: IISD.

ENB. 1997h. *Highlights from the Meetings of the FCCC Subsidiary Bodies: 22 October 1997*. Vol. 12 (59). 23 October. New York: IISD.

ENB. 1997i. *Highlights from the Meeting of the FCCC Subsidiary Bodies: 23 October 1997*. Vol. 12 (60). 24 October. New York: IISD.

ENB. 1997j. *Highlights from the Meetings of the FCCC Subsidiary Bodies: 27 October 1997*. Vol. 12 (62). 28 October. New York: IISD.

ENB. 1997k. *Report of the meetings of the FCCC Subsidiary Bodies: 20–31 October 1997*. Vol. 12 (66). 3 November. New York: IISD.

ENB. 1997l. *Highlights from the Third Conference of the Parties to the UNFCCC: 30 November 1997*. Vol. 12 (67). 1 December. New York: IISD.

ENB. 1997m. *Highlights from the Third Conference of the Parties to the UNFCCC: 1 December 1997*. Vol. 12 (68). 2 December. New York: IISD.

ENB. 1997n. *Highlights from the Third Conference of the Parties to the UN Framework Convention on Climate Change: 2 December 1997*. Vol. 12 (69). 3 December. New York: IISD.

ENB. 1997o. *Highlights from the Third Conference of the Parties to the UNFCCC: 3 December 1997*. Vol. 12 (70). 4 December. New York: IISD.

ENB. 1997p. *Highlights from the Third Conference of the Parties to the UNFCCC: 5 December 1997*. Vol. 12 (72). 6 December. New York: IISD.

ENB. 1997q. *Highlights from the Third Conference of the Parties to the UNFCCC: 6–7 December 1997.* Vol. 12 (73). 8 December. New York: IISD.

ENB. 1997r. *Highlights from the Third Conference of the Parties to the UNFCCC: 8 December 1997.* Vol. 12 (74). 9 December. New York: IISD.

ENB. 1997s. *Highlights from the Third Conference of the Parties to the UNFCCC: 9 December 1997.* Vol. 12 (75). 10 December. New York: IISD.

ENB. 1997t. *Report of the Third Conference of the Parties to the UNFCCC: 1–11 December 1997.* Vol. 12 (76). 13 December. New York: IISD.

ENB. 1998. *Report of the Fourth Conference of the Parties to the UNFCCC.* Vol. 12 (97). 16 November. New York: IISD.

ENB. 1999. *Summary of the Fifth Conference of the Parties to the UNFCCC.* Vol. 12 (123). 8 November. New York: IISD.

ENB. 2000. *Summary of the Sixth Conference of the Parties to the UNFCCC.* Vol. 12 (163). 27 November. New York: IISD.

ENB. 2001a. *Summary of the Resumed Sixth Conference of the Parties to the UNFCCC.* Vol. 12 (176). 30 July. New York: IISD.

ENB. 2001b. *Summary of the Seventh Conference of the Parties to the UNFCCC.* Vol. 12 (189). 12 November. New York: IISD.

ENB. 2002. *Summary of the Eighth Conference of the Parties to the UNFCCC.* Vol. 12 (209). 4 November. New York: IISD.

ENB. 2003. *Summary of the Ninth Conference of the Parties to the UNFCCC.* Vol. 12 (231). 15 December. New York: IISD.

ENB. 2004a. *Summary of the Tenth Conference of the Parties to the UNFCCC.* Vol. 12 (260). 20 December. New York: IISD.

ENB. 2004b. *UNFCCC COP-10 Highlights: 13 December.* Vol. 12 (256). 14 December. New York: IISD.

ENB. 2005a. *Summary of the UNFCCC seminar of Governmental Experts: 16–17 May 2005.* Vol. 12 (261). 19 May. New York: IISD.

ENB. 2005b. *Summary of the Eleventh Conference of the Parties to the UNFCCC and First Meeting of the Parties to the Kyoto Protocol (MOP).* Vol. 12 (291). 20 Dec. New York: IISD.

ENB. 2006. *Summary of the Twelfth Conference of the Parties to the UNFCCC and Second MOP: 6–17 November 2006.* Vol. 12 (318). 20 December. New York: IISD.

ENB. 2007a. *Twenty-Sixth Sessions of the Subsidiary Bodies of the UNFCCC and associated meetings.* Vol. 12 (333). 21 May 2007. New York: IISD.

ENB. 2007b. *AWG 4 and Dialogue 4 highlights: 30 August 2007.* Vol. 12 (338). 31 August. New York: IISD.

ENB. 2007c. *Fourth sessions of the AWG-LCA and Convention Dialogue: 27–31 August 2007.* Vol. 12 (339). 3 September. New York: IISD.

ENB. 2007d. *Thirteenth Conference of the Parties to the UNFCCC and Third MOP.* Vol. 12 (343). 3 December. New York: IISD.

ENB. 2007e. *COP 13 and COP/MOP 3 highlights: 5 December 2007*. Vol. 12 (346). 6 December. New York: IISD.

ENB. 2007f. *COP 13 and COP/MOP 3 highlights: 6 December 2007*. Vol. 12 (347). 7 December. New York: IISD.

ENB. 2007g. *COP 13 and COP/MOP 3 highlights: 7 December 2007*. Vol. 12 (348). 8 December. New York: IISD.

ENB. 2007h. *COP 13 and COP/MOP 3 highlights: 8 December 2007*. Vol. 12 (349). 10 December. New York: IISD.

ENB. 2007i. *COP 13 and COP/MOP 3 highlights: 11 December 2007*. Vol. 12 (351). 12 December. New York: IISD.

ENB. 2007j. *Summary of the Thirteenth Conference of Parties to the UNFCCC and Third MOP*. Vol. 12 (354). 18 December. New York: IISD.

ENB. 2008a. *First session of the AWG-LCA and Fifth Session of the AWG-KP*. Vol. 12 (357). 31 March. New York: IISD.

ENB. 2008b. *Summary of the First Session of the AWG-LCA and the Fifth Session of the AWG-KP*. Vol. 12 (362). 7 April. New York: IISD.

ENB. 2008c. *SB 28 and AWG highlights: 9 June 2008*. Vol. 12 (371). 10 June. New York: IISD.

ENB. 2008d. *Twenty-Eighth Sessions of the UNFCCC Subsidiary Bodies, Second Session of the AWG-LCA, and Fifth Session of the AWG-KP: 2–13 June 2008*. Vol. 12 (375). 16 June. New York: IISD.

ENB. 2008e. *AWG-LCA 3 and AWP-KP 6 highlights: 23 August 2008*. Vol. 12 (380). 25 August. New York: IISD.

ENB. 2008f. *AWG-LCA 3 and AWP-KP 6 highlights: 25 August 2008*. Vol. 12 (381). 26 August. New York: IISD.

ENB. 2008g. *Summary of the Third Session of the AWG-LCA and Sixth Session (Part One) of the AWG-KP*. Vol. 12 (383). 30 August. New York: IISD.

ENB. 2008h. *COP 14 highlights: 2 December 2008*. Vol. 12 (387). 3 December. New York: IISD.

ENB. 2008i. *COP 14 highlights: 4 December 2008*. Vol. 12 (389). 5 December. New York: IISD.

ENB. 2008j. *COP 14 highlights: 9 December 2008*. Vol. 12 (392). 10 December. New York: IISD.

ENB. 2008k. *COP 14 highlights: 11 December 2008*. Vol. 12 (394). 12 December. New York: IISD.

ENB. 2008l. *Summary of the Fourteenth Conference of Parties to the UNFCCC and Fourth MOP*. Vol. 12 (395). 15 December. New York: IISD.

ENB. 2009a. *AWG-LCA 5 and AWG-KP 7 highlights: 29 March 2009*. Vol. 12 (398). 30 March. New York: IISD.

ENB. 2009b. *AWG-LCA 5 and AWG-KP 7 highlights: 30 March 2009*. Vol. 12 (399). 31 March. New York: IISD.

ENB. 2009c. *AWG-LCA 5 and AWG-KP 7 highlights: 31 March 2009*. Vol. 12 (400). 1 April. New York: IISD.

ENB. 2009d. *AWG-LCA 5 and AWG-KP 7 highlights: 1 April 2009.* Vol. 12 (401). 2 April. New York: IISD.

ENB. 2009e. *SB 30 and AWG highlights: 1 June 2009.* Vol. 12 (411). 2 June. New York: IISD.

ENB. 2009f. *SB 30 and AWG highlights: 2 June 2009.* Vol. 12 (412). 3 June. New York: IISD.

ENB. 2009g. *SB 30 and AWG highlights: 4 June 2009.* Vol. 12 (414). 5 June. New York: IISD.

ENB. 2009h. *SB 30 and AWG highlights: 5 June 2009.* Vol. 12 (415). 6 June. New York: IISD.

ENB. 2009i. *SB 30 and AWG highlights: 6 June 2009.* Vol. 12 (416). 8 June. New York: IISD.

ENB. 2009j. *SB 30 and AWG highlights: 8 June 2009.* Vol. 12 (417). 9 June. New York: IISD.

ENB. 2009k. *Summary of the Bonn Climate Change Talks.* Vol. 12 (421). 15 June. New York: IISD.

ENB. 2009l. *Bonn Climate Change Talks.* Vol. 12 (422). 10 August. New York: IISD.

ENB. 2009m. *AWG-LCA 7 and AWG-KP 9 highlights: 28 September 2009.* Vol. 12 (429). 29 Sept. New York: IISD.

ENB. 2009n. *AWG-LCA 7 and AWG-KP 9 highlights: 2 October 2009.* Vol. 12 (433). 3 Oct. New York: IISD.

ENB. 2009o. *AWG-LCA 7 and AWG-KP 9 highlights: 5 October 2009.* Vol. 12 (435). 6 Oct. New York: IISD.

ENB. 2009p. *Summary of the Bangkok Climate Change Talks.* Vol. 12 (439). 12 Oct. New York: IISD.

ENB. 2009q. *AWG-LCA 7 and AWG-KP 9 highlights: 2 November 2009.* Vol. 12 (443). 3 Nov. New York: IISD.

ENB. 2009r. *AWG-LCA 7 and AWG-KP 9 highlights: 3 November 2009.* Vol. 12 (444). 4 Nov. New York: IISD.

ENB. 2009s. *Summary of the Barcelona Climate Change Talks.* Vol. 12 (447). 9 Nov. New York: IISD.

ENB. 2009t. *Copenhagen highlights: 7 December 2009.* Vol. 12 (449). 8 Dec. New York: IISD.

ENB. 2009u. *Copenhagen highlights: 8 December 2009.* Vol. 12 (450). 9 Dec. New York: IISD.

ENB. 2009v. *Copenhagen highlights: 9 December 2009.* Vol. 12 (451). 10 Dec. New York: IISD.

ENB. 2009w. *Copenhagen highlights: 10 December 2009.* Vol. 12 (452). 11 Dec. New York: IISD.

ENB. 2009x. *Copenhagen highlights: 12 December 2009.* Vol. 12 (454). 14 Dec. New York: IISD.

ENB. 2009y. *Copenhagen highlights: 14 December 2009.* Vol. 12 (455). 15 Dec. New York: IISD.

ENB. 2009z. *Copenhagen highlights: 15 December 2009.* Vol. 12 (456). 16 Dec. New York: IISD.

ENB. 2009aa. *Copenhagen highlights: 16 December 2009.* Vol. 12 (457). 17 Dec. New York: IISD.

ENB. 2009bb. *Copenhagen highlights: 17 December 2009.* Vol. 12 (458). 18 Dec. New York: IISD.

ENB. 2009cc. *Summary of the Copenhagen Climate Change Conference: 7–19 December 2009.* Vol. 12 (459). 22 Dec. New York: IISD.

ENB. 2010a. *Summary of the Bonn Climate Change Talks.* Vol. 12 (460). 14 April. New York: IISD.

ENB. 2010b. *Final Summary of the Bonn Climate Change Talks.* Vol. 12 (472). 14 June. New York: IISD.

ENB. 2010c. *Summary of the Bonn Climate Talks.* Vol. 12 (478). 9 August. New York: IISD.

ENB. 2010d. *Summary of the Tianjin Climate Change Talks.* Vol. 12 (485). 12 October. New York: IISD.

ENB. 2010e. *Summary of the Cancun Climate Change Conference.* Vol. 12 (498). 13 December. New York: IISD.

ENB. 2011a. *Summary of the Bangkok Climate Talks.* Vol. 12 (499). 11 April. New York: IISD.

ENB. 2011b. *Summary of the Bonn Climate Change Conference.* Vol. 12 (513). 20 June. New York: IISD.

ENB. 2011c. *Summary of the Panama City Climate Change Talks.* Vol. 12 (521). 10 October. New York: IISD.

ENB. 2011d. *Summary of the Durban Climate Change Conference.* Vol. 12 (534). 13 December. New York: IISD.

ENB. 2012a. *Summary of the Bonn Climate Change Conference.* Vol. 12 (546). 28 May. New York: IISD.

ENB. 2012b. *Summary of the Bangkok Climate Change Conference.* Vol. 12 (555). 8 Sept. New York: IISD.

ENB. 2012c. *Summary of the Doha Climate Change Conference.* Vol. 12 (567). 11 Dec. New York: IISD.

EPL. 1999. Climate Change: Plan of Action Adopted. *Environmental Policy and Law* 29 (1): 2–8.

EUObserver. 2008. *Climate deal slammed as a 'mirage'.* 12 December. (available at http://euobserver.com/9/27297/?rk=1, last accessed 9 March 2009).

Eurobarometer. 2008. *Europeans' attitudes towards climate change.* Special Report 300. Brussels, September.

Falleti, T., and J. Lynch. 2009. Context and Causal Mechanisms in Political Analysis. *Comparative Political Studies* 42 (9): 1143–1166.

Fermann, G. 1993. Japan's 1990 Climate Policy under Pressure. *Security Dialogue* 24 (3): 287–300.

Fisher, R., W. Ury, and B. Patton. 1991. *Getting to Yes*. New York: Houghton Mifflin.

Flavin, C. 1998. Last Tango in Buenos Aires. *World Watch* (Nov./Dec.): 11–18.

Fodella, A. 2004. The Ninth Session of the Conference of the Parties. *Environmental Policy and Law* 34 (1): 24–26.

Friedman, L. 2011. *Nations Heading to Durban Climate Talks Remain Deeply Divided*. NY Times. October 10 (available at http://www.nytimes.com/cwire/2011/10/10/10climatewire-nations-heading-to-durban-climate-talks-remai-1993.html, last accessed 20 December 2011).

Garber, K. 2008. *The U.N. on the State of Climate Change. Q & A: Yvo de Boer discusses progress toward a global treaty*. 22 April (available at http://www.usnews.com/articles/news/world/2008/04/22/the-un-on-the-state-of-climate-change_print.htm, last accessed 15 October 2009).

Geden, O. 2011. *Was an der EU-Strategie falsch ist*. Berlin: German Institute for International and Security Affairs.

Germanwatch. 2006. *Results of the UN Climate Summit in Montreal*. Bonn: Germanwatch.

Germanwatch. 2007. *EU threatens MEM boycott*. Eco Newsletter. 14 Dec.

Gerring, J. 2007. *Case study research. Principles and Practice*. Cambridge: Cambridge UP.

Giddens, A. 1984. *The Constitution of Society. Outline of the Theory of Structuration*. Cambridge: Polity Press.

Ginsberg, R. 2001. *The European Union in International Politics. Baptism by Fire*. Boston: Rowman and Littlefield Publishers.

Goertz, G. 2006. *Social science concepts: A user's guide*. Princeton: Princeton UP.

Goldenberg, S. 2009a. *Great clean-up – can economic rescue plans also save planet?* The Guardian. 24 February (available at http://www.guardian.co.uk/environment/2009/feb/24/obama-environment-economic-rescue, last accessed 26 February 2009).

Goldenberg, S. 2009b. *US and India pledge common action on climate change*. The Guardian. 24 Nov. (available at http://www.guardian.co.uk/environment/2009/nov/24/climate-change-india-barack-obama, last accessed 14 February 2010).

Goldenberg, S., and A. Stratton. 2009. *From dinner to desperation: The 24-hour race for a deal in Copenhagen*. The Guardian. 18 Dec. (available at http://www.guardian.co.uk/environment/2009/dec/18/copenhagen-race-for-a-deal, last accessed 24 February 2010).

Goldenberg, S., and J. Vidal. 2009. *US scales down hopes of global climate change treaty in Copenhagen*. The Guardian. 4 Nov. (available at http://www.guardian.co.uk/environment/2009/nov/04/us-climate-change-copenhagen-treaty, last accessed 23 March 2010).

Goldenberg, S., and J. Watts. 2009. *US aims for bilateral climate change deals with China and India*. The Guardian. 14 Oct. (available at http://www.guardian.co.uk/environment/2009/oct/14/obama-india-china-climate-change, last accessed 14 March 2010).

Goldenberg, S., J. Vidal, and J. Watts. 2009. *Copenhagen draft text reveals deal is still out of reach.* The Guardian. 18 Dec. (available at http://www.guardian.co.uk/environment/2009/dec/18/copenhagen-draft-text, last accessed 24 March 2010).

Goldenberg, S., and M. Elder. 2011. *Japan nuclear crisis escalates.* The Guardian. 16 March. (available at http://www.guardian.co.uk/world/2011/mar/16/japan-nuclear-crisis-escalates, last accessed 22 April 2011).

Goldstein, J., and R.O. Keohane. eds. 1993. *Ideas and Foreign Policy: Beliefs, Institutions and Political Change.* Ithaca, NY: Cornell UP.

Graham-Harrison, E. 2007. *Bali breakthrough launches historic climate talks.* Yahoo News. 15 December (available at http://news.yahoo.com/s/nm/2007 1215/wl_nm/bali_saturday_dc_9, last accessed 20 February 2010).

Greenpeace. 2007. *Hiding behind the poor. A report by Greenpeace on climate injustice* (available at http://www.greenpeace.org/raw/content/india/press/reports/hiding-behind-the-poor.pdf, last accessed 23 October 2009).

Groen, L., and A. Niemann 2012. *EU actorness and effectiveness under political pressure at the Copenhagen climate change negotiations.* Mainz: University of Mainz.

Groenleer, M.L.P., and L.G. van Schaik. 2007. United We Stand? The European Union's International Actorness in the Cases of the International Criminal Court and the Kyoto Protocol. *Journal of Common Market Studies* 45 (5): 969–998.

Grubb, M. 2001. The UK and the European Union: Britannia Waives the Rules? In *Climate Change After Marrakech: The Role of Europe in the Global Arena. German Foreign Policy in Dialogue*, ed. D. Sprinz, 9–12. University of Trier.

Grubb, M., and F. Yamin. 2001. Climatic Collapse at The Hague: What happened, why, and where do we go from here? *International Affairs* 77 (2): 261–276.

Grubb, M., and J. Gupta. 2000. Towards a theoretical analysis of leadership. In *Climate Change and European leadership: A sustainable role for Europe*, ed. J. Gupta, and M. Grubb, 15–24. Dordrecht: Kluwer Academic Publishers.

Grubb, M., C. Vrolijk, and D. Brack. 2001. *The Kyoto Protocol. A Guide and Assessment.* London: The Royal Institute of International Affairs.

Gupta, J. 1998. Leadership in the Climate Regime: Inspiring the Commitment of Developing Countries in the Post-Kyoto Phase. *Review of European Community and International Environmental Law* 7 (2): 180–190.

Gupta, J. 2000. North-South Aspects of the Climate Change Issue: Towards a Negotiating Theory and Strategy for Developing Countries. *International Journal of Sustainable Development* 3 (2): 115–135.

Gupta, J., and L. Ringius. 2001. The EU's Climate Leadership: Reconciling Ambition and Reality. *International Environmental Agreements* 1 (1): 281–299.

Gupta, J., and M. Grubb. 2000. *Climate Change and European Leadership: A sustainable role for Europe?* Dordrecht: Kluwer Academic Publishers.

Gysen, J., H. Bruyninckx, and K. Bachus. 2006. The Modus Narrandi. A Methodology for Evaluating Effects of Environmental Policy. *Evaluation* 12 (1): 95–118.

Habermas, J. 1981. *Theorie des kommunikativen Handelns 2*. Frankfurt am Main: Suhrkamp.

Haigh, N. 1996. Climate Change Policies and Politics in the European Community. In *Politics of Climate Change. A European Perspective*, ed. T. O'Riordan, and J. Jäger, 155–185. London: Routledge.

Hallding, K., M. Olsson, A. Atteridge, A. Vihma, M. Carson, and M. Román. 2011. *Together Alone. BASIC countries and the climate change conundrum*. Copenhagen: Nordic Council of Ministers.

Harris, P.G. 2000. International Norms of Responsibility and U.S. Climate Change Policy. In *Climate Change and American Foreign Policy*, ed. P.G. Harris, 225–240. New York: St. Martin's Press.

Harris, P.G. 2001. International Environmental Affairs and U.S. Foreign Policy. In *The Environment, International Relations, and U.S. Foreign Policy*, ed. P.G. Harris, 3–42. Washington: Georgetown UP.

Harris, P.G. 2007. Europe and the politics and foreign policy of global climate change. In *Europe and Global Climate Change: Politics, Foreign Policy, and Regional Cooperation*, ed. P.G. Harris, 3–37. Cheltenham: Edward Elgar.

Harris, P.G., and Y. Hongyuan. 2009. Climate Change in Chinese Foreign Policy: Internal and External Responses. In *Climate Change and Foreign Policy. Case Studies from East to West*, ed. P.G. Harris, 53–67. London: Routledge.

Harrison, N.E. 2000. From the Inside Out: Domestic Influences on Global Environmental Policy. In *Climate Change and American Foreign Policy*, ed. P.G. Harris, 89–110. New York: St. Martin's Press.

Harvey, F. 2009. *Views diverge on cuts by 2020*. Financial Times. 8 April (available at http://www.ft.com/cms/s/0/50a7f2de-23d3-11de-996a-00144feabdc0.html?nclick_check=1, last accessed 22 March 2010).

Hasenclever, A., P. Mayer, and V. Rittberger. 1996. Interests, Power, Knowledge: The Study of International Regimes. *Mershon International Studies Review* 40 (2): 177–228.

Hasselmann, K., and T. Barker. 2008. The Stern Review and the IPCC Fourth Assessment Report: implications for interaction between policymakers and climate experts. An editorial essay. *Climatic Change* 89: 219–229.

Hedegaard, C. 2011. *Europe has brought about a new phase in global climate policy*. The Guardian. 13 December.

Hedegaard, C. 2012. *Why the Doha climate conference was a success*. The Guardian. 14 December.

Held, D., A. McGrew, D. Goldblatt, and J. Perraton. 1999. *Global Transformations, Politics, Economics and Culture*. Cambridge: Polity Press.

Herold, A., M. Cames, V. Cook, and L. Emele. 2012. *The Development of Climate Negotiations in View of Doha (COP 18)*. Study. Brussels: European Parliament.

Holsti, K.J. 1995. *International Politics: A Framework for Analysis*. Englewood Cliffs, NJ: Prentice Hall.

Holzinger, K. 2004. Bargaining through Arguing: An Empirical Analysis Based on Speech Act Theory. *Political Communication* 21 (2): 195–222.

Hood, M. 2009. *US won't speed up emissions cuts: top climate negotiator.* 24 May (available at http://www.google.com/hostednews/afp/article/ALeq M5gJP2lqVLhpEPLc0Spl8dd7OOD6Zw, last accessed 24 February 2010).

Huberman, A.M., and M.B. Miles. eds. 2002. *The qualitative researcher's companion.* Thousand Oaks, CA: Sage.

Huberts, L.W.J.C. 1994. Intensieve procesanalyse. In *Methoden van invloedsanalyse*, ed. L.W.J.C. Huberts, and J. Kleinnijenhuis, 38–60. Amsterdam: Boom.

Hudson, V., and C. Vore. 1995. Foreign policy analysis yesterday, today and tomorrow. *Mershon International Studies Review* 39 (2): 209–238.

Hultman, N. 2011. *The Durban Platform.* Brookings. 12 Dec. (available at http://www.brookings.edu/opinions/2011/1212_durban_platform_hultman.aspx, last accessed 4 February 2012).

Hunter, D., J. Salzman, and D. Zaelke. 2002. *International Environmental Law and Policy.* 2nd ed. New York: Foundation Press.

Imhasly, B. 2008. Ein reiches Land mit armen Menschen. *Aus Politik und Zeitgeschichte* 22: 13–19.

International Center For Trade and Sustainable Development (ICTSD). 2010. *Tianjin Climate Meeting Delivers Little.* Geneva: ICTSD.

Jachtenfuchs, M. 1996. *International Policy-Making as a Learning Process: the European Union and the Greenhouse Effect.* Hampshire: Avebury Studies in Green Research.

Jacoby, H., and D. Reiner. 2001. Getting climate policy on track after The Hague. *International Affairs* 77 (2): 297–312.

Jakobson, L. 2009. China. In *Towards a new climate regime? Views of China, India, Japan, Russia and the United States on the road to Copenhagen*, ed. A. Korppoo, and A. Luta, 22–46. Helsinki: The Finnish Institute of International Affairs (FIIA).

Johnston M., and P. Hudson. 2011. *Protesters disrupt Question Time after carbon bills pass lower house.* Herald Sun. 12 October. (available at http://www.heraldsun.com.au/news/more-news/carbon-tax-bills-pass-lower-house-of-federal-parliament/story-fn7x8me2-1226164570957, last accessed 14 November 2011).

Jordan, A., and T. Rayner. 2010. The evolution of climate policy in the European Union: an historical overview. In *Climate Change Policy in the European Union*, ed. A. Jordan, D. Huitema, H. van Asselt, T. Rayner, and F. Berkhout, 52–80. Cambridge: Cambridge UP.

Jørgensen, K.E. 2007. The European Union and the World. In *Handbook of European Union Politics*, ed. K.E. Jørgensen, M.A. Pollack, and B. Rosamond, 507–544. London: Sage.

Jørgensen, K.E., S. Oberthür, and J. Shahin. 2011. Introduction: Assessing the EU's Performance in International Institutions – Conceptual Framework and Core Findings. *Journal of European Integration* 33 (6): 599–620.

Kameyama, Y. 2004: Evaluation and Future of the Kyoto Protocol: Japan's perspective. *International Review for Environmental Strategies* 5 (1): 71–81.

Kanie, N. 2003. Leadership in Multilateral Negotiation and Domestic Policy: The Netherlands at the Kyoto Protocol Negotiations. *International Negotiation* 8 (4): 339–365.

Kasa, S., A.T. Gullberg, and G. Heggelund. 2008. The Group of 77 in the international climate negotiations: recent developments and future directions. *International Environmental Agreements* 8 (1): 113–127.

Kennedy, M. 2009. *India's familiar stance on climate change*. Gulf News. 30 Nov. (available at http://gulfnews.com/opinions/columnists/india-s-familiar-stance-on-climate-change-1.541953, last accessed 29 March 2010).

Keohane, R.O. 1984. *After Hegemony: Cooperation and Discord in the World Political Economy*. Princeton: Princeton UP.

Keohane, R.O. 1989. The Demand for International Regimes. In *International Institutions and State Power*, ed. R.O. Keohane, 101–131. Boulder, CO: Lynne Rienner.

Keohane, R.O. 2001. Governance in a Partially Globalized World. Presidential Address, American Political Science Association 2000. *American Political Science Review* 95 (1): 1–13.

Keukeleire, S., and H. Bruyninckx. 2011. The European Union, the BRICs and the emerging new world order. In *International Relations and the European Union*, 2nd ed., ed. C. Hill, and M. Smith, 380–403. Oxford: Oxford UP.

Keukeleire, S., and J. MacNaughtan. 2008. *The Foreign Policy of the European Union*. Basingstoke: Palgrave Macmillan.

Kjellen, B. 1994. A Personal Assessment. In *Negotiating Climate Change: The Inside Story of the Rio Convention*, ed. I.M. Mintzer, and J.A. Leonard, 149–174. Cambridge: Cambridge Studies in Energy and Environment.

Kleine, M., and T. Risse. 2005. *Arguing and Persuasion in the European Convention*. Berlin: Center for Transatlantic Foreign and Security Policy.

Knocke, D. 1990. *Political Networks: The Structural Perspective*. Cambridge: Cambridge UP.

Korppoo, A. 2009a. Japan. In *Towards a new climate regime? Views of China, India, Japan, Russia and the United States on the road to Copenhagen*, ed. A. Korppoo, and A. Luta, 66–80. Helsinki: FIIA.

Korppoo, A. 2009b. Russia. In *Towards a new climate regime? Views of China, India, Japan, Russia and the United States on the road to Copenhagen*, ed. A. Korppoo, and A. Luta, 81–99. Helsinki: FIIA.

Korppoo, A. 2009c. *The Russian Climate Doctrine. Emerging Issues on the Road to Copenhagen*. Briefing Paper. Helsinki: FIIA.

Korppoo, A., and A. Luta. 2009. India. In *Towards a new climate regime? Views of China, India, Japan, Russia and the United States on the road to Copenhagen*, ed. A. Korppoo, and A. Luta, 47–65. Helsinki: FIIA.

Krasner, S. 1983. Structural Causes and Regime Consequences: Regimes as Intervening Variables. In *International Regimes*, ed. S. Krasner, 1–21. Ithaca: Cornell UP.

Krasner, S. 1991. Global Communications and National Power: Life on the Pareto Frontier. *World Politics* 43 (2): 336–366.

Kratochwil, F.V. 1989. *Rules, Norms, and Decisions: On the Conditions of Practical and Legal Reasoning in International Relations and Domestic Affairs.* Cambridge: Cambridge UP.

La Viña, A.G.M., and L.G. Ang. 2010. *From Copenhagen to Cancun: Challenges and Prospects for the UNFCCC Negotiations.* London: FiELD.

Lacasta, N. 2008. *The EU as an Actor in International Climate Policy: External Competence, Internal Procedure and Actual Practice.* Lecture slides. VUB Brussels, 26 November.

Lakshmanan, I. 2009. *India Refuses to Bend to U.S. Pressure on Carbon Caps.* Bloomberg. 19 July (available at http://www.bloomberg.com/apps/news?pid=20601090&sid=aZcBZXJ9RZ1Y, last accessed 25 June 2010).

Landler, M. 2009. *Clinton Paints China Policy With a Green Hue.* NY Times. 21 February (available at http://www.nytimes.com/2009/02/22/world/asia/22diplo.html, last accessed 25 Nov. 2009).

Lenaerts, U. 2009. *Climate change – EU internal coordination and outreach.* Presentation at the workshop "EU foreign policy and global climate change: towards a comprehensive European climate diplomacy?" Leuven. 13 Oct.

Lescher, T.B. 2000. *Die EU als eigenständiger Akteur in der Entstehung des internationalen Klimaregimes.* Studien zur Deutschen und Europäischen Aussenpolitik. Trier: Universität Trier.

Lewis, J.I. 2007/2008. China's Strategic Priorities in International Climate Change Negotiations. *The Washington Quarterly* (Winter 2007/2008) 31 (1): 155–174.

Lieb, J., and A. Maurer. eds. 2009. *Der Vertrag von Lissabon. Kurzkommentar.* 3rd ed. Berlin: Stiftung Wissenschaft und Politik.

Lindenthal, A. 2009. *Leadership im Klimaschutz. Die Rolle der Europäischen Union in der internationalen Umweltpolitik.* Frankfurt am Main: Campus.

Little, R. 2008. International regimes. In *The Globalization of World Politics. An introduction to international relations.* 4th ed., ed. J. Baylis, S. Smith, and P. Owens, 296–311. Oxford: Oxford UP.

Lucarelli, S., and L. Fioramati. eds. 2010. *External Perceptions of the European Union as a Global Actor.* London: Routledge.

Luta, A. 2009. *Climate Sudoku. Japan's bumpy ride towards a post-2012 target.* Briefing Paper. Helsinki: The Finnish Institute of International Affairs.

Lynas, M. 2009. *How do I know China wrecked the Copenhagen deal? I was in the room.* The Guardian. 22 Dec. (available at http://www.guardian.co.uk/environment/2009/dec/22/copenhagen-climate-change-mark-lynas, last accessed 25 June 2010).

Mackie, J.L. 1974. *The Cement of the Universe: A Study of Causation.* Oxford: Oxford UP.

Mahoney, J. 2003. *Tentative Answers to Questions about Causal Mechanisms.* Paper presented at the Annual Meeting of the American Political Science Association. Philadelphia, PA. 28 August.

Manners, I., and R. G. Whitman. eds. 2000. *The foreign policies of European Union Member States*. Manchester: Manchester UP.

Marcu, A. 2012. *Doha/COP 18: Gateway to a New Climate Change Agreement*. Brussels: Centre for European Policy Studies.

Marini, M., and B. Singer. 1988. Causality in the Social Sciences. *Sociological Methodology* 18 (2): 347–409.

Matsumura, H. 2000. *Japan and the Kyoto Protocol. Conditions for Ratification*. London: The Royal Institute of International Affairs.

Meilstrup, P. 2010. *The Runaway Summit: The Background Story of the Danish Presidency of COP15, the UN Climate Change Conference* (available at http://www.fao.org/fileadmin/user_upload/rome2007/docs/What%20really%20 happen%20in%20COP15.pdf, last accessed 3 February 2012).

Meltzer, J. 2011. *After Fukushima: What's Next for Japan's Energy and Climate Change Policy?* Washington, DC: Brookings Institute.

Menon, S. 2009. *What the US has done is too little*. Business Standard. 19 April (available at http://www.business-standard.com/india/news/whatus-has-done-is-too-little/355512/, last accessed 24 February 2010).

Metz, B. 2000. International equity in climate change policy. *Integrated Assessment* 1 (1): 111–126.

Meunier, S. 2000. What Single Voice? European Institutions and EU-US Trade Negotiations. *International Organization* 54 (1): 103–135.

Michaelowa, A. 2003. Editorial. *Climate Policy* 2 (1): 1–2.

Michaelowa, K., and A. Michaelowa. 2011. *India in the international climate negotiations: from traditional nay-sayer to dynamic broker*. Zurich: University of Zurich.

Miles, M., and A. Huberman. 1994. *Qualitative data analysis*. 2nd ed. Thousand Oaks, CA: Sage.

Minten, D. 2009. *Europa in tranen*. De Standaard. 21 December 2009. P. 15.

Mintzer, I.M., and J.A. Leonard. eds. 1994. *Negotiating Climate Change: The Inside Story of the Rio Convention*. Cambridge: Cambridge Studies in Energy and Environment.

Missbach, A. 2000. Regulation Theory and Climate Change Policy. In *Climate Change and American Foreign Policy*, ed. P.G. Harris, 131–150. New York: St. Martin's Press.

Mohiuddin, Y. 2009. *India to rein in emissions, but at minimal cost: analysts*. 3 Dec. (available at http://news.yahoo.com/s/afp/unclimatewarmingindia, last accessed 29 March 2010).

Morgera, E., K. Kulovesi, and M. Muñoz. 2011. Environmental integration and multi-faceted international dimensions of EU law: Unpacking the EU's 2009 climate and energy package. *Common Market Law Review* 48(3): 829–891.

Mrusek, K. 2009. *Der Zwei-Grad-Kippschalter*. Frankfurter Allgemeine Zeitung. 9 July.

Müller, B. 2006. *Montreal 2005: What Happened and What It Means*. Oxford: Oxford Institute for Energy Studies.

Müller, B. 2008. *Bali 2007: on the road again*! Oxford: Oxford: Climate Policy Paper.

Müller, B. 2010. *Copenhagen 2009. Failure or final wake-up call for our leaders*. Oxford: Oxford Institute for Energy Studies.

Müller, H. 1994. Internationale Beziehungen als kommunikatives Handeln. Zur Kritik der utilitaristischen Handlungstheorien. *Zeitschrift für Internationale Beziehungen* 1 (1): 15–44.

Murphy, D. 2009. *Status of UNFCCC Negotiations. Outcomes of COP 14 in Poznan*. Winnipeg: International Institute for Sustainable Development.

Murray, J. 2009. India and China ink climate pact, as Copenhagen speculation rumbles on. Business Green. 21 Oct. (available at http://www.businessgreen.com/business-green/news/2251737/india-china-ink-climate-pact, last accessed 22 March 2010).

Muthoo, A. 2000. A Non-Technical Introduction to Bargaining Theory. *World Economics* 1 (2): 145–166.

Nagel, J. 1975. *The Descriptive Analysis of Power*. New Haven, CT: Yale UP.

Najam, A., S. Huq, and Y. Sokona. 2003. Climate negotiations beyond Kyoto: developing countries concerns and interests. *Climate Policy* 3 (2): 221–231.

Nazret.2010.*Ethiopia–MelesZenawiwasbothpraisedandslammedforhisroleinCopenhagen*. 20 December (available at http://nazret.com/blog/index.php?title=ethiopia_meles_zenawi_was_both_praised_a&more=1&c=1&tb=1&pb=1, last accessed 10 May 2010).

Nitze, W.A. 1994. A Failure of Presidential Leadership. In *Negotiating Climate Change: The Inside Story of the Rio Convention*, ed. I.M. Mintzer, and J.A. Leonard, 187–200. Cambridge: Cambridge Studies in Energy and Environment.

Nordhaus, T., and M. Shellenberger. 2009. *Apocalypse Fatigue: Losing the Public on Climate Change*. E360.yale.edu. 16 November (available at http://e360.yale.edu/content/feature.msp?id=2210, last accessed 18 Nov. 2009).

Nuttall, S. 2005. Coherence and Consistency. In *International Relations and the European Union*, ed. C. Hill, and M. Smith, 91–112. Oxford: Oxford UP.

O'Riordan, T., and J. Jäger. eds. 1996. *Politics of Climate Change. A European Perspective*. London: Routledge.

Oberthür, S. 1993. *Politik im Treibhaus: Die Entstehung des internationalen Klimaschutzregimes*. Berlin: Sigma.

Oberthür, S. 1994. Climate Change Convention: Preparations for the First Conference of the Parties. *Environmental Policy and Law* 24 (6): 299–303.

Oberthür, S. 2007. The European Union in international climate policy: The prospect for leadership. *Intereconomics* 42 (2): 77–83.

Oberthür, S. 2009. The role of the EU in global environmental and climate governance. In *The European Union and Global Governance*, ed. M. Telo, 192–209. London: Routledge.

Oberthür, S., and C. Roche Kelly. 2008. EU Leadership in International Climate Policy: Achievements and Challenges. *The International Spectator* 43 (3): 35–50.

Oberthür, S., and H.E. Ott. 1999. *The Kyoto Protocol. International Climate Policy for the 21st Century*. Berlin: Springer.

Oberthür, S., M. Pallemaerts, and C. Roche Kelly. eds. 2010. *The New Climate Policies of the European Union: Internal Legislation and Climate Diplomacy*. Brussels: VUB Press.

Ochs, A. 2008. *The Good, the Bad, and the Ugly? Europe, the United States, and China at the World Climate Conference*. Facet Analysis. February.

Okereke, C., P. Mann, H. Osbahr, B. Müller, and J. Ebeling. 2007. *Assessment of key negotiating issues at Nairobi climate COP/MOP and what it means for the future of the climate regime*. Tyndall Working Paper 106. Norwich: Tyndall Centre for Climate Change Research.

Ott, H.E. 1998. Report on developments in international climate policy and law. *Yearbook of International Environmental Law* 9: 183–189.

Ott, H.E. 2001a. Climate change: An important foreign policy issue. *International Affairs* 7 (1): 277–296.

Ott, H.E. 2001b. The Bonn Agreement to the Kyoto Protocol: Paving the way for ratification. *International Environmental Agreements* 1 (3): 469–476.

Ott, H.E. 2003. *Warning signs from Delhi: Troubled waters ahead for global climate policy*. Paper (available at http://www.wupperinst.org/download/Warning-Signs-Ott.pdf, last accessed 24 July 2009).

Ott, H.E., B. Brouns, W. Sterk, and B. Wittneben. 2005. It takes two to tango. Climate policy at COP 10 in Buenos Aires and beyond. *Journal for European Environmental and Planning Law* 2 (1): 84–91.

Ott, H.E., W. Sterk, and R. Watanabe. 2008. The Bali roadmap: new horizons for global climate policy. *Climate Policy* 8 (1): 91–95.

Paarlburg, R. 1997. Earth in Abeyance: Explaining Weak Leadership in US International Environmental Policy. In *Eagle Adrift: American Foreign Policy at the End of the Century*, ed. R.J. Lieber, 137–159. New York: Longman.

Pallemaerts, M. 2004. Le cadre international et européen des politiques de lutte contre les changements climatiques. *Courrier hébdomadaire No. 1858–1859*. Bruxelles: CRISP.

Pallemaerts, M., and R. Williams. 2006. Climate change: The international and European policy framework. In *EU climate change policy: the challenge of new regulatory initiatives*, 22–50, ed. M. Peeters, and K. Deketelaere. Cheltenham: Edward Elgar.

Parker, C.F., and C. Karlsson. 2010. Climate Change and the European Union's Leadership Moment: An Inconvenient Truth? *Journal of Common Market Studies* 48 (4): 923–953.

Pataki, G.E., and T.J. Vilsack. eds. 2008. *Confronting Climate Change: a strategy for U.S. foreign policy. Report of an independent task force*. New York: Council on Foreign Relations.

Paterson, M. 1996. *Global Warming and Global Politics*. London: Routledge.

Paterson, M., and M. Grubb. 1992. The International Politics of Climate Change. *International Affairs* 68 (2): 293–310.

Pawlak, J. 2009. *EU calls for more U.S. involvement in climate works*. Reuters. 22 Dec. (available at http://www.reuters.com/article/idUSTRE5BL21F2009 1222, last accessed 25 June 2010).

PBL (Netherlands Environmental Assessment Agency). 2008. *Global CO_2 emissions: increase continued in 2007*. Report released on 13 June 2008 (available at http://www.pbl.nl/en/publications/2008/GlobalCO2emissions through2007.html, last accessed 9 March 2009).

Peake, R. 2007. *Labor ratifies Kyoto Protocol Climate of change as Kevin Rudd takes over reins of power*. Canberra Times. 4 Dec. (available at http://www.canberratimes.com.au/news/local/news/general/labor-ratifies-kyoto-protocol-climate-of-change-as-kevin-rudd-takes-over-reins-of-power/463389.aspx, last accessed 23 February 2010).

Pearce, F. 2009. *Did Ed Miliband save the Copenhagen summit from complete failure?* The Guardian. 23 December.

Petersen, F.A. 2009. *Danish Ambassador: 'We Will Seal the Deal in Copenhagen'*. Green Inc. 23 Nov. (available at http://greeninc.blogs.nytimes.com/2009/11/23/danish-ambassador-we-will-seal-the-deal-in-copenhagen/, last accessed 23 March 2010).

Phillips, L. 2008. *Parliament negotiators sign off on climate deal*. EUObserver. 15 Dec (available at http://euobserver.com/9/27300/?rk=1, last accessed 29 March 2010).

Phillips, L. 2009a. *UN climate chief: 'The world is waiting for the EU'*. EUObserver. 23 June (available at http://euobserver.com/19/29040, last accessed 23 March 2010).

Phillips, L. 2009b. *Recession and hot weather push CO_2 emissions down in EU*. EUObserver. 2 Sept. (available at http://euobserver.com/885/28609, last accessed 26 June 2010).

Phillips, L. 2009c. *EU pessimistic about Copenhagen climate change deal*. EUObserver. 6 Nov. (available at http://euobserver.com/9/28950/?rk=1, last accessed 15 March 2010).

Phillips, L. 2010. *Battling the 'Multilateral Zombie' – EU climate strategy after Copenhagen*. EU Observer. 3 February (available at http://www.euobserver.com/19/29354, last accessed 11 February 2010).

Pidwirny, M. 2006. Greenhouse effect. In *Encyclopedia of Earth*, ed. C.J. Cleveland. Washington, D.C.: Environmental Information Coalition, National Council for Science and the Environment.

Pinholt, K. 2004. *Influence through arguments? A study of the Commission's influence on the climate change negotiations*. Oslo: Arena.

Pop, V. 2010. *Van Rompuy and Barroso to both represent EU at G20*. EUObserver. 19 March. (available at http://euobserver.com/9/29713, last accessed 29 March 2010).

Purvis, N., and A. Stevenson. 2010. *Rethinking Climate Diplomacy. New ideas for transatlantic cooperation post-Copenhagen.* Washington: The German Marshall Fund of the United States.

Rabe, B.G. 2004. *State and greenhouse: the emerging politics of American climate change policy.* Brookings: Washington.

Rajamani, L. 2000. The Principle of Common but Differentiated Responsibility and the Balance of Commitments under the Climate Regime. *Review of European Community and International Environmental Law* 9 (2): 120–131.

Rajamani, L. 2008. Indiens internationale Klimapolitik. *Aus Politik und Zeitgeschichte* 22: 19–25.

Rajamani, L. 2009. *The Copenhagen Agreed Outcome: Form, Shape and Influence.* New Delhi: Centre for Policy Research.

Rajamani, L. 2010a. *Neither fish nor fowl.* New Delhi: Centre for Policy Research.

Rajamani, L. 2010b. *The can-can't at Cancun.* Indian Express. 14 December. (available at http://www.indianexpress.com/story-print/724374, last accessed 16 January 2012).

Rajan, M.G. 1997. *Global Environmental Politics. India and the North-South Politics of Global Environmental Issues.* Oxford: Oxford UP.

Rettmann, A. 2009. *Russia makes surprise CO_2 pledge at summit.* EUObserver. 18 Nov. (available at http://euobserver.com/24/29014/?rk=1, last accessed 28 February 2010).

Rettmann, A. 2010. *Ashton names EU foreign-service priorities at low-key launch event.* EUObserver. 2 Dec. (available at http://euobserver. com/18/31413, last accessed 19 December 2010).

Reuters. 2007a. *Ban Ki-moon's speech at UN Bali Climate Talks,* 12 December. (available at http://www.alertnet.org/thenews/newsdesk/L1248845.htm, last accessed 12 May 2008).

Reuters. 2007b. *China to reject binding emissions caps, Europe says.* 7 Nov. 2007 (available at http://www.alertnet.org/thenews/newsdesk/PEK359325.htm, last accessed 30 July 2009).

Reuters. 2007c. U.S. *U-turn brings Bali climate deal.* 15 December (available at http://www.guardian.co.uk/feedarticle?id =7155608, last accessed 15 January 2008).

Reuters. 2009a. *EU wants more on climate from China, lauds Japan.* 1 Sept. (available at http://www.reuters.com/article/latestCrisis/idUSL1608967, last accessed 13 March 2010).

Reuters. 2009b. *Rich nations trying to kill Kyoto pact, says China.* 5 Oct. (available at http://www.reuters.com/article/latestCrisis/idUSSP280493, last accessed 10 February 2010).

Reuters. 2010. *Snubbed in Copenhagen, EU weighs climate options.* 13 Jan. (available at http://www.busrep.co.za/ind ex.php?fSectionId=552&fArticleI d =5311734, last accessed 22 March 2010).

Reuters. 2011a. *Euro zone crisis to widen climate fund gap: report.* 17 November (available at http://www.reuters.com/article/2011/11/17/us-climate-finance-idUSTRE7AG0L720111117, last accessed 16 December 2011).

Reuters. 2011b. *Kanadas Kyoto-Wende erzürnt China*. 13 December. (available at http://www.spiegel.de/wissenschaft/natur/0,1518,803367,00.html, last accessed 20 December 2011).

Reuters.2011c.*U.N.climatetalksreachmodestdeal*.11December(availableathttp://thestar.com.my/news/story.asp?file=/2011/12/11/worldupdates/20111211T053801Z_17_TRE7B909W_RTROPTT_0_UKCLIMATE&sec=Worldupdates, last accessed 28 December 2011).

Rhinard, M., and M. Kaeding. 2006. The International Bargaining Power of the European Union in 'Mixed' Competence Negotiations: The Case of the 2000 Cartagena Protocol on Biosafety. *Journal of Common Market Studies* 44 (5): 1023–1050.

Ricard, P. 2009. *L'Europe divisée sur les négociations climatiques*. Le Monde. 29 Sept.(availableathttp://www.lemonde.fr/planete/article/2009/09/29/l-europe-divisee-sur-les-negociationsclimatiques_1246660_3244.html, last accessed 29 March 2010).

Risse, T. 2002. Constructivism and International Institutions: Towards Conversations Across Paradigms. In *Political Science: State of the Discipline*, ed. I. Katznelson, and H. Milner, 597–629. New York: Norton.

Risse, T. 2004. Global Governance and Communicative Action. *Government and Opposition* 39 (2): 298–302.

Romàn, M., and M. Carson. 2009. *Sea Change: US Climate Policy Prospects Under the Obama Administration*. Stockholm: The Commission on Sustainable Development.

Romero, J. 2004. *An assessment of the Swiss experience with the Environmental Integrity Group in the UNFCCC process*. Presentation. Ankara Climate Change Conference 2004. (available at http://www.iklim.cevreorman.gov.tr/sunumlar/romero.pdf, last accessed 24 July 2009).

Rowbotham, E. 1996. Legal obligations and uncertainties in the climate change convention. In *Politics of Climate Change. A European Perspective*, ed. T. O'Riordan, and J. Jäger, 32–50. London: Routledge.

Rowlands, I. H. 2001. Classical theories of international relations. In *International relations and global climate change*, ed. U. Luterbacher and D.F. Sprinz, 43–65. Cambridge and London: The MIT Press.

Runge-Metzger, A. 2008. *The link between EU domestic policies and international negotiations or "The Power of the Example"*. Informal EU Workshop on Climate Change. Brdo, Slovenia, 20–22 January.

Santarius, T., C. Arens, U. Eichhorst, D. Kiyar, F. Mersman, H.E. Ott, F. Rudolph, W. Sterk, and R. Watanabe. 2009. *Pit Stop Poznan. An Analysis of Negotiations on the Bali Action Plan at the Stopover to Copenhagen*. Wuppertal: Institute for Climate, Environment and Energy.

Santini, J.-L. 2009. *US pressed over climate leadership role*. AFP. 4 March (available at http://news.yahoo.com/s/afp/20090304/pl_afp/climateenergy useurope_20090304153150, last accessed 21 February 2010).

Saretzki, T. 1996. Wie unterscheiden sich Verhandeln und Argumentieren? In *Verhandeln und Argumentieren. Dialog, Interessen und Macht in der Umweltpolitik*, ed. V. Prittwitz, 19–39. Opladen: Leske und Budrich.

Sbragia, A. 1998. Institution-Building from below and above: The European Community in Global Environmental Politics. In *European Integration and Supranational Governance*, ed. W. Sandholtz, and A. Stone Sweet, 283–303. Oxford: Oxford UP.

Sbragia, A. 2000. Environmental Policy: Economic Constraints and External Pressures. In *Policy-making in the European Union*, 4th ed., ed. H. Wallace, and W. Wallace, 293–316. Oxford: Oxford UP.

Scheipers, S., and D. Sicurelli. 2007. Normative Power Europe: A credible Utopia? In *Journal of Common Market Studies* 45 (2): 435–457.

Schneider, F., and A.F. Wagner. 2002. *What happened in Marrakech? COP-7 and recent agreements on the implementation of the Kyoto Protocol*. Online Paper (available at http://www.economics.uni-linz.ac.at/members/Schneider/files/publications/MarrakechJune4.PDF, last accessed 3 August 2009).

Schreurs, M.A., and Y. Tiberghien. 2007. Multi-Level Reinforcement: Explaining European Union Leadership in Climate Change Mitigation. *Global Environmental Politics* 7 (4): 19–46.

Schröder, H. 2001. *Negotiating the Kyoto Protocol. An analysis of negotiation dynamics in international negotiations*. Münster: LIT Verlag.

Schumer, S. 1996. *Die Europäische Union als Akteur in der internationalen Klimapolitik: das Beispiel des Ozon- und Klimaregimes*. Mosbach: AFES-PRESS-Report 55.

Schunz, S. 2009. The European Union's external climate policy: a foreign policy analysis. *CFSP Forum* 7 (2): 1–6.

Schunz, S. 2010. *European Union foreign policy and its effects: a longitudinal study of the EU's influence on the United Nations climate change regime (1991–2009)*. Ph.D. dissertation. Leuven: Katholieke Universiteit Leuven.

Schunz, S. 2011. *Beyond leadership by example*. Discussion Paper. Berlin: German Institute for International and Security Affairs.

Schunz, S., S. Basu, H. Bruyninckx, S. Keukeleire, and J. Wouters. 2012. Analysing the position of the EU in the United Nations system: analytical framework. In *The European Union and Multilateral Governance: Assessing EU Participation in United Nations Human Rights and Environmental Fora*, ed. J. Wouters, H. Bruyninckx, S. Basu, and S. Schunz, 25–48. Houndmills, Basingstoke: Palgrave Macmillan.

Schunz, S., S. Happaerts, and K. Van den Brande. 2009. *European Union foreign policy and global climate change: towards a comprehensive European climate diplomacy?* Leuven: Centre for Global Governance Studies.

Sebenius, J.K. 1991. Designing Negotiations Toward a New Regime: The Case of Global Warming. *International Security* 15 (4): 110–148.

Shanka Jha, P. 2009. *Indian public perceptions of the international climate change negotiations* (available at http://www.stimson.org/rv/pdf/Indian_Climate_Change/Jha_Indian_Public_Perceptions.pdf, last accessed 23 March 2010).

Sjöstedt, G. 1977. *The external role of the European Community*. Farnborough: Hants.

Sjöstedt, G. 1998. The EU Negotiates Climate Change. External Performance and Internal Structural Change. In *Cooperation and Conflict* 33 (3): 227–256.

Sjöstedt, G. 1999. Leadership in Multilateral Negotiations: Crisis or Transition? In *International Negotiation: Actors, Structure/Process, Values*, ed. P. Berton, H. Kimura, and I.W. Zartman, 223–253. New York: St. Martin's Press.

Skjaerseth, J. 1994. Climate Policy of the EC: Too Hot to Handle. *Journal of Common Market Studies* 32 (1): 25–45.

Skjærseth, J., and J. Wettestad 2008. *EU Emissions Trading: Initiation, Decision-making and Implementation*. Farnham: Ashgate.

Smith, K.E. 2003. *European Union Foreign Policy in a Changing World*. Cambridge: Polity Press.

Smith, K.E. 2007. *The EU in the World: Future Research Agendas*. European Foreign and Security Policy Studies Programme. Anna Lindh award lecture. Brussels. 19 October.

Smith, K.E. 2010. The European Union in the World: Future Research Agendas. In *Research Agendas in EU Studies. Stalking the Elephant*, ed. M. Egan, N. Nugent, and W. Paterson, 329–353. Houndmills, Basingstoke: Palgrave.

Somers, M.R. 1998. "We're No Angels": Realism, Rational Choice, and Relationality in Social Science. *The American Journal of Sociology* 104 (3): 722–784.

Spence, C., K. Kulovesi, M. Gutiérrez, and M. Muñoz. 2008. Great Expectations: Understanding Bali and the Climate Change Negotiation Process. *Review of European Community and International Environmental Law* 17 (2): 142–153.

Spencer, T., K. Tangen, and A. Korppoo. 2010. *The EU and the Global Climate Regime. Getting Back in the Game*. Helsinki: The Finnish Institute of International Affairs.

Spiegel. 2008a. *Geheimpapier belegt Blockadehaltung der USA*. Spiegel online. 2 June (available at http://www.spiegel.de/wissenschaft/natur/0,1518,557 167,00.html, last accessed 25 June 2010).

Spiegel. 2008b. *US-Senat blockiert Klimaschutzgesetz*. Spiegel online. 6 June (available at http://www.spiegel.de/wissenschaft/mensch/0,1518,558211,00.html, last accessed 25 October 2009).

Spiegel. 2009a. *Japans neue Regierung will CO_2-Ausstoß drastisch senken*. Spiegel online. 7 Sept. (available at http://www.spiegel.de/wissenschaft/natur/0,1518,647364,00.html, last accessed 10 April 2010).

Spiegel. 2009b. *Asien-Gipfel torpediert Klimakonferenz in Kopenhagen*. Spiegel online. 15 Nov. (available at http://www.spiegel.de/politik/ausland/0,1518,661350,00.html, last accessed 4 February 2010).

Spiegel. 2010. *Kopenhagen-Protokolle: Geheimer Mitschnitt enthüllt Eklat bei Klimagipfel*. Spiegel online. 2 May (available at http://www.spiegel.de/wissenschaft/natur/0,1518,692471,00.html, last accessed 5 May 2010).

Spillmann, C. 2009. *Climate heating up at EU global warming talks*. AFP. 27 Sept. (available at http://www.google.com/hostednews/afp/article/ALeqM5ibLioxT2NPbGrjYcuXEa61I0Vtwg, last accessed 25 February 2010).

Sprinz, D. ed. 2001. *Climate Change After Marrakech: The Role of Europe in the Global Arena. German Foreign Policy in Dialogue*. Trier: University of Trier.

Stearns, J., and A. Morales. 2009. *U.S. Climate Plan Threatens EU Goal for Global Accord*. Bloomberg. 22 May (available at http://www.bloomberg.com/apps/news?pid=20601090&sid=aACte9HS5us8, last accessed 22 March 2010).

Steffek, J. 2005. Incomplete agreements and the limits of persuasion in international politics. *Journal of International Relations and Development* 8 (2): 229–256.

Sterk, W., C. Arens, U. Eichhorst, F. Mersmann, and H. Wang-Helmreich. 2011a. *Processed, Refried – Little Substance Added: Cancún Climate Conference Keeps United Nations Process Alive but Raises More Questions than It Answers*. Wuppertal: Institute for Climate, Energy, and Environment.

Sterk, W., C. Arens, F. Mersmann, H. Wang-Helmreich, and T. Wehnert. 2011b. *On the road again: progressive countries score a realpolitik victory in Durban while the real climate continues to heat up*. Wuppertal: Institute for Climate, Environment and Energy.

Stern, N. 2007. *The Economics of Climate Change*. Cambridge: Cambridge UP.

Stern, T., and W. Antholis. 2007/2008. A Changing Climate: The Road Ahead for the United States. *The Washington Quarterly* 2007/2008 31 (3): 175–188.

Tangen, K. 1999: The climate change negotiations: Buenos Aires and beyond. *Global Environmental Change* 9 (1): 175–178.

Ten Kate, D. and C. Airlie. 2011. *EU May Extend Kyoto Climate Treaty if Conditions Are 'Right'*. Bloomberg. 8 April (available at http://www.bloom berg.com/news/2011-04-08/eu-may-extend-kyoto-climate-treaty-if-condition s-are-right-1-.html, last accessed 18 May 2011).

The Guardian. 2009. *Copenhagen climate change summit – final day live blog*. 18 Dec. (available at http://www.guardian.co.uk/environment/blog/2009/dec/18/copenhagen-climate-change-summit-liveblog/print, last accessed 22 March 2010).

Torney, D. 2013. *European Climate Diplomacy. Building capacity for external action*. Briefing Paper 141. Helsinki: FIIA.

Torvanger, A. 2001. *An analysis of the Bonn Agreement: Background information for evaluating business implications*. Oslo: CICERO.

Toth, F. 2008. *Adaptation to climate change*. Presentation. Central European University Budapest. 1 July.

Ulbert, C., T. Risse, and H. Müller. 2004. *Arguing and Bargaining in Multilateral Negotiations*. Paper presented at the conference "Empirical Approaches to Deliberative Politics". European University Institute Florence. 21–22 May.

Urpelainen, J. 2009. The United States of America. In *Towards a new climate regime? Views of China, India, Japan, Russia and the United States on the road to Copenhagen*, ed. A. Korppoo, and A. Luta, 100–117. Helsinki: The Finnish Institute of International Affairs.

Van Asselt, H., N. Kanie, and I. Masahiko. 2009. Japan's position in international climate policy: navigating between Kyoto and the APP. *International Environmental Agreements* 9 (2): 319–336.

Van Schaik, L., and C. Egenhofer. 2003. *Reform of the EU Institutions: Implications for the EU's Performance in Climate Negotiations*. Brussels: Centre for European Policy Studies.

Van Schaik, L., and K. van Hecke. 2008. *Skating on thin ice. Europe's internal climate policy and its position in the world*. Brussels: Egmont Institute.

Van Schaik, L., and S. Schunz. 2012. Explaining EU Activism and Impact in Global Climate Politics: Is the Union a Norm- or Interest-Driven Actor? *Journal of Common Market Studies* 50(1): 169–186.

Verdonk, M., C. Brink, H. Vollebergh, and M. Roelfsema. 2013. *Evaluation of policy options to reform the EU Emissions Trading System*. The Hague: PBL.

Verolme, H. 2012. *European Climate Leadership, Durban and beyond*. Brussels: Heinrich Böll Foundation.

Victor, D.G., and Salt, J.E. 1994. From Rio to Berlin. Managing Climate Change. *Environment* 36 (10): 7–32.

Vidal, J. and F. Harvey. 2011. *COP17 climate talks: Durban text follows EU roadmap for new global deal*. The Guardian. 9 December (available at http://www.guardian.co.uk/environment/2011/dec/09/un-climate-talks-durban-text, last accessed 26 January 2012).

Vidal, J., and J. Watts. 2009. *Copenhagen: The last-ditch drama that saved the deal from collapse*. The Guardian. 20 Dec. (available at http://www.guardian.co.uk/environment/2009/dec/20/copenhagen-climate-global-warming, last accessed 25 February 2010).

Vogler, J. 2005. The European contribution to global environmental governance. *International Affairs* 81 (4): 835–850.

Vogler, J. 2008. *Climate Change and EU Foreign Policy: The Negotiation of Burden-Sharing*. Dublin: University College.

Von Seth, H. 1999. Global climate. *Yearbook of International Environmental Law* 10: 223–233.

Watanabe, R. 2003. Less than Satisfactory Results. *Environmental Policy and Law* 33 (1): 18–20.

Watts, J. 2009. *China sets first targets to curb world's largest carbon footprint*. The Guardian. 26 November.

Watts, J. 2010. *Climate deal is closer, says UN envoy, despite China and US locking horns*. The Guardian. 9 October. (available at http://www.guardian.co.uk/environment/2010/oct/09/climate-talks-china-america-clash, last accessed 22 October 2010).

Webber, M., and M. Smith. eds. 2002. *Foreign Policy in a Transformed World*. London: Prentice Hall.

Werksman, J. 1998. The Clean Development Mechanism: Unwrapping the 'Kyoto Surprise'. *Review of European Community and International Environmental Law* 7 (2): 147–158.

Wettestad, J. 2000. The complicated development of EU climate policy. In *Climate Change and European leadership: A sustainable role for Europe*, ed. J. Gupta, and M. Grubb, 25–47. Dordrecht: Kluwer Academic Publishers.

White, B. 2001. *Understanding European Foreign Policy*. Basingstoke: Palgrave.

Wigley, T.M.L. 1999. *The science of climate change*. Arlington, VA: Pew Center on Global Climate Change.

Wilkenfeld, J., G. Hopple, P. Rossa, and S. Andriole. 1980. *Foreign Policy Behavior. The Interstate Behavior Analysis Model*. Beverly Hills: Sage.

Willis, A. 2010a. *France, UK and Germany call for 30 percent CO_2 cut*. EUObserver. 15 July (available at http://euobserver.com/884/30482, last accessed 18 September 2010).

Willis, A. 2010b. *Cancun climate deal restores faith in UN process*. EUObserver. 12 Dec. (available at http://euobserver.com/9/31482/?rk=1, last accessed 22 March 2012).

Wittneben, B., W. Sterk, H.E. Ott, and B. Bround. 2006. *In from the Cold: The Climate Conference in Montreal Breathes New Life into the Kyoto Protocol*. Wuppertal: Institute for Climate, Environment and Energy.

Wong, P. 2008. *Australia's goal of cutting emissions by 15% in 2020*. New Europe. Issue 815. 5 January.

World Bank. 2008. *World Bank Ranking Table. 2008 GDP*.

Wurzel, R.K.W., and J. Connelly. 2010. Conclusion: the European Union's leadership role in international climate politics reassessed. In *The European Union as a Leader in International Climate Change Politics*, ed. R.K.W. Wurzel, and J. Connelly, 271–290. London: Routledge.

Xinhua. 2008a. *Meeting on climate change ends in Paris without tangible agreement*. 19 April (available at http://news.xinhuanet.com/english/2008-04/19/content_8008112.htm, last accessed 22 March 2010).

Xinhua. 2008b. *World major economies wrap up Hawaii climate meeting*. 1 February (available at http://english.peopledaily.com.cn/90001/90777/63495 02.html, last accessed 13 March 2010).

Yamin, F. 1998. The Kyoto Protocol: Origins, Assessment and Future Challenges. *Review of European Community and International Environmental Law* 7 (2): 113–127.

Yamin, F. 2000. The role of the EU in climate negotiations. In *Climate Change and European leadership: A sustainable role for Europe*, ed. J. Gupta, and M. Grubb, 47–66. Dordrecht: Kluwer Academic Publishers.

Yamin, F., and J. Depledge. 2004. *The International Climate Change Regime: A Guide to Rules, Institutions and Procedures*. Cambridge: Cambridge UP.

Yin, R. 2003. *Case Study Research*. 3rd ed. Thousand Oaks, CA: Sage.

Young, O.R. 1989a. *International Cooperation. Building Regimes for Natural Resources and the Environment*. Ithaca: Cornell UP.

Young, O.R. 1989b. The Politics of International Regime Formation: Managing Natural Resources and the Environment. *International Organization* 43 (3): 349–376.

Young, O.R. 1994. *International Governance: Protecting the Environment in a Stateless Society*. Ithaca: Cornell UP.

Young, T. 2010. *Australia ditches carbon cap-and-trade plans: Australian Prime Minister drops plans for climate bill*. Businessgreen. 27 April (available at http://www.businessgreen.com/bg/news/1804424/australia-ditches-carbon-cap-trade-plans, last accessed 23 June 2010).

Yuxia, J. 2008. *Barroso: Financial crisis no excuse for ignoring climate change*. Chinaview. 25 Oct. (available at http://news.xinhuanet.com/english/2008-10/25/content_10250642.htm, last accessed 25 March 2009).

Zammit Cutajar, M. 2009. *Presentation: Fulfilment of the Bali Action Plan and components of the agreed outcome*. Bonn, 27 March.

Zhang, Z. 2010. *Copenhagen and Beyond: Reflections on China's Stance and Responses*. Washington, DC: East-West Center.

Zito, A. 2005. The European Union as an Environmental Leader in a Global Environment. *Globalizations* 2 (3): 363–375.

Annexes

ANNEX I – Overview of Research Interviews Conducted

Interviewee (affiliation) and location	Date
Director of Climate Policy, WWF Europe, Brussels	30/10/2007
Secretariat to the Temporary Committee on Climate Change, European Parliament, Brussels	19/12/2007
Policy Coordinator, International Climate Negotiations, DG Environment, European Commission, Brussels	28/1/2008
Director, Climate and Energy Policy, Greenpeace Europe, Brussels	18/3/2008
Desk Officer Climate Change, DG Environment, Council of the European Union, Brussels	20/5/2008
Policy Officer Energy/Climate Change, Bond Beter Leefmilieu, Brussels	20/5/2008
Policy Officer Climate Change Post-2012, Climate Action Network Europe, Brussels	22/5/2008
Policy Advisor Climate Negotiations, Unit Air Pollution, Risk Management, Environment and Health, Flemish Department of Environment, Nature and Energy, Brussels	15/12/2008
Advisor on environmental issues, Green/FEA Group, European Parliament, Brussels	9/1/2009
DG External Relations, European Commission, Brussels	14/1/2009
Climate Change Commission, Lisbon (telephone)	16/1/2009
Delegation of the European Commission, Beijing (telephone)	5/2/2009
Member of G-77/China negotiation team (telephone)	10/2/2009
Advisor, Federal Public Service for Health, Food Chain Safety and the Environment, Brussels (telephone)	25/2/2009
Delegation of the European Commission, Tokyo (telephone)	27/2/2009
Policy Officer, DG Environment, European Commission, Brussels (telephone)	13/3/2009
Federal Ministry for the Environment, Nature Conservation and Nuclear Safety, Berlin (telephone)	13/3/2009
Climate Change responsible, WWF-USA, Washington, DC (telephone)	24/3/2009
Department for Environment, Food and Rural Affairs, London (telephone)	25/3/2009
State Department, Washington, DC	17/4/2009
Associate Professor, Department of International Organizations and Law, Shanghai Institute for International Studies, China (telephone)	28/4/2009

Interviewee (affiliation) and location	Date
Policy Advisor, Coordination of post-2012 negotiations, Ministère de l'Ecologie, du Développement durable et de l'Aménagement du territoire, Paris, France (telephone)	28/4/2009
Former Chief Negotiator on Climate Change, Ministry of the Environment, Stockholm (telephone)	7/5/2009
Delegation of the European Commission, Washington, DC	8/5/2009
Senior Professional Staff for US Senator, US Senate Foreign Relations Committee, Washington, DC	8/5/2009
Senior Policy Advisor, Embassy of the United Kingdom, Washington, DC	13/5/2009
Counselor, Environment and Energy, Deputy Head Economic and Commercial Section, Embassy of Germany, Washington, DC	15/5/2009
Center for Clean Air Policy, Washington, DC	19/5/2009
Federal Ministry for the Environment, Nature Conservation and Nuclear Safety, Berlin (telephone)	24/2/2010
Attaché, Sustainable Development and Environment, Ministry for Foreign Affairs, Foreign Trade and Development Cooperation of Belgium, Brussels (and telephone)	26/2 + 2/3/2010
Policy Officer, DG Environment, European Commission, Brussels (telephone)	2/3/2010
Climate Change Commission, Lisbon (telephone)	10/3/2010
State Department, Washington, DC (telephone)	22/3/2010

ANNEX II – UN Climate Regime Negotiations between 1991 and 2012

Period	Meeting	Dates	Place
1991–1995	INC 1	4–14/2/1991	Chantilly, Virginia
	INC 2	19–28/6/1991	Geneva
	INC 3	9–20/9/1991	Nairobi
	INC 4	9–20/12/1991	Geneva
	INC 5.1/5.2	18–28/2, 30/4–8/5/1992	New York
	INC 6	7–10/12/1992	Geneva
	INC 7	15–20/3/1993	New York
	INC 8	16–27/8/1993	Geneva
	INC 9	7–18/2/1994	Geneva
	INC 10	22/8–2/9/1994	Geneva
	INC 11	6–17/2/1995	New York
1995–1997	COP 1	28/3–7/4/1995	Berlin
	AGBM 1	21–25/8/1995	Geneva
	AGBM 2	30/10–3/11/1995	Geneva
	AGBM 3	5–8/3/1996	Geneva
	COP 2/AGBM 4	8–19/7/1996	Geneva
	AGBM 5	9–13/12/1996	Geneva
	AGBM 6	3–7/3/1997	Bonn
	AGBM 7	28/7–7/8/1997	Bonn
	AGBM 8	20–31/10, 30/11/1997	Bonn/Kyoto
	COP 3	1–11/12/1997	Kyoto
1998–2007	COP 4	2–13/11/1998	Buenos Aires
	COP 5	25/10–5/11/1999	Bonn
	COP 6	13–25/11/2000	The Hague
	COP 6bis	16–27/7/2001	Bonn
	COP 7	29/10–10/11/2001	Marrakech
	COP 8	23/10–1/11/2002	New Delhi
	COP 9	1–12/12/2003	Milan
	COP 10	6–18/12/2004	Buenos Aires
	COP 11/MOP 1	28/11–10/12/2005	Montreal
	AWG-KP 1/Dialogue 1	17–26, 15–16/5/2006	Bonn
	COP 12/MOP 2 – AWG-KP 2/Dialogue 2	5–17/11/2006	Nairobi
	AWG-KP 3/Dialogue 3	7–18, 16–17/5/2007	Bonn
	AWG-KP 4.1/Dialogue 4	27–31/8/2007	Vienna
2007–2012	COP 13/MOP 3/AWG-KP 4.2	3–15/12/2007	Bali
	AWG-KP 5.1/AWG-LCA 1	31/3–4/4/2008	Bangkok
	AWG-KP 5.2/AWG-LCA 2	2–13/6/2008	Bonn
	AWG-KP 6.1/AWG-LCA 3	21–23/8/2008	Accra
	COP 14/MOP 4 – AWG-KP 6.2/AWG-LCA 4	2–13/12/2008	Poznan
	AWG-KP 7/AWG-LCA 5	29/3–8/4/2009	Bonn

Period	Meeting	Dates	Place
	AWG-KP 8/AWG-LCA 6	2–13/6/2009	Bonn
	AWG-KP 8/AWG-LCA 6	10–14/8/2009	Bonn
	AWG-KP 9.1/AWG-LCA 7.1	28/9–9/10/2009	Bangkok
	AWG-KP 9.2/AWG-LCA 7.2	2–6/11/2009	Barcelona
	COP 15/MOP 5 – AWG-KP 10/AWG-LCA 8	7–19/12/2009	Copenhagen
	AWG-KP 11/AWG-LCA 9	9–11/04/2010	Bonn
	AWG-KP 12/AWG-LCA 10	31/5–11/6/2010	Bonn
	AWG-KP 13/AWG-LCA 11	2–6/8/2010	Bonn
	AWG-KP 14/AWG-LCA 12	4–9/10/2010	Tianjin
	COP 16/MOP 6 – AWG-KP 15/AWG-LCA 13	29/11–10/12/2010	Cancun
	AWG-KP 16.1/AWG-LCA 14.1	3–8/4/2011	Bangkok
	AWG-KP 16.2/AWG-LCA 14.2	6–17/6/2011	Bonn
	AWG-KP 16.3/AWG-LCA 14.3	1–7/10/2011	Panama City
	COP 17/MOP 7 – AWG-KP 16.4/AWG-LCA 14.4	28/11–9/12/2011	Durban
	AWG-KP 17/AWG-LCA 15/ ADP 1	14–25/5/2012	Bonn
	AWG-KP 17/AWG-LCA 15/ ADP 1 (informal)	30/8–5/9/2012	Bangkok
	COP 18/MOP 8 – AWG-KP 17.2/AWG-LCA 15.2/ADP 1.2	26/11–7/12/2012	Doha

Index

A

Actorness 34

Actor capacity 25, 34–5, 38–9, 46, 51–3, 60, 73–5, 111, 116, 125, 135, 149, 154, 166, 247–9, 254, 274, 278, 282–3, 290–2, 294–5, 298, 302–3, 306–8

Adaptation, *see* Climate change

Ad Hoc Working Group on Long-Term Cooperative Action 44, 173, 178, 180–4, 188, 193, 198–9, 201, 205, 209–10, 222, 225, 233–5, 258–63, 265–8, 277

Ad Hoc Group on the Berlin Mandate 67, 70, 76, 78, 80–90, 92, 98–102, 104, 116, 277

Ad Hoc Working Group on the Durban Platform for Enhanced Action 262–3, 265, 267–8, 273, 277

Ad Hoc Working Group on the Kyoto Protocol 138–44, 167, 171–3, 175, 178, 180–1, 184–5, 188–9, 191, 193–6, 198–9, 201–2, 206, 209–11, 222, 225, 233, 242, 258, 260–3, 265–8, 277

African Group 163, 201, 211–3, 240, 304

African Union 163, 182, 190, 222

Alliance of Small-Island States 44, 53, 55, 57, 66, 72, 77–9, 81–2, 84, 89, 95, 101–6, 116–7, 127, 131, 133, 143, 145, 159, 163, 172, 183, 189, 193, 205, 213, 216, 220, 223–4, 234–5, 238, 254, 258, 260–2, 270–1, 274, 277, 284–6, 304, 308

Argentina 79, 90, 116–8, 132, 152

Asia-Pacific Economic Cooperation 153, 202, 231–2, 236–7, 249

Asia-Pacific Partnership on Clean Development and Climate Change 134–5, 147, 149, 153, 155, 157, 161–2, 279

Assessment Report of the IPCC

Fifth 21, 276

Fourth 21, 25, 135, 153, 165, 170, 173, 223

Third 114, 127, 130

Second 68, 82

First 50, 52

Australia 24, 44, 50, 53, 55, 66, 68, 71, 78, 81, 84–5, 87, 90, 94, 96, 98, 114–5, 118–9, 122, 126, 133–5, 138, 140, 143, 149, 152, 159, 161, 177, 186, 189, 192, 202, 205, 213, 229, 237–8, 245, 247, 253–4, 259, 265, 268, 285, 298

B

Bali Roadmap 151, 172, 175–6, 178, 184, 192, 196, 201, 232–5, 239, 260, 267–8, 289

BASIC 159, 163, 200, 220, 224, 232, 238–9, 242–4, 247, 254, 270–1, 277–8, 281, 284–6

Berlin Mandate 24, 67, 78, 80–3, 85–7, 89, 91–2, 101–2, 105–6, 232, 289

Brazil 56, 71–2, 90, 92, 134, 152, 159, 162–3, 168, 178, 180, 204, 211, 213, 226, 237, 254, 304

BRIC 152

Buenos Aires Plan of Action 117–8, 121–2

Burden-sharing 84–5, 87, 91, 100, 106–7, 115, 128, 228

Byrd-Hagel resolution 89, 109

C

Canada 40, 50, 53, 61, 66, 68, 71, 81, 84, 90–1, 94, 96, 114, 119, 121, 126–7, 130, 136, 140, 142–3, 158–9, 172–4, 234, 253, 260–1, 268, 272

Cancun Agreements 17, 20, 25, 251, 253, 260–1, 265–8, 273

Cartagena Group 254, 259–62, 264, 274, 308

Case study 23, 27, 41

Methodology 20, 42, 205, 305

Causal mechanism 35–6, 42, 110, 112, 148, 244, 287–8, 298

Arguing 25, 31–2, 35–6, 46, 53, 63, 72, 80, 82, 89, 92, 100, 109, 115, 120, 129, 140, 145, 169, 181, 230, 243–5, 247, 258, 260, 270, 275–7, 279–81, 285, 287–91, 293, 297–9, 301, 303

Bargaining 25, 31–2, 35–6, 38, 46, 59, 61, 63–4, 92, 100–1, 107, 109–12, 125, 127, 133, 147–9, 200, 228, 230, 242, 244, 247–8, 270, 275–7, 279, 281, 284–5, 287–91, 293–7, 299, 303

Causality 25

Conditionality 229, 288, 296, 309

Conjunctive causality 25, 294, 296, 298, 303

China 25, 44, 52–3, 55–7, 61, 68, 71–2, 78–9, 81–3, 85, 89–90, 92, 95, 98, 102–5, 108–12, 115, 117–23, 127, 129–36, 138–40, 142–5, 149, 152, 155–7, 159–63, 168, 171–6, 178–81, 186–90, 193, 197–200, 202–7, 209–14, 216, 218, 220, 223, 226–7, 229, 232–4, 236–40, 243–4, 246–7, 253–4, 257–8, 260–4, 267, 271, 280–1, 284, 294–7, 304, 308–9

Clean Development Mechanism 97, 113, 117, 129–30, 132, 138, 178, 262, 286

Climate change

Adaptation 22, 40, 99, 121, 126–7, 129, 132, 135–9, 141, 143, 152, 163, 173, 181, 183–6, 190, 192, 196–7, 199, 201, 207, 212, 229, 261, 266–7, 286, 293, 307

Definition 21–23

Mitigation 22, 40, 55, 57–8, 88, 99, 103, 108–11, 127, 129, 132, 135–6, 140–1, 144, 152, 157, 163, 173–4, 176–8, 180–1, 183, 186, 188–91, 196, 198–9, 201–3, 205, 208–12, 214–19, 223–4, 228–9, 231–4, 236–9, 241, 243, 245–7, 253–4, 257–9, 261, 263–5, 268, 281, 284–5, 287, 294, 298

Common but differentiated responsibilities 23, 40–1, 58, 60, 62, 76, 87, 102–3, 115, 144, 148, 159, 161–2, 171, 173, 176–7, 180, 182, 229, 232–3, 293, 302

Common Foreign and Security Policy 165, 255

Conference of the parties
  Bali 25, 134, 141–5, 171–176, 260, 265, 267–8, 289
  Berlin 24, 64, 77–80, 133, 232, 289
  Cancun 17, 20, 25, 251–274, 306
  Copenhagen 18–19, 25, 43, 151–243
  Delhi 129–132, 191, 200
  Doha 20, 25, 251, 256, 263–5, 265–74
  Durban 20, 25, 251, 254, 256–7, 261–74
  Geneva 48, 50, 54–5, 65–7, 80–1, 83, 103
  Kyoto 67–112
  Marrakech 24, 113–150, 279
  Milan 130–2
  Montreal 48, 96, 138, 140, 142, 172–3
  Nairobi 139–41
  Poznan 180–6, 226–7, 231, 286
  The Hague 49, 116–26, 147

Copenhagen Accord 19, 25, 43, 222–4, 251–60, 266–70, 279, 291

Council of the EU
  Environment 51, 65, 75–6, 84, 99, 128, 165–70, 178–9, 182, 187, 200, 204, 226, 248, 255–7, 262, 307
  Finance 187
  Presidency, *see* Presidency of the Council of the EU

D

Delhi Declaration 129–30
Doha Gateway 20, 251, 265–8

Durban Package 20, 251, 262, 265–8, 271–2

E

Earth Summit 23, 47, 50, 57, 86, 117
Emissions Trading 18, 82, 84, 90, 92, 95, 97–8, 101, 106, 108, 113, 128, 177–9, 185–6, 228, 230, 237, 242, 253, 264, 270, 285–7, 298, 306, 309
European Commission 34, 51, 88, 90, 165, 170, 182–3, 186–7, 196, 213, 254, 259, 264, 309
European Council 34, 51, 99, 137, 141–2, 147, 165, 167–70, 185, 187, 200–1, 216, 221, 227, 231, 256
European Emissions Trading System, *see* Emissions Trading
European External Action Service 255–6, 307
European Union
  Actor capacity, *see* Actor capacity
  Ad Hoc Working Group on Climate Change 75, 84, 128
  Climate and Energy Package 18, 170, 177–9, 182–3, 185, 227, 245
  External representation 35, 74, 248, 254–5, 307
  Leadership 18, 20, 24, 28, 86, 124, 157, 169, 175, 181, 227, 229–30, 247, 273–4, 304
  Legal status 35, 54, 191
  Troika, *see* Troika

F

Flexible mechanism 23–4, 29, 82, 84, 89, 92, 95, 97, 103–9, 112–9, 122–5, 128, 155, 158, 178, 180, 186, 189, 193, 206–7, 225, 280

see Clean Development Mechanism
see Emissions Trading
see Joint Implementation
Foreign policy
  Analysis 34–37
  Definition 17–18, 34
  European Union 17–25, 34–37
  Implementation 29
  Instruments 34–37
France 54–6, 60, 63, 75, 78, 86, 90, 120, 164, 167, 190, 208, 212–3, 216, 226, 240, 252, 256, 259, 308

G

G-7 49, 52
G-7+1 86, 98
G-8 86, 134–5, 142, 144, 146–7, 149, 152–3, 157, 168, 180, 194–5, 202, 216–7, 220, 225–6, 231, 234–5, 238, 276
  Gleneagles 134, 145, 177, 226
  Heiligendamm 142, 144, 180–1, 195, 226, 235
  L'Aquila 231, 238
G-8+5 135, 147, 177, 181, 194–5, 225, 276
G-20 152–3, 168, 182, 190, 197–8, 200, 225, 245, 256, 259, 276, 307
Geneva Ministerial Declaration 50
Germany 51, 55–6, 63, 65–6, 73, 75, 81, 84, 86, 90, 120, 142, 164–5, 167–8, 179, 185, 190, 200, 208, 210, 213, 226, 252, 256, 259
Goodness of fit 283–6, 290, 292–3, 303, 306
Green Diplomacy Network 136–7

Greenhouse effect 22, 51–2, 68
Greenhouse gas 22, 25, 41, 54–8, 65–72, 80–1, 83, 85, 88–9, 91, 96, 99–100, 104, 107, 114, 118, 127, 129, 134–5, 143–4, 151–5, 159–62, 164, 169, 177, 189–90, 192, 196, 200, 208, 223, 229, 243–5, 264
Greenland Dialogue 194–5
Group of 77 and China 44, 53, 55, 57, 61, 68, 71–2, 78–9, 82–3, 85, 89–90, 92, 95, 102–5, 108–12, 115, 117, 119–23, 127, 129–33, 135, 138, 140, 144, 149, 159, 161, 163, 171–6, 178, 180–1, 186, 188–9, 198–9, 205–7, 209–14, 223, 229, 234, 240, 243, 253–4, 258, 260, 271, 281, 284, 294–7

H

High Representative for Foreign and Security Policy 34, 165, 255, 307

I

India 44, 55, 71, 79, 90, 92, 117–8, 120, 129–31, 134–6, 139, 145–7, 149, 152, 155–63, 168, 172, 178, 180, 182, 184, 187, 189–91, 200, 204, 211, 213, 222, 226, 232–7, 243, 246, 254, 258, 260–2, 267, 271, 280, 295, 304, 308
Influence
  Analysis 27–33
  Constitutive dimensions 30–31
  Definition 30
  Degrees 33
  Typology 33
  Constitutive dimensions 29
  Definition 29
  Typology 33

Influence attempt

Intergovernmental Negotiating Committee 23, 50, 52, 54–66, 72, 76–9

Intergovernmental Panel on Climate Change 21–2, 48, 50, 52, 68, 71–2, 78, 81–2, 104, 110, 135, 144–6, 153, 165, 172–4, 170, 188–9, 191, 193–4, 224, 229, 233–5, 238, 245, 257, 267, 269, 280–1, 298–9

Assessment Report, *see* Assessment Report

International relations theory 28, 30, 289, 299

Italy 75, 86, 90, 164, 168, 185, 190, 194, 216

J

Japan 24, 44, 50, 53, 56, 66, 68, 70, 81, 84–5, 88, 90–4, 96, 98, 104–8, 110–1, 114, 117–8, 120–2, 124–7, 130, 134, 136, 138–9, 142, 145, 156–7, 160, 168, 171, 173, 177, 179, 180, 183, 189–90, 192, 194, 196, 199, 202–6, 212–3, 222, 228–9, 234, 237–8, 246, 252–3, 257, 260–1, 268, 272, 280–1, 283

Joint Implementation 59, 84, 97, 113, 138

JUS(S)CA(N)NZ 68, 71–2, 78–80, 82, 84–6, 88, 90–1, 100, 102, 104–5, 107–8, 114–5

see also Umbrella Group

K

Kyoto Protocol 17, 19, 24, 43, 65, 67–8, 70, 74, 76–7, 95–7, 101–9, 113–8, 120, 122–34, 138, 142, 147–9, 124, 156–9, 161–2, 165, 167–72, 177, 179–81, 183–5, 190, 192–3, 196, 198, 200–2, 206, 208–10, 231–2, 234, 238–9, 242–3, 246, 249, 253, 257–8, 260, 262–4, 268–70, 272, 279, 282, 291, 294–6, 304

L

Land-use, Land-use change and forestry 118, 122, 130, 132, 143, 178–80, 189, 193, 206, 209, 225

Least Developed Countries 71–2, 121, 163, 191, 216

Lisbon Treaty 34, 166, 255–6, 269, 273, 307

M

Maastricht Treaty 17, 282

Major Economies Forum 190, 194–7, 200, 213, 225–6, 231, 234–5, 238, 240, 259, 261, 264, 279, 308

Major Economies Meeting 145, 153, 155, 168, 177, 179, 181

Marrakech Accords 24, 113–132, 138, 148, 279

Meeting of the parties 43, 97, 113–4, 117, 127, 130, 132, 135, 138–46, 151, 171–2, 175, 182–3, 185, 194, 202, 205–7, 209–13, 221–2, 225, 231–2, 238–9, 245, 251, 254, 258–60, 262–8, 270–2

Mitigation, *see* Climate change

MRV 178, 184, 191–2, 197, 209, 215–6, 218–20, 224, 238, 259

N

Nationally appropriate mitigation actions 55, 174, 176, 181, 184, 189, 191–2, 200, 202–3, 207–9, 215, 218, 223–4, 232–3, 242, 266

New Zealand 44, 53, 66, 68, 71, 78, 85, 87, 90, 92, 96, 98, 102, 114, 127, 254, 259, 268

Non-governmental organization 43, 89, 102–3

North American Free Trade Agreement 91

Norway 44, 68, 71, 78, 114, 177, 213, 254, 259, 268

O

Organisation for Economic Cooperation and Development 52–3, 55–6, 58, 61–3, 65–6, 83, 98, 162, 178, 281

Organisation of Petroleum Exporting Countries 44, 53, 72, 78–82, 89, 102–3, 131, 159, 163, 205, 213

P

Petersberg Dialogue 259, 261, 264

Poland 164, 169, 185, 187, 216, 256–7, 264–5, 273

Power 21, 27, 39–40, 52, 64, 113, 125–7, 149, 228, 247, 275, 281–4, 290–7, 303, 305

Presidency of the Council of the EU 34, 54, 60, 65, 75–8, 80–1, 84, 93, 98–9, 116, 120, 125, 128, 167–8, 179–80, 185, 187, 196, 198, 202, 205–6, 210–11, 213, 223, 226, 229, 236, 238, 247, 259, 273, 307

Belgian 256

British, *see also* United Kingdom 75, 226

Cypriot 256

Czech 167, 187

Dutch 54, 75, 84

French, *see also* France 75–6, 78, 125, 167, 180, 185

German, *see also* Germany 75–6, 142, 167, 226

Irish 84, 128

Italian, *see also* Italy 81, 84

Polish, *see also* Poland 167, 179

Portuguese 54, 167

Slovenian 167, 179

Spanish 80, 167

Swedish 120, 167, 198, 229

Principles 40–1, 57–8, 63, 117, 159, 197, 265, 299–300

see also Common but differentiated responsibilities

Equity 40, 58, 71–2, 79, 111, 161, 246

Precautionary 40, 53, 58, 73, 111, 165, 245, 280

Process-tracing 42–46

Q

Quantified Emission Limitation and Reduction Objectives 87, 91, 93, 176, 185, 241, 258, 260, 266–9, 283, 294

R

Regime

Analysis 38–42

Definition 40

Theory 38–9

Regional Economic Integration Organisation 65, 72, 74, 106, 166

Russia 24, 44, 52–3, 56, 68, 72, 78, 82–4, 90–1, 96, 98, 103, 114,

Index

120–2, 124, 126–7, 130–2, 143, 147, 152, 156, 158, 172–3, 190, 202, 204, 206, 213, 229, 234–5, 237, 253, 260–1, 265, 268, 272, 278, 280, 285–7

S

Saudi Arabia 72, 78, 83, 90, 143, 152, 163, 213

Single European Act 51, 133

South Africa 25, 79, 134, 143, 152, 159, 162–3, 175, 180, 183, 188, 193, 213, 226, 254, 262, 267, 270, 304

Sustainable development 58, 77, 80, 127, 129, 130–1, 134, 174, 177, 263

T

Troika 75, 93, 98, 106, 116, 119–20, 136, 149, 167–8, 177, 182, 187, 190, 196, 212–3, 220–1, 225, 278, 282, 294, 307

U

Ukraine 44, 72, 114, 265, 268

Umbrella Group 44, 114, 119–27, 130–31, 135, 144, 147–9, 154, 158, 172, 174–6, 180–1, 186, 195, 198, 229, 232–3, 244–5, 252, 258, 265, 285–6

United Nations, *see also* United Nations Framework Convention on Climate Change 17, 50, 52, 128, 139, 145, 151, 197, 204, 304

United Nations Conference on the Human Environment 48, 162

United Nations General Assembly 23, 49–50, 54, 86, 142, 177

United Nations Security Council 141

United Nations Framework Convention on Climate Change 17, 19, 23–4, 40–1, 43, 47, 52–3, 56–9, 62–6, 68–74, 76–80, 83, 86, 90–5, 97–103, 106, 110, 114, 118–9, 121, 123, 126, 129, 134, 138–9, 141, 143, 166, 171, 176, 180–1, 183, 192, 198–9, 202, 207, 223, 236, 243, 257, 270, 279, 282, 295–6, 301

Secretariat 59, 83, 86, 98–9, 121, 143, 181, 193–4, 223, 257

United Kingdom 51, 54–65, 73, 76, 84, 86, 90, 94, 96, 98, 106, 111, 126, 134–5, 139, 164–70, 187, 190, 200, 209–13, 216, 221–2, 226, 240, 260

United States 24, 44, 47–8, 52–3, 55–64, 66, 68–71, 78–96, 98, 100–27, 129, 131, 133–9, 144–9, 151–62, 168, 171, 173–7, 179, 187–94, 196–206, 209, 212–6, 218, 220–4, 226–8, 231–47, 253, 257, 260–2, 264–5, 271, 277–81, 283–4, 294–6, 304, 309

House of Representatives 69, 190, 194, 214

Senate 69, 88, 109–10, 115, 117, 154–55, 193, 205, 253

W

Working Party on International Environmental Issues-Climate Change 134, 137, 166–68, 261, 288, 307

World Climate Conference 48, 50

World Event Interaction Coding Scheme 30, 99, 228

College of Europe Studies

Europe is in a constant state of flux. European politics, economics, law and indeed European societies are changing rapidly. The European Union itself is in a continuous situation of adaptation. New challenges and new requirements arise continually, both internally and externally. The College of Europe Studies series exists to publish research on these issues done at the College of Europe, both at its Bruges and its Warsaw campus. Focused on the European Union and the European integration process, this research may be specialised in the areas of political science, law or economics, but much of it is of an interdisciplinary nature. The objective is to promote understanding of the issues concerned and to make a contribution to ongoing discussions.

Series Editors
Professors D. Hanf, D. Mahncke, I. Govaere, J. Pelkmans, E. Lannon and J. Monar for the College of Europe (Bruges and Warsaw)

Series Titles

No. 18 – Simon SCHUNZ, *European Union Foreign Policy and the Global Climate Regime*, 2014, 371 p., ISBN 978-2-87574-134-9

No. 17 – Inge GOVAERE & Dominik HANF (eds.), *Scrutinizing Internal and External Dimensions of European Law/Les dimensions internes et externes du droit européen à l'épreuve.* Liber Amicorum *Paul Demaret*, 2013, 882 p., ISBN 978-2-87574-085-4

No. 16 – Michele CHANG & Jörg MONAR (eds.), *The European Commission in the Post-Lisbon Era of Crises. Between Political Leadership and Policy Management*, 2013, 298 p., ISBN 978-2-87574-028-1

No. 15 – Dieter MAHNCKE & Sieglinde GSTÖHL (eds.), *European Union Diplomacy. Coherence, Unity and Effectiveness*, 2012, 273 p., ISBN 978-90-5201-842-3

No. 14 – Erwan LANNON (ed.), *The European Neighbourhood Policy's Challenges / Les défis de la politique européenne de voisinage*, 2012, 491 p., ISBN 978-90-5201-779-2

No. 13 – Marise CREMONA, Jörg MONAR & Sara POLI (eds.), *The External Dimension of the European Union's Area of Freedom, Security and Justice*, 2011, 434 p., ISBN 978-90-5201-728-0

No. 12 – Jing MEN & Giuseppe BALDUCCI (eds.), *Prospects and Challenges for EU-China Relations in the 21st Century. The Partnership and Cooperation Agreement*, 2010, 262 p., ISBN 978-90-5201-641-2

No. 11 – Jörg MONAR (ed.), *The Institutional Dimension of the European Union's Area of Freedom, Security and Justice*, 2010, p., ISBN 978-90-5201-615-3

N° 10 – Dominik HANF, Klaus MALACEK & Elise MUIR (dir.), *Langues et construction européenne*, 2010, 286 p., ISBN 978-90-5201-594-1

No. 9 – Jacques PELKMANS, Dominik HANF & Michele CHANG (eds.), *The EU Internal Market in Comparative Perspective. Economic, Political and Legal Analyses*, 2008, 314 p., ISBN 978-90-5201-424-1

No. 8 – Inge GOVAERE & Hanns ULLRICH (eds.), *Intellectual Property, Market Power and the Public Interest*, 2008, 315 p., ISBN 978-90-5201-422-7

No. 7 – András INOTAI, *The European Union and Southeastern Europe. Troubled Waters Ahead?*, 2007, 414 p., ISBN 978-90-5201-071-7

No. 6 – Inge GOVAERE & Hanns ULLRICH (eds.), *Intellectual Property, Public Policy, and International Trade*, 2007, 234 p., ISBN 978-90-5201-064-9

No. 5 – Dominik HANF & Rodolphe MUÑOZ (dir.), *La libre circulation des personnes. États des lieux et perspectives*, 2007, 329 p., ISBN 978-90-5201-061-8

No. 4 – Dieter MAHNCKE & Sieglinde GSTÖHL (eds.), *Europe's Near Abroad. Promises and Prospects of the EU's Neighbourhood Policy*, 2008, 318 p., ISBN 978-90-5201-047-2

No. 3 – Dieter MAHNCKE & Jörg MONAR (eds.), *International Terrorism. A European Response to a Global Threat?*, 2006, 191 p., ISBN 978-90-5201-046-5

No. 2 – Paul DEMARET, Inge GOVAERE & Dominik HANF (eds.), *European Legal Dynamics – Revised and updated edition of 30 Years of European Legal Studies at the College of Europe / Dynamiques juridiques européennes – Édition revue et mise à jour de 30 ans d'études juridiques européennes au Collège d'Europe*, 2007, 571 p., ISBN 978-90-5201-067-0

No. 1 – Dieter MAHNCKE, Alicia AMBOS & Christopher REYNOLDS (eds.), *European Foreign Policy. From Rhetoric to Reality?*, 2004, 2nd printing/2e tirage 2006, 381 p., ISBN 978-90-5201-247-6

www.peterlang.com